Anne Riley

DINA

Nature's Case for Democracy

Created Independently

Published Independently

DINA

Nature's Case For Democracy

This is a work of fiction.
The scientific discussions are based on fact and are annotated in the Sources Section.
The characters and events of the story are the products of my own mind with parts stolen from real life.
The theory is an extension of the scientific facts but is not provable. Please read with an open mind. Your thoughts and opinions are welcome.

Cover Art:
Nature's Creatures in Support of Democracy
Generated by Shutterstock AI
Licensed Image: 2401579419 to Anne Riley

ISBN: 9798875700873

To Jane and Goldie

And your future

The difficulty lies not so much in developing new ideas, as in escaping from old ones.

~~John Maynard Keynes~

Chapter 1

DINA

The pungent fragrance of the Mortgage Lifter spiced the May morning. Sara Wallace closed her eyes and breathed in the scent of tomato leaves that foretold of summer to come.

"Hey Beautiful. You look like have a busy day planned."

She smiled up at Randy, her husband of many years. "I do indeed. I'm going to plant my first victims, I mean, tomatoes, of the season."

"What do you have there?"

"Ooh, Early Girl, Amana Orange, and, my personal favorite, Mortgage Lifter."

"Hmm. Did you pick them by name or . . ."

"Of course. And the fact they can be planted early. I'm not a complete sucker for clever names disguised as savvy marketing tactics."

Randy rolled his eyes. "News to me. Couldn't find an Alexander the Great or a Charles Darwin in the bunch?"

"Funny man. Where are you off to this morning?"

"Hardware store. I thought I'd repaint my office."

"Nice. What color?"

"White. Off-white. I'm still thinking about it."

"Really going out on a limb there, aren't you?"

"I need one room not blasting with color, my dear. I'm okay with you painting every room in the house in your own special Sara style, but my office is mine and mine alone."

"Fair enough." She tilted her head and gave him a rueful smile. "I must admit some of my ideas haven't worked out as well as I thought. Maybe sometime you could help . . . er . . . temper some of my more audacious choices?"

Randy wrapped his arms around Sara. "That's what I like about you, sweetheart. Not afraid to adjust when things don't go as planned. But I hesitate to ask; does this mean you're going neon?"

"I might need to go the other direction a teensy-weensy tiny bit. A touch more subtle."

"Well, it's taken nigh forty years for me to hear those words. A red-letter day, or should I say, a delicate shade of coral, day?"

"Ha! I earned that one."

"Not to worry. I'll be glad to lend a hand." He released her with a kiss. "Enjoy the Mortgage Girls."

"Correction: Mortgage Lifter and Early Girl. But you knew that, didn't you? Anyway, Josh is coming over around noon to help me dig an extension to the garden in exchange for half the tomato plants."

"Oh, good. Say hi for me." He kissed her again. "Later, sweetheart."

She was starting to dig the third hole when she saw it. It looked like a blob, a little ball of jelly hiding in the dirt. What is that? A bird's egg? A slew of slugs? Eww, ick. She did not like slugs. She dug down and lifted out the blob. Beautiful. Prismatic rainbows pulsed inside as she rotated the soft sphere in the light. She turned it over carefully, but as she did, the outer shell tore. She half expected a thick liquid to spill out, like an egg yolk, but to her surprise, a fine gold mist emerged. She caught her breath. "Oh, how lovely."

Frantic barking interrupted her. Shadow must have gotten out and found a squirrel. She stood up and followed the sound. "Shadow. Shadow. Come, boy. Where are you, sweet pup?" How did he get out, she wondered? He was whining now. He seemed to be in a panic. "Shadow? Where are you, boy?" She ran toward the noise, worried now. The barking went silent, and she found herself in a clearing surrounded by a thicket of trees. Where am I, she thought? Was she in a forest? The trees stood huge and dark. Pines and oaks and aspens were interlaced with tropical vines and bushes. In the middle of the clearing rested a stone bench. She peered through the mass of green trees. "Shadow. Shadow, where are you, boy?"

"Shadow is not here," a voice said.

Sara whirled around "Who's there?" No one answered. She was alone in this strange forest, with only a golden haze hanging in the air. She sat down. Well, the bench was real. She had half expected to fall through the air and land on her tailbone. I'm dreaming, she thought. She pinched her arm hard. "Ouch." Nothing changed, except now her arm stung and glowed with a deep yellow tinge around the pinch mark. "Shadow? Shadow?"

"I told you. Shadow is not here."

She scanned the forest. "Who is that? Where are you? Come out so I can see you."

"That will be difficult. I am here, next to you."

"Where?"

"You cannot see me."

The voice was soft and quiet, like a whisper on a breeze, but so clear she could understand every word. "Who are you?"

"Your question is irrelevant."

Sara closed her eyes. What is wrong with me? She didn't feel sick or dizzy, but something was clearly off. "Where is Shadow? Is he all right?"

"Shadow is in your house, where he always is at this time of day."

"But I heard him."

"To be accurate, you experienced a memory of your canine companion that appeared quite real."

She stood up. "What a relief. I want to go make sure."

"Wait, Sara. I want to speak with you."

She stopped. "Who are you? How do you know my name?"

"It is interesting how important this identifying information is to you. I did not expect that. You may call me . . . Dina."

"Dina. Then, you're a girl? A woman?"

"Irrelevant. Again. I should have expected you would continue to ask questions. After all, that is why we chose you."

"Chose me? Who is 'we?' Chose me for what?"

"Sit down, Sara. I will explain." She sat. "You cannot see me . . . or, us, to be more precise."

"Oh, so you're invisible. Of course. How the heck am I supposed to believe that?"

"Not invisible, but too small to see."

Sara shuddered. "Are you slugs? Am I dreaming? Are you slugs haunting me because I think you're yucky and gross?"

"You humans are curious beings. You can easily veer from the well-used road.'"

"What? Veer from the . . . Do you mean we. . . go off the beaten path?'"

"Yes, that is the phrase. We have observed humans ever since you appeared on Earth, but we still do not understand how you put words together to produce atypical meanings. It is quite mysterious."

"Blame Shakespeare. He started it."

"No, it began long before Mr. Shakespeare, but your attribution contains some veracity. We thought he was one of your better hopes."

"Better hopes for what?"

"For understanding."

"Understanding *what*?"

"That is why we are here, Sara. That is why we chose you. We have watched humans from their earliest development. We expected you would survive as all other species have. We assumed you needed time to acquire your survival skills, but we were wrong. You are too different. You do not act the way other life forms do. Humans are an example of a binary tipped knife."

"What?" Sara tilted her head. "Oh, I get it. A double-edged sword."

"As I said, some of your sayings make little sense."

"What do you mean? About the double-edged sword?"

"It means humans are the most extraordinary beings on the planet. And the most dangerous. Right now, we have assessed you to be far more dangerous than extraordinary. Which is why we are seeking your assistance."

"Assistance? From me?" Sara scoffed. "Why me? I'm nobody in this world. I'm not a leader or celebrity or preacher or CEO or social media influencer. I'm nothing. I'm a retired wife and Mom who never had much of a career to speak of, only a lot of different jobs."

"True. However, you possess a combination of characteristics we think are important."

"Such as?"

"Experience. You have worked across many industries and in many positions."

"That's a failure, not an accomplishment."

"That depends on the angle from which you choose to observe. From the human perspective of specialized expertise, indeed, you have failed. But if one values multiple experiences which allows consideration of problems from different viewpoints, then your personal history is widely prized."

"Hmm. I always just thought I was a failure."

"You humans. For creatures with such expansive minds, you close them so easily. However, you are not being entirely forthcoming, Sara. You willingly sought out new experiences, did you not?"

Sara stopped. "You're right, now that you mention it."

"Indeed. You ignored the message when others labelled you as a failure. Why?"

"Because I was . . ."

"Curious," both said at the same time.

"Yes. Your curiosity is what we noted. You possess a mind open to new ideas and a variety of experiences. Plus, your strong excrement indicator. That is why we chose you."

"W-w-w-wait a super sanctimonious second. What did you say? Excrement Indicator? What are you taking about?"

"That is the term you use, is it not? The mental measurement device that helps you remain skeptical of assertions of truth?"

Sara laughed out loud. "Oh, I'm with you now. My bullshit meter."

"*That* is the term. My apologies. I often confuse your maxims."

"How do you know about my bullshit meter? That's private. I don't share that with just anyone."

"I am not just anyone, Sara. I told you. We have been watching you."

"And my bullshit meter, too, obviously."

"Yes, that should be clear from our discussion. Your . . . bullshit meter is what kept you from pursuing a particular focus."

"How did you kn. . .? Who are you again?"

Dina ignored her question. "Your bullshit meter makes itself known and tells you when something does not seem truthful. It is why you are hesitant to commit to a specific path. Think about your past. Your response to certain religious teachings-"

"Wait a second. Not all of them. But some of the things the Church tries to pass off as true are perfectly ridic-"

"To certain fundamentals of economics-"

"Hold on. Even Keynes agreed with me. Technology does not remain fixed. Especially nowadays-"

"To the veracity of accounting."

"W-w-w-wait a miniscule minute. I have a point, there. Do you realize how many ways there are to account for inventory alone?"

"I put my argument to sleep."

Sara stopped and considered Dina's words. "Oh. I rest my case. Dang, Dina creature. You make me laugh, even if you do hoist me with my own petard."

"I do not under-"

"Never mind. Blame Shakespeare for that one."

"I see we have veered off course. Let us return to our discussion of your bullshit meter. The point is, you sense when ideas do not . . . shall I say, fit?"

"You're right. Though a fat lot of good it has done over the years," Sara agreed. "I can just feel when some 'supposed' solution is not really true. Then I can't accept it, no matter how hard I try."

"And then you search elsewhere for . . . the truth? And you continue to search, do you not? With even more zeal since you lost Maggy."

"How do you know about Maggy?" Sara asked sharply.

"You had a strong affection for your child. Her loss overwhelmed you for a long time."

"Yes. I still miss her. Every. Single. Day."

"Her loss has motivated you to search for a world that makes sense, but still, understanding eludes you."

"Yes. No. Well, maybe. I've learned a few things along the way, but I don't like to talk about them. People think I've gone loco. And at my age, they start throwing around terms like dementia and Alzheimer's when all I am doing is . . . trying to put ideas together in a way that makes sense."

"Will you tell me about your thoughts? I pledge not to mock them."

"Well, okay." Sara remained silent for a long moment. "I think everything should make sense, but it doesn't. At least not to me."

"I am not sure I understand."

"Well, everything we see around us: the stars, the Earth, all the things on Earth, they're connected. But somehow humans seem disconnected from it all, from the rest of the universe."

"That is a broad statement. Would you please explain?"

"Well, that's just it. I'm not sure. Before humans, everything sort of . . . flowed. Everything co-existed. I know that's a gross simplification. I know animals preyed on each other, and meteors smashed into the Earth and destroyed things. I'm not saying Earth was a perfect Eden, but it was a system where all the pieces worked in sync with each other. Then humans came along. Now things don't seem to flow. Somehow, we've changed the basic connection that existed before. Or maybe we broke it. I don't know. But nothing makes sense anymore. Humans don't make sense. Humans don't fit. And I don't know why. But deep down, I think it should all make sense; everything should fit."

"Sara, would you give me an example of how humans do not fit?"

"Sure. Here's one that always ties me up in knots. Animals kill things so they can eat. It's necessary for an animal to survive. That's a perfectly understandable behavior. But humans? We kill *each other*. And sometimes for the stupidest reasons. Because we want more land. Or because we believe in a different god. Or because our skin is a different color. It's like humans came along and slipped off the wheel of life everyone else had been riding on for billions of years. We don't seem to understand the basic 'rules of engagement' that every other creature has known for billions of years. And sometimes I think we are going to destroy ourselves if we keep going like this."

"Some animals do kill their own kind, Sara, but your point is generally true. Do you think it is possible for humans to exist on this Earth the way other creatures do?"

"Well, if it is, we haven't figure out how."

"Let me ask the question this way. Is it possible for people to build societies that operate like the natural ecosystem?"

"Hmm. That's an interesting thought. Are you saying our societies don't work like the natural world, but they could? Or should?"

"We believe this to be true."

"Hold the super cellular telephone. I need to wrap my head around that idea."

"That would be an impossible feat, Sara. Ideas are not physical in nature."

"Ha! That won't stop me from trying. But tell me, Dina, is that why you're here? To convince people to be more like animals in nature?"

"It has taken us many years to understand how your species operates. We now understand you have failed to learn the successful survival lessons of your predecessors. We have decided to step in before catastrophe occurs."

"What do you mean, 'catastrophe?'"

"This is not a human axiom, Sara. Catastrophe. Harm. Disaster. Ruin."

"Okay, Webster. I get your message. And you need my help?"

"Yes. We want you to learn how humans have veered away from the natural path and share this information with others."

"What? I told you I'm nobody here. No one's going to listen to me."

"No one who is living is a nobody, Sara. Your statement is nonsensical."

"Still. Dina, there are thousands, millions of experts who strive every day to learn about the mysteries of Earth. Scientists, doctors, engineers, poets-"

"True, and your experts are necessary to find the solutions you need to survive. But those who are adept at understanding the details, often miss the woodlands for the pines and aspens."

"What? Oh, the forest for the trees. Gotcha."

"People understand a great deal about how the universe works, and they are learning more every day. However, as people grow in their specialties, the gaps between them widen. Opportunities for uniting disparate findings grow fewer. We fear people are now so specialized they have lost the ability to understand how all the elements of life work together as a whole."

"Aha. So, there is such a thing."

"Yes."

"But we don't have this knowledge now?"

"Your situation is ironic. The knowledge is all around you. It exists in every person and every object people have produced on this Earth. It lies in pieces everywhere. All you need to do is put the pieces together, but you have not done so."

"And you're saying we can't put the pieces together without your help?"

"We have reluctantly arrived at this conclusion. We thought you might achieve this knowledge by yourselves. It has not happened."

Sara looked at the mosaic of trees crowding around her. That's what's odd, she thought. The trees. They don't belong together.

Oaks and aspens and pines don't grow in the same area. And the tropical vines? No way. What is this place?

"We are here to explain, Sara. We hope you will understand our message and share your knowledge with others before time runs out."

"Time runs out?"

"Yes, or more precisely, before you humans kill yourselves. And take many other creatures with you."

Sara caught her breath. "Are we really in trouble? Is our situation that dire? Is it about what we are doing to the climate?"

"That is of immediate concern, but our alarm goes far beyond the climate. We are troubled by how you operate in general. People move quickly and unpredictably. You are following a path toward destruction and do not seem to be able to stop yourselves."

"So, you are saying that you can help us change our self-destructive ways?"

"We debated this for a long time, but yes, we believe we must act. Which is why we seek your assistance. We need someone who will learn, understand, and communicate this knowledge to others."

Sara sighed. She understood what was happening. She was dreaming. Shadow's barking. The forest. The golden glow. She shouldn't have thrown so much turmeric on her eggs this morning. The spice made her woozy, and now she was in full-fledged hallucination mode.

"It might be dangerous, Sara."

"What? Learning? Ha. Learning is always dangerous. That's why it's fun."

"No. Seriously. Dangerous. To your body. To your brain."

"How so?"

"We have never interfered so directly, and we are not sure of the outcome of our efforts. We believe we must try. You may choose to help or not. We will honor your choice, but you must make the decision. After all, you are human; this is what humans do."

She looked down. Her arm glowed. No, not just her arm. Her whole body seemed to be pulsating with a golden turmeric glow. "I . . . I don't . . . I don't even know how to think about this."

"Yes, you do. Ponder. Is that the right word? Think. Take time to think. If you decide to assist us, we will find you."

At that moment, she heard a strange noise, like a sharp gust of wind at the top of the trees. "Dina, what is that noise?"

"Mom! Mom! Are you all right?" Josh ran across the yard. "Dad, come quick; something's happened to Mom." She lay prone in the grass, the tomato plants neatly stowed in their planters, untouched. He bent down and turned Sara over. Her skin was jaundiced, her breathing shallow. Josh felt the pulse at the base of her neck. Rapid and light, but still beating.

Randy hurried up behind him. "What's happened?"

"I don't know. Call 911."

Chapter 2

ENERGY MATTERS

Dr. Victoria Chen stepped into the hospital waiting room. "Mr. Wallace?"

Randy hurried over. "How is she? What's wrong with her?"

The doctor removed her mask. "Well, we don't know. She's stable. She shows no signs of stroke or heart attack, or any type of catastrophic event for that matter. She's in a coma right now. Can you tell me what happened?"

He shook his head. "I don't know. She went out to plant tomatoes around ten this morning. I left to run a few errands. I got back at 11:30 or so. I saw Josh in the back yard."

"She was just lying there," Josh said. "On her side with her knees still tucked under her. It looked like she collapsed in the middle of planting. She hadn't even put the plants in the holes yet. I turned her over. Her skin was yellowish. Dad came, and we called 911."

"Well," Dr. Chen sighed. "I've ordered a full blood panel. If that doesn't tell us anything, I'll order an MRI to scan her brain. You may go in to see her, now. We'll keep you posted on any developments."

Father and son walked into the hospital room. Sara lay quietly. No tubes. No ventilator. No . . . consciousness. "Hey sweetheart," Randy whispered. No response. Not a twitch.

Sara was floating. Aware, but not aware. Awake, but not awake. Thinking, thinking, thinking.

She had struggled, like every human being, to figure out what she was doing on this Earth. As a child, she thought she knew the answers. Her folks sent her to Church, and she learned all the stories about Heaven and Hell, God, and Satan. Heaven, she envisioned, was knowledge. Perfect knowledge. She pictured herself sitting back on a fluffy white cloud while God rolled the tape of the history of the universe. She would be able to soak it all in and learn exactly how it all happened. Maybe eat some popcorn too. Was there popcorn in heaven?

Somewhere along the line, as the joys and tragedies of her life strung out like pearls on a string, she lost sight of God. Now, she knew better. She had absolutely no answers. This world didn't make any sense.

Could this Dina creature, this figment of her dreams, be right? Was the world really a sensible place whose fundamental elements were just hidden from humans? Was it possible to construct the pieces of this puzzle into a coherent whole? That would be some kind of amazing. Maybe as good as the version of heaven from her childhood? Hmm.

Thoughts of Dina drifted through her mind. Dina was a dream. A soft voice without a body. A gold glow in the air. With ideas she had never considered before. Maybe she was unconsciously trying to work out her frustrations over the state of this mixed-up world. What if she gave in?

She had finally, painstakingly, learned that the secret to her own happiness required her to accept things as they were. Should she try the same tactic with this dream? Give in. Go wherever it took her. Listen to the voice. Let her dream-laden brain make the connections she could not make while she was awake. Why not?

"You have made a decision." Dina's voice sounded soft. Almost whispery. Gold mist suddenly surrounded Sara.

"I couldn't help it. You probably knew that."

"Your curiosity is why we picked you."

"So, where do we begin?"

"To answer your question, I need to make the basement first."

"Make what?" Sara paused. "Do you mean. . . 'lay the groundwork?'"

"That is an equivalent term. I must explain the context so you will understand how humans came to be. If I do not start at the beginning, the story will not make sense and you will not fully understand."

"Okay. What do I need to do?"

"Listen and do what you do best; ask questions."

Sara felt herself smile. "Do you have a cloud I could sit on?"

"No." Dina paused. "You humans. You will always be a mystery to me."

"Ha! To me too. I guess I'm ready, though a cloud would have been a nice touch."

"To understand how all things work, we must start at the very onset of time. In the beginning there was . . . energy. The universe was, and still is, composed entirely of energy. As time passed, energy transformed into innumerable forms. But it is, and always will be, energy. Time and energy are the basic drivers of change in our universe.

"Energy does only one thing. It moves. Energy moves in whatever way it can. At the beginning of time, energy within a very small space expanded to form the universe."

"The Big Bang!"

"Indeed. Energy in its purest form moved outward. At first, the temperatures were so hot only pure energy existed. As the universe expanded, the speed of the movement began to slow. Within one ten billionth of a second, some of the energy changed into particles containing the forces of our universe; what your scientists call the strong interaction, the weak interaction, electromagnetism, and gravity. These forces still exist today.

"Within one ten millionth of a second, the universe cooled more, and energy collided with the force particles and converted them into tiny particles scientists call 'subatomic.' Each of these particles was still energy, but now it was in the form of matter, meaning it possessed tangible physical properties."

"W-w-wait a sacrilegious second. You mean to tell me forces are stuffed inside matter?"

"Energy exists in many forms, Sara. Some energy particles escaped too far and too quickly during the Big Bang, and still float across the universe. And some energy particles are held captive inside matter."

"Dina, can I ask a silly question?"

"No questions are silly, Sara. Go ahead."

"So, if some force particles are caught up in an atom, and force particles are made up of energy that is always moving, does that mean an atom is always moving?"

"Your question is quite astute. What do you think?"

"Well, it would have to. Energy moves. Energy is in the atom. Therefore, the atom must move."

"You are correct. The particles inside an atom do move, but the movement varies. The electromagnetic force inside an electron moves quite freely around the nucleus, but the forces trapped inside the nucleus cannot move nearly as easily. But that slight ability of the nucleus to move gives each element its particular characteristics."

"No way!"

"Yes, way. Let us stop for a moment and consider this early universe at the age of one ten millionth of a second. The energy emitted at high speed had slowed and transformed into at least two general types of particles: force particles that take up no space but have many ways to move, and matter particles that exist in space but have fewer ways to move. These two groups represent only six percent of the known universe."

"Six percent? Hmm. What's the rest?"

"Can you determine the answer to your question, Sara?"

"Hmm. Well, it's energy. I can figure that much out. Maybe it's energy we can't detect. Or we don't understand. Or can't measure. Honestly, Dina, I don't know."

"You are quite close, Sara. Scientists call the unknown ninety-four percent of the universe dark energy and dark matter."

"There's both?"

"Indeed. Ninety-four percent of all energy in the universe is not yet accessible to humans. Sixty-nine percent of all the substance in the universe is dark energy, and twenty-five percent is what scientists call dark matter. But it is all in one's perspective. Your scientists call it dark because they do not understand what it is or how it works."

"Well, we're six percenters. Of course, we don't understand it."

"Indeed. How could humans, as children of force and matter combined into atoms, possibly understand what exists outside their own physical environment? Dark energy and dark matter exist outside current human understanding. Scientists are studying these phenomena right now, and they will learn more about the dark universe in time. Because these substances are part of the universe, they are subject to the same forces as we are, so they will succeed in this endeavor. Of course, they may need to expand their understanding of physics."

"What do you mean?"

"Think, Sara. You are in the six percent group. Everything you know is based on what exists in the six percent of the universe you understand, including the laws of physics. It stands to reason there are other laws of physics humans do not know."

"My head is exploding," Sara said.

"I will leave, then, Sara. I cannot afford to endanger you."

She laughed. "No, it's an expression. It means . . . you've given me a lot to think about."

"Remember this is only background information, Sara. We have many more ideas to discuss."

"Hoo boy. Are you sure I'm the right person for this job?"

"Yes. Let us summarize. Everything existing in the universe is made of energy that has been converted into many different forms. The more time passes, the more types of forms energy can create."

"And we are only a measly six percent of the whole picture."

"Indeed. For the next approximately three minutes and forty-five seconds in the life of the universe, the force and subatomic matter particles collided at high speeds. During this time, the universe expanded and cooled rapidly. After the 3 minute, 45 second point, some of the larger subatomic particles that slammed into each other did not possess enough energy to break apart. That is when the first nuclei appeared. A nucleus is a combination of a proton and a neutron. Protons and neutrons are combinations of matter and force particles stable enough to stay together.

"Remember this concept, Sara. It is fundamental to your understanding as we go forward. Energy moves in whatever direction it can and will create any possible form it can. Energy does not stop. Not all the forms energy creates are stable. Only those that fit the conditions in which they find themselves will remain. Do you understand?" Dina stopped. "Is your brain blasting again? Should I stop?"

"My brain . . . Oh, no. But I do like the term 'brain blasting' a lot more than 'head exploding.' Dina, you do have a way with words."

"Let us continue. For the next 380,000 years or so, our six percent of the universe was a swirling combination of nuclei, electrons, and force particles. If you were able to see the universe at this time, it would look like a hazy bluish-orange orb with a temperature of about one billion degrees Kelvin. As the universe continued to expand and cool, these particles began to adhere to one another. This is how the first element, hydrogen, formed. The universe never looked the same after hydrogen emerged. The opaque haze that filled the universe disappeared, and the electromagnetic particles, called photons, trapped inside the haze, began to flow across the universe.

"Light," Sara said. "Photons. Light. And they moved fast. At the speed of light. Ha!"

"They are light particles, Sara, so of course they moved at the speed of light."

“Dina, I remember reading the matter in the universe is made up mostly of hydrogen. Is that true?"

"Yes. The universe continued to cool, but it was still hot enough for some of the hydrogen atoms to move fast enough to bump into other hydrogen atoms. These combinations created helium. But the universe cooled too quickly for that process to go on for too long. Only 6% of the matter in the universe is made up of helium atoms, and 92% is hydrogen."

"That's only 98%. What is the other two percent?"

"All other elements.”

“What?!? You're kidding. All the other elements: oxygen, gold, uranium, einsteinium. . . They are only 2% of all the elements in the universe?”

“They are 2% of your 6%, Sara.”

“Damn,” Sara exclaimed. I don't think I really understand how big the universe is.”

“No, you do not. Right now, it is more important you understand the concept I am trying to impart to you. The hydrogen atom, indeed, all atoms, are just energy stored in an efficient manner. The amount of energy the universe started with is the same amount that exists today. From the earliest moments of the universe, energy has merely changed forms. Matter is one of those forms. In fact, matter is an extremely efficient way to store vast quantities of energy. Albert Einstein was a human who made this important discovery."

"Oh! E=mc2. I remember that equation from high school physics."

"Indeed. This classic equation shows how mass and energy are related. A massive amount of energy is stored in every single atom.

In fact, a kilogram of matter contains about 25 billion kilowatt hours of energy."

"25 billion? Hmm. That's a lot of energy. My electric bill last month showed I used 100 kilowatt hours." Sara stopped. "Wow. Billions of galaxies are filled with matter, and each bit of matter stores a shitload of energy. It's kind of incredible when you think about it."

"What is this? Shitload? Excrement cargo? I do not understand."

"Don't tell me you don't understand what 'shit' is?

"I have heard this word, but in too many contexts for it to be meaningful to me. For example, the excrement indicator indicates one meaning as a measure of veracity. Excrement cargo is new to me."

"Ha! It means a lot of something."

"It is not related to excrement?"

"No. It means a lot, a large quantity."

"Then I would agree with you. The universe contains an energy . . . shitload."

Sara giggled. "I'm not sure what kind of slug creature you are, Dina, but you make me laugh."

"You humans are curious creatur-." Dina stopped. "I must leave, Sara. Immediately. I will return when I can."

"But where are you goin-"

There was a sudden pull on Sara. It felt as if her body was being slowly hoisted into the air. What's happening, she wondered frantically?

"All right, Mrs. Wallace. We're underway." The MRI technician checked the images as the scanner rotated and the radio waves pulsed across the huge magnetic ring with Sara's head positioned inside. "Let's see what's going on inside that brain of yours, Mrs. Mystery Coma Woman."

Floating. She was floating again. All hazy and dreamy and hovering over . . . what? It was the universe, she kept seeing. A hazy blue cloud spreading out across her vision like an impressionist painting. A golden orb with rainbow prisms floating inside the haze.

The soft voice humming in her head. Where was she? What was happening? Suddenly, the floating sensation stopped.

"All done, Mrs. Wallace." The technician wheeled the gurney away from the MRI machine. "Perfect timing. Your team is here to take you back to your room." Two nurses lifted Sara off the gurney and onto the bed. As one of them placed a blanket over Sara, she noticed something strange. She peered closely at Sara's right foot. There, between the toes. A deep yellow area. Almost, gold. She touched the spot gently. The glow faded and spread up Sara's leg before disappearing. The nurse examined Sara's other foot. It was clear. No discoloration showed anywhere. She turned to the right foot once more. She sighed. "That's weird. I must have been seeing things." The foot looked perfectly normal.

Chapter 3

CHANGE = MC^2

"How are you feeling, Sara?"

"Dina? Is that you?" The gold mist appeared. "Why did you leave so suddenly? What happened?"

"I am here now. Do you feel well enough to continue?"

"It's funny because I don't feel . . . anything. I can't explain it." She looked around. "Oh, we're in the forest again. I'm sitting on the bench. Wait, how did-'"'

"It is not important, Sara. Your situation will become clear soon. Now, let us continue."

"Okay. We left off when the universe started making hydrogen."

"Yes. The universe expanded and cooled with the passing of time. At the age of one billion years, hydrogen was abundant, but it was not the only amusement in the city."

"Wait. Amusement in the city? You mean, the only game in town?"

"Yes, Sara. Going forward, please do not correct my idiomatic attempts. Your communication mechanism is difficult, and I do my superlative."

"Bes-" She bit her tongue. "Okay."

"Thank you."

"So, you were saying hydrogen wasn't the only knife in the drawer."

"Sara-"

"I'm teasing you, Dina. Hydrogen wasn't the only. . . amusement in the city."

"Correct. All energy forms, including matter, interact. Do you recall that forces are contained within an atom?"

"Yes. And they can move around inside the atom, too."

"Indeed. Within the atom, the strong and weak interaction forces only work over short distances, which is why they can hold a hydrogen atom together. But gravity, is a weak force. Weak as it is, it can work at any distance."

"Why is that important?"

"The forces inside a hydrogen atom have no effect on other hydrogen atoms that are far away, but gravity does. As the universe expanded, gravity pulled hydrogen atoms towards each other across vast distances. Gravity shaped the early universe, Sara.

"When gravity draws atoms very close together, the atoms' boundaries are compressed. The strong and weak interaction forces inside the nucleus then push against gravity, and an impasse ensues. When so many hydrogen atoms are drawn together, gravity can overwhelm the interaction forces and strip the electrons from their nuclei. Since electrons are bound to the nucleus by the weak force, that force is stripped away too, releasing photon particles, or light, in the process. The light escapes into space, and the now electron-less hydrogen nuclei fuse with each other to form helium. Do you know what I am describing, Sara?"

"Hmm. A hydrogen party?"

"Please be serious, Sara."

"Okay, okay, hydrogen is squeezed together to produce helium and a bunch of energy in the form of light." She stopped. "I'm a doofus. It's a star."

"Correct. Stars are made by the interaction of hydrogen and gravity. In a star, the hydrogen fuses into helium until all the hydrogen runs out. Then gravity proceeds to press in on the helium atoms until they give up their electrons just like the hydrogen atoms before them. Heavier atoms like carbon and oxygen are formed this way."

"Oh. That's how the other elements came about."

"Partly. In large stars, this process continues and creates heavier elements until they produce iron. Iron is a heavy atom. Unlike all the previous elements that give off energy when they fuse, iron absorbs energy when it fuses. So, when iron forms in a star, no energy is emitted outward, and the impasse is broken."

"What happens?"

"Gravity wins the battle. With nothing pushing against it, gravity squeezes the star until the atoms explode in a giant detonation. Humans call this a supernova. The power of the supernova explosion spreads the matter outward and causes the elements to collide and create larger elements."

"Wow! Did all the elements form this way?"

"Indeed. Not just elements, but some simple molecules, like water and ammonia and methane."

"I had no idea. So, every piece of matter larger than an iron atom is made up of blown-up stars."

"Yes. Some remains of an exploding star get blown so far away they drift out into space. Gravity still pulls on them, and, depending on how far they float from the original star, different things can occur. Some may form clusters and move whatever way gravity pulls them. These objects become comets or asteroids. The matter remaining closer to the explosion may coalesce. Lighter elements may form into a new star. The heavier ones may form clusters that orbit the new star. Does this sound familiar, Sara?"

"It does! It's a solar system. Matter floating around a star. Planets. Tell me, Dina, are there a lot of solar systems in the universe?"

"More than you can count."

"Is there life on them too?"

"Always to the core of the issue, Sara. I do appreciate your directness. What do you think?"

"There would have to be," she said. "It stands to reason. If all matter is the result of energy moving in certain conditions, then as long as the conditions are similar in any corner of the universe, the same general outcome would occur. Right?"

"Interesting. Go on."

"Well, if the same conditions that created our solar system and our Earth existed all over the universe, many solar systems would contain Earth-like planets. If life could form in our solar system, then it could form any place with similar conditions."

"You are correct. You are connecting these ideas much more quickly than I expected. When conditions similar to life's formation exist elsewhere, similar outcomes will also result. This is a concept known as convergent evolution. I would add one thing. In many places in the universe where different conditions exist, life may emerge as well, but in a form unfamiliar to you."

"Oh. I never thought of that."

"Indeed. Before we delve into the solar system in more detail, I want to impart some important concepts about energy."

"Okay. I'm all ears."

"This is not a true statement, Sara."

She giggled. "Sorry. I'm paying attention."

"We discussed how, with time, energy has changed into many different forms. Correct?"

"Yes."

"So, can we equate energy and change?"

"I'm not sure. What do you mean?"

"Try to reason it out, Sara."

"Okay." She took a deep breath. "Here we go. The universe is composed of energy. What is energy? From what you've described, energy in its most basic sense, is movement. When energy moves it can collide with other energy particles and produce new forms. It's all still energy, but they're different types of energy. So, yes, I suppose we can think of energy as change. Energy is the cause of change, and change is the result of the movement of energy. So, if we say the universe is made of energy, we must also say the universe is made of change."

"Yes. This is a fundamental truth of the universe. Now, let us go a little further. In the universe, change generally moves from simple to complex. Sooner or later, stable energy forms will collide with other stable energy forms. If the amount of energy it takes to join them remains in the environment, there is a chance those combinations will fly apart. But in a cooling universe, combined energy forms can remain fixed in their conjoined structures."

"Oh. So, it looks like energy moves from simple to complex, but really, it's just the energy in the area has spread too far to break the combined form apart."

"Yes. Complex forms resulting from a supernova, for example, will remain because the energy from the supernova dissipates, leaving the heavy elements unable to revert to their pre-supernova structures."

"Oh, I get it. So, when we see gold or silver, or any element on Earth heavier than iron, what we are seeing is the piece of an exploding star that cooled fast."

"Indeed. Given this fact, Sara, can you determine if change is random?"

"Yes. No. Hmm. I don't know."

'I believe you have stopped on all your recreational landing pads, Sara. Please explain any one of your positions."

"Ha! Covered my bases. Well, at first glance, it seems random. In the early universe energy bits just smacked into other energy bits in whatever way they could."

"Let me ask the question this way. Do you think the universe could produce a complex living unit, like a dinosaur in its first million years?"

"Ooh, that would be so cool, dinosaurs flying around in space breathing fire on everything."

"Sara, dinosaurs do not breathe fire. Furthermore, there is no fire in space."

"Still, it would be cool."

"I disagree. Fire is quite warm. Now, please answer my question."

Sara was quiet for several moments. "Well, energy moves in whatever way its surroundings allow, and energy bits can crash into each other. The combined units that were created, like hydrogen made from subatomic particles, can only remain as a unit if the environment allows it. In this case, hydrogen stuck around because the energy required to pull it apart had spread too far apart. Then hydrogen atoms floated around and coalesced into stars. The environment changed again, and when the stars exploded, they produced enough energy to create larger atoms and molecules."

"Do we have a dinosaur, yet Sara?"

"Hold your horses, Dina. I've only just gotten to heavy elements."

"My point exactly. You are already billions of years into the life of the universe, and there is no sun, no Earth, no life, no dinosaur anywhere."

"Ouch. I was still thinking."

"In this litany of events you just recited, can you say change is random, Sara?"

"No, because we have dinosaurs."

"Your answer is true, but nonsensical."

"I know. I'm buying time while I think."

"Let me provide a clue. Remember that energy tries every possible way to produce any and every form possible. Do you agree?"

"I do."

"So why did it not produce a dinosaur?"

"But . . . it did! Only not right away . . ." Sara stopped. "Wait a miraculous muddled minute. Let me think about this question from the other direction." She pondered for several moments. "I think I see now. As random as the efforts of early energy particles appeared to be, they could only make what they could make. They were limited by their own characteristics. Energy could only make forces. Forces could only make subatomic particles. Subatomic particles could only make hydrogen. As powerful as they were, they couldn't possibly make a dinosaur right away, although that would have been exceedingly cool. Even with all their energy, they wouldn't be able to manage that feat. Each change must go in order, like a stairstep. You can get to the second floor, but only if you go one step at a time. You can't leap up from the first floor to the second floor unless you're Superman. Energy can only create a dinosaur if it goes step by step and makes all the other things it needs first, like large atoms and complex molecules and planets and cells and bodies. Hmm. Which means all forms only develop incrementally."

"That is the concept I wanted you to understand, Sara. The reality is far more complicated than your description. Each of those 'stairsteps' includes many events and interactions. Only a few of the countless forms created are ever stable enough to remain to become the basis for further changes."

"Hmm. Dina, are you saying small, step-by-step changes occurring over billions of years is what finally created dinosaurs . . . and humans?"

"Yes. Let me provide another perspective on this phenomenon. If we start with life forms that we know exist today, like humans or bees, you can look backward and trace the changes that produced them all the way back to the beginning of life. When you see history laid out like this, it makes perfect sense that humans or bees formed. If you turn the direction around, the sense of perfection disappears."

"What do you mean?"

"If we started at the beginning of time and attempted to look forward, it would be impossible to predict bees or humans would ever emerge, much less the Earth and the Solar System and-"

"You're right," Sara said in wonder. "How could we know? Too many variables occur along the way. Too many possibilities. Too many things change all the time. Hmm. Conditions had to be just right to create the Solar System, and this Earth, and life as we know it-"

"No, Sara, you are looking at the idea backwards, as humans tend to do. The conditions are always right. Conditions rule over all forms of life. Conditions determine what energy units remain. Humans exist because they fit the conditions of the planet on which they live, and because every single one of their ancestors fit into their respective environments, going all the way back to the emergence of life."

"Oh," she exclaimed. "Then we aren't special. We only exist because we fit the conditions around us. Like every other living thing on the planet."

"That does not make you 'special' or 'non-special.' Sara. Humans exist because the corner of this universe you call Earth, maintained a series of conditions, changeable though they were and still are, for a long enough time to produce creatures like you. And me, I might add. We are all shaped to match the conditions of the Earth, not the other way around."

"But we are so complicated. It's distressing to think of us as a product of random changes. This is the basic argument of most religions, you know; humans are too intricate and complicated to be randomly generated."

"Not randomly, Sara. Step by logical step. Each of your ancestors, starting with the simplest ones from billions of years ago, managed to survive in the Earth's environment in which they found themselves. Do you understand how long a billion years is, let alone 13.8 billion years, which is the age of the universe according to your human scientists? It is difficult for humans to grasp this important idea. Think about one second of time." Dina paused for a moment, which Sara guessed was exactly one second. "One second passes by quickly. A billion seconds? That takes 31.7 years. Think about a billion *years*. It is very difficult to comprehend. Change may be slow, but with enough time, energy can produce very complex forms, even humans."

"I never thought about it that way. I guess when we don't notice change happening, we think no change is happening at all."

"That is a good way to express your perspective, Sara, however erroneous it is. Now, there is one last thing about change I want you to understand."

"Oh? Only one more? Good. Because my brain is filling up and it's going to top out soon."

"That is not possible, Sara. Brains do not work this way."

"She laughed. You're probably right."

"Yes, I am."

"Okay, what's this last thing?"

"Change does not always move at the same pace. Shortly after our universe came into being, energy flowed in every direction. As the universe cooled, it began to behave differently. Matter attracted other matter and formed galaxies and stars with abundant energy. Other parts of the universe became free of matter, what we call space. Over time, stars exhibited a certain pattern of behavior. They would spend a lot of time burning hydrogen in a relatively steady manner. When all the fuel was consumed, the star would explode and create heavy elements and molecules. This pattern of short bursts of energy followed by long periods of relative stability is important and occurs repeatedly in the universe. It is during the long periods of stability that the environment determines which forms remain."

She stopped for a long moment. "I think my head is exploding again."

"Sara, are you feeling all right?"

"Yes. No. My head hurts a little."

"I will leave you then. It is important you do not feel pain when we talk."

"It's because you give me so much to think about. I'm getting overwhelmed."

"Then I shall go."

"Where do you go? When you leave? Do you hide in the trees?"

"It is too early to discuss this, Sara, but I am not far. We shall speak again soon. Now, rest."

"Dina?"

"Yes?"

"I want to hug you. How do I hug you when I can't see you?"

"Your request is strange and yet predictably human." Dina paused. "I deem to have received your embrace. Thank you, Sara."

"I like that." Sara smiled as the gold mist disappeared. "I deem to have received your embrace," she repeated as she closed her eyes.

"What's wrong with her, Doctor?" Randy asked, cradling Sara's hand in his own. "Why won't she wake up?"

Dr. Chen shook her head. "We are still unsure, Mr. Wallace. Her blood panel was normal. She has some elevated proteins and bacteria, but nothing so far outside the normal ranges that would result in a comatose state. Her MRI showed no trauma anywhere. Blood flow is strong. It's odd. The EEG displayed normal brain activity, which is unusual. Typically, in comatose patents, brain activity is low. She scores only a 3 on the Glasgow Coma scale."

"What does that mean?"

"She is unresponsive to stimulus. She is in a deep coma." Dr. Chen sighed. "Yet her brain activity is still normal. Nothing adds up. I have ordered a PET scan to find out more."

"What is that, and what will it do?"

"It's a Positron Emission Tomography scan. It will give me a more detailed view of her brain activity. Something is going on, but I can't pinpoint it, yet." She squeezed Sara's hand. "Good night, Mrs. Wallace. We will solve the mystery of your condition."

"Thank you, Dr. Chen," Randy said as she left the room. He turned once more to Sara. "Wake up, Sara. Please, sweetheart. After everything we've been through, you can't leave yet. We have so much more to do in this world." He leaned over and kissed her gently. He did not see the golden glow gather under the skin of her cheeks.

Chapter 4

THE GREAT AND NATURAL OZ

"Hey, Mom." Josh leaned over and kissed his mother's cheek. "I got all those tomato plants in the ground. I added a few green peppers and some lettuce. You better wake up soon because they're growing fast in this warm weather."

His mother lay still in the bed. She looked peaceful. What could be wrong with her? She had always been healthy. Except after Maggy died. That had nearly destroyed her. But his sister had been gone five years now, and Mom seemed to have made it through the worst. They still cried on June 5th, Maggy's birthday. They texted each other during the day, ate dinner at her favorite restaurant, and polished her memory into bittersweet lore. Could stress, long suppressed, finally emerge and cause something like this? It seemed so unlikely. The doctors could not explain this . . . coma. He hated to use the word. It felt hopeless. "Come on, Mom. Wake up. Nina and I have some good news for you."

Sara looked around. She was in the forest again. She was getting the hang of this dream now. If she's in the fake forest filled with trees that don't belong together, then Dina must be close. Plus, the weird gold mist just descended from nowhere. "Dina? Are you there?"

"Hello, Sara. How are you feeling?"

She sighed. "I guess this is my new normal. Floating in Nothing Land. Why do we always come to this forest?"

"Do you not like it?"

"Oh, yes, I do. It reminds me of something, but I can't quite recall what. It's strange, that's all. It's not real, but it's familiar."

"Would you like to go somewhere else?"

"How is that possible? I don't seem to be able to go anywhere on my own. In fact, I'm not sure what's happening with me; where I am, what I'm doing. It feels like I'm in an extended dream."

"Think of another place, Sara. One significant to you."

She took a deep breath. The breeze was cool, but the sun warmed her. She opened her eyes. She was on a cliff, sitting on a bench. Another bench. But below her the view could not have been more different. A long sweep of sand and ocean curved around an outcrop covered with pines. The sun glittered on the waves, and the grasses on the dune below her swayed in the breeze. "I know this place, but I've never been here."

"It is named Ecola Beach in-"

"In Oregon. Northern Oregon. I remember now. My daughter and I found this picture years ago. We looked at it for a long time. We vowed we would visit one day." Sara paused. "We never had the chance. She died . . . at nineteen."

"Yes. She encountered fatal damage while in her automobile."

"She was hit by a drunk, driving a sports car the wrong way on a narrow road. Such a waste. The creep walked away without a scratch," Sara said bitterly.

"Would you like to return to the forest?"

She gazed out at the waves rolling against the shore, just a whisper from her perch here on the cliff. It was so lovely. It reminded her of Maggy when she was young and full of life and wanted to try everything. "No, I'd like to stay here for a while."

"All right. Are you ready to proceed?"

"I am." She gathered a deep breath. "The last time we spoke, we talked about dinosaurs emerging from the Big Bang, all fire breathing and huge in outer space."

"No, Sara, we did not. I am now quite concerned you are not grasping the concepts I am imparting to you."

"Don't worry, Dina, I'm teasing you. We talked about energy and change and the step-by-step process it takes to make dinosaurs."

"I am relieved you remember our conversation. Before we talk about Earth, I must discuss one more important concept. I want to talk about 'conditions.'"

Sara thought for a moment. "We talked about conditions. Conditions are the great Oz of the universe."

"Oz? I do not comprehend this reference."

"Oz as in the Wizard of Oz. The great and all-powerful Oz. Conditions are like that. All powerful."

"Perhaps, but Oz in that story was a fraud. Not all-powerful at all. Conditions are the opposite. By their very existence, they exert control over all energy forms within their confines. Let us call conditions by the common human term, 'environment.' Humans use this term in many contexts, so I will define it for our discussion. An environment is any area in which some of the energy inside remains stable long enough to change form. We can also describe an environment as a collection of energy existing in a defined location."

"W-w-w wait a melting macaroni minute. Let me make sure I understand. Are you saying the universe became an environment after the Big Bang?"

"Yes, Sara. At first it was hot and full of energy particles moving at high speeds. As time went on, the universe cooled and the forms inside changed to fit within those changed conditions. We can think of the 'stability' of the environment itself by considering the relative permanence of the forms inside. As the universe expanded, it formed sub-areas containing different amounts of matter and energy. These areas became environments of their own. The smaller environments each affected the forms inside in a different way."

"I'm not sure I understand. Would you give me an example?"

"How about billions of examples? Every galaxy is a collection of matter and energy distributed into stars and gases. The energy forms that exist in stars are far different from the energy forms that exist in relatively empty space. And the ability for the energy forms to change is different in each of these areas."

"Right. Stars can hatch new elements. Empty space can't. Hmm. So, the universe is a huge environment that holds a lot of smaller environments. Can we say environments are layered, like nesting dolls?"

"I am not sure of the relationship to brooding figurines, but yes, the universe is composed of successively smaller divisions. From open space to galaxies, to solar systems, to planets, to oceans, to forests or deserts, down to a drop of water, all of these can be considered environments. Each environment, no matter the layer, can be viewed as a unit. The energy present in each environment is unique, and though relatively stable, is always changing."

"That makes so much sense."

"We took a rather circular method of approaching the idea that the Earth itself is an environmental unit, but I believe we have arrived. We can think of the Earth as a separate environment."

"Hmm. From what you said, we fall smack in the middle of the nesting doll set."

"Yes. There are many environments beyond the Earth and many within. The Earth too, is constantly changing. As in the wider universe, the movement of energy irrevocably alters every environment."

"What do you mean?"

"Let me tell you a brief story. Early in Earth's existence, a planet called Theia crashed into Earth, tilting it onto its current axis. The crash also pulverized part of the Earth into pieces. Do you know what happened to those pieces, Sara?

"No. I didn't know anything about this Theia or that Earth was never more than a hot blob that slowly cooled down. What happened?"

"Two important things. The pieces flying farthest away in the explosion formed the moon."

"The Moon? You mean, the moon is made up of parts of Earth?"

"Parts of the early Earth and parts of Theia. Because pieces of the crust were very light, they flew far during the explosion. Before gravity could pull them back to the planet, they coalesced into the moon and began to rotate around the Earth instead. When it first formed, the moon hovered only 25,000 miles away from Earth."

"Twenty-five thousand? But it's a lot further away now, isn't it?"

"Correct Sara. The moon today is nearly 240,000 miles away from Earth, almost ten times the distance from which it started. When the moon first formed, its strong gravitational pull caused Earth to spin rapidly. In fact, the Earth-day lasted only four hours."

"Wow, that would mean, I would be like, over 300 years old in early Earth years."

"I do not think your calculation is meaningful."

She giggled. "Hmm, I wonder how old Shadow would be. Dog years times early Earth years times-"

"Veering, Sara."

"Oops."

"Since the crash with Theia, Earth and the moon have steadily moved away from each other. The moon's gravity pulls at the Earth's oceans and slows down its spin. The slower the Earth's spin, the less pull it exerts on the moon, and the further away the moon moves. Today, the moon travels away from Earth at a rate of 1.5 inches per year."

"I had no idea. So, Theia blasts earth, the Earth tilts and suddenly has a big fat moon as its neighbor. What was Earth like after the crash?"

"Well, exceedingly hot for one thing. The collision heated Earth and caused pieces of Theia to melt into Earth's core and mantle. It is this abundance of heavy metals inside that gives Earth its strong magnetic field."

"I read about the magnetic field. It's what causes the beautiful auroras at the poles."

"Yes, but it plays a more critical role. The magnetic field acts as a deflection mechanism for objects approaching Earth. It keeps the planet protected from forces of change, particularly radiation from the sun."

"Which means the Earth is more stable. Aha!"

"Precisely. Less energy injected into a system results in more stability."

"I see. The crash contained a lot of energy and changed the Earth irrevocably, as in, it would never be the same again."

"Exactly. And every change following the collision occurred in that altered environment. What might have developed before the crash was never to be."

"Hmm. So, we can say humans exist today because Earth and Theia collided."

"Not only from this one change, Sara. Life on earth resulted from millions upon millions of environment-changing events from the time the universe began. The Big Bang, the formation of hydrogen, the development of galaxies and solar systems. Each planet throughout this universe has experienced a unique set of events that created its own unique environment. Earth is merely one of millions."

"And humans? But, Dina, we're unique-"

"As is every life form, Sara. You only exist because countless changes have occurred to alter the conditions on the planet and enable humans to emerge. Meteors have collided with Earth from when it first formed. Some of these collisions have profoundly affected the planet."

"Meteors? Like the one that killed the dinosaurs?"

"Indeed. Chicxulub is one of many, but it is one that proved to be important to your species."

"Well, of course. Chick-sa-bob, or whatever you called it, killed the dinosaurs. They'd been around a long time."

"Yes, they existed for approximately 165 million years, far longer than humans. But that is not why this meteor impact was so significant."

"What do you mean?"

"Chicxulub's impact changed the temperature of the Earth for a long enough time to kill the dinosaurs. That left space for smaller creatures to develop, creatures that were previously dining options for dinosaurs. One of those creatures was a small rodent-like entity that produced its offspring from within its body and nourished them with food they produced within their body. That is the branch from which humans descended. Without the unpredictable crash of a meteor on Earth at that particular time, you humans might never have formed the way you did."

"Wow, what a lucky break for us. It's too bad about the dinosaurs, but…"

"Sara, you continually miss the point. There is no luck involved. Energy moves wherever it can. Matter flows wherever it can. Change happens. Results occur. There is no luck, good or bad. There is only change and the consequence of change."

"Still. I guess I'm kind of glad we got the chance to… exist."

"I hope you will assist us now, so we all can continue to exist."

"Are things really so bad, Dina? I've asked you this before, but you haven't answered me directly."

"Yes, Sara, they are, but I need to explain more about the early Earth before we delve further into the problems of today."

"Okay. I'm getting a bit tired anyway."

"Close your eyes and rest while I continue."

She closed her eyes. "Wow, it's been a long time since I've heard a bedtime story."

"Sara, you have been in bed this entire time."

"What are you talking about? Look around at this beautiful view. The ocean, the breeze-"

"Of course." Dina paused for a long moment. "I am mistaken."

She sighed. "I know I'm dreaming, but I do love this view. Go on, Dina. Tell me your story."

"Very well. Even as the early Earth cooled, the metallic outer core of the planet was compressed so tightly it stayed hot, as it is to this day. Heat slipped out of the core into the Earth's mantle, making it hot and viscous."

"What's viscous? Sara murmured.

"Sticky. Glue like."

"Oh. Ick."

"No judgments, Sara. That hot and sticky substance gave you life. When the material from the mantle found a weak point in the Earth's crust, it would force its way to the surface, where it would be released onto the Earth's surface. You know what those are, don't you, Sara?"

"Of course. They're volcanoes. Volcanoes gave us life?"

"Sara, you are intent on seeking easy answers. Humans possess a complicated history. You did not develop quickly or via a short cut. We must proceed step by step."

"Step by step. Just like the universe."

"The mantle-vents were indeed volcanoes. During Earth's early years, volcanoes ejected significant amounts of hot liquids and gases. When energy and matter collide at high speeds as in a volcanic explosion, elements and molecules bind together in different shapes and structures or blow apart into components that then combine into new forms."

"Hmm. That sounds like an exploding star."

"Correct. Because the Earth is a subunit of the universe, explosions on Earth act the same way as explosions all around the universe."

"That makes sense."

"Except, the energy levels are different, and the forms being changed vary. Do you understand the significance of this, Sara?"

"What? The process is basically the same, but energy levels are different, and the forms being changed vary. Sure. Yep. I get it."

"I believe you are repeating my words back to me. Please, Sara, tell me if you do not understand my explanation. These topics are quite important."

She opened her eyes. "All right. Let me see if I can say it in my own words." She paused, gathering her thoughts. "The action in the universe is generally the same everywhere. Energy moves in whatever way it can. Whatever it hits, it changes in some way, including creating new forms like elements and molecules."

"Indeed. The Earth contains far more complex matter than a star or the wider universe, so the changes that occur when a volcano erupts are, by definition, more complex than those in a simple star. The pattern is the same, however. Once an energy burst subsides, forms that collided in the heat of the explosion may not be able to separate."

"Like the heavy elements. Hmm. Same pattern. They have different starting points, so, of course, different outcomes will result."

"Quite so, Sara. Before we continue, I want to make sure you understand the key characteristics of environments."

"Okay."

"Would you describe change in an environment as predictable or unpredictable?"

"Hmm. Both, I guess. I think you can predict what generally may happen, but not exactly what will happen."

"Explain, please."

"I used to do this activity with my kids. We would put two twigs into a stream at the same time, right next to each other, and follow them downstream for two minutes. They rarely ended up in the same place. We could predict the twigs would flow downstream at the pace of the current, but we couldn't be sure exactly where they would end up. Sometimes a twig would run into plants growing on the bank and stop. Another twig may sail down the strongest part of the current and be out of sight within the two-minute timeframe. Even in a small stream, there were so many variables we couldn't predict exactly where the twigs would end up."

"This is quite true. A small difference in an initial condition can cause large differences in the result. When a change occurs, a general range of outcomes can be predicted, but an exact outcome cannot.

Specific predictability and change in a complex environment are not compatible."

"And if that's true in a tiny stream, I can just imagine how unpredictable changes would be on a planet like Earth. Sheesh, who can tell what would happen?"

"Indeed. Humans exist because unpredictable and unrepeatable changes occurred that enabled the human life form to develop. One small change along the way, like your twig running into the bank of the stream, and humans would not exist."

"Ooh, that's humbling. I guess it's pretty darn unlikely we would even form if you turned back the tape recorder of time to the day when the meteor exploded and ran it forward again."

"Perhaps if the dinosaurs did not die from the meteor collision, they would have evolved into human-like creatures. We shall never know because we are not able to turn back time."

"Human dinosaurs. That would be so cool!"

"One last word about environments, Sara, and it is the most important one of all. In the end, environments are collections of energy separate from other collections of energy. Though environments always change, they remain stable over a long enough time to produce more complex forms. If an environment receives a burst of energy from outside its boundaries, the forms inside will be affected, either by getting squeezed together into new forms, or by being broken up and then reforming in different ways. Humans have a term for this process."

"We do? What is it? 'Squeeze or be squeezed'?"

"Survival of the fittest. It means those forms that best fit the environment are more likely to survive."

"That's funny. I always thought it meant survival of the strongest. But now you've explained it, I understand its true meaning. 'Fit' means the forms match with the conditions of the environment, so they are more likely to remain. They don't need to be strong; they just need to mesh with the conditions around them."

"Correct."

"Dina, I think we should change this term. It's confusing."

"Possibly, but Mr. Spencer and Mr. Darwin chose this phrase precisely. It is others who have twisted its definition."

"You know about Charles Darwin?"

"Of course, Sara. We observed him with great interest."

"You watched him, too?"

"Not me. Others like me."

"Well, why didn't you visit him? He would have been a 'more fit' person. . . ha! . . . to hear your message and tell it to the world. He was pretty much a rock star in his day."

"Mr. Darwin did well enough on his own. He changed the world without any assistance from us. Besides, we did not feel the need to interfere at that time. We thought humans were well on the way to learning how life operates, but . . ."

"But . . . we didn't? Where did we go wrong, Dina? What did we miss? How did we-"

"Not now, Sara. It is not time for that discussion. You must be patient. We must go step by-"

"Step. I know, but I'm upset. I'm really starting to worry about our future."

"I want to make sure you understand the fundamentals of this world before we delve into the complexities of humans. I will explain. Trust me. Is that what you say when you want one human to believe another without knowing all the facts?"

For some reason, Dina's words made Sara feel warm inside. "All right, Dina. I trust you."

"Good. Then rest now. You will need your strength for our coming conversations."

She closed her eyes. The cool breeze was dying. The sweet scent of pines was dissipating. She could no longer hear the whisper of the waves.

Chapter 5

LIPID-RNA-DNA-PROTEIN CLUMP

"Sara? Are you awake?"

Sara opened her eyes. "Oh, hello, Dina. Yes, I'm awake. I was just thinking."

"About what?"

"About how it all started. Dina, where did the energy that made the universe come from?"

"Sara, this is an important question but not of immediate importance to us. We must turn our focus to the environment we call Earth."

"Maybe it's not important to you, but it is to me and probably every human being that ever lived on this planet."

"It is logical for humans to want to understand their origin, but we do not have time for that discussion, right now. Besides, you already know the answer."

"W-w-w-wait a murky meritorious minute. Me? I know?"

"Yes."

"So, if I know the answer, why don't I tell myself?"

"You ask an illogical question, Sara."

"Aha! Because I don't know. I can't know and not know at the same time-"

"This is not the time, Sara. We must discuss more pressing matters. If we are able to complete our discussion, I will aid you in answering this question."

Sara sighed. "Fair enough. I can wait to find out the most basic question of all mankind. What's the hurry?"

"Precisely. The answer will not solve our current problem."

"If you say so." Sara took a deep breath. "So, where are we going today?"

"Where would you like to go?"

"Back to the beach." And there they were. The sun shone down on the glistening grains of sand. The waves rolled in, frothy and white over the deep teal of the water. Sara breathed in. "It's lovely. All right, Dina. I'm ready. Let's continue with the story of Earth."

"When we last left early Earth, volcanoes were emitting lava and gases from the inner depths of the planet. When the hot materials met the cool temperatures at the surface, the gases formed Earth's atmosphere. As the Earth cooled further, water vapor precipitated into liquid and created the oceans. Liquid water is a solvent. Do you know what a solvent is?"

"I do. It means it has more assets than liabilities."

"I do not understand."

"Stupid accounting joke. I'm teasing you, Dina. A solvent is something that dissolves things."

"Precisely. Water can dissolve many kinds of molecules into their components. Water is the best solvent on Earth. Once materials are dissolved, the components can drift along in the water and are able to combine with other materials. Does this sound familiar, Sara? Do you recognize the pattern?"

"Yes!" Sara exclaimed. "This is how energy operates. Forms move around, collide into each other, then break apart or stick together depending how much energy is in the environment. It's the same process, isn't it?"

"The very same."

"W-w-w wait a sneaky sagacious second. How do they stick together?"

"You know the answer to this question, Sara. Atoms contain forces and energy. The forces within the hydrogen and oxygen atoms interact to create the molecule of water. The interaction is called bonding. In water, the hydrogen and oxygen atoms bond by sharing electrons."

"Ooh, so that's how they are held together."

"Exactly. Energy is stored in the bonds between the atoms. Now, what do you think happens when something like salt is deposited into water?"

"It dissolves."

"How does it dissolve?"

"Um, the sodium and chlorine get divorced?"

"I am impressed you know salt is composed of sodium and chlorine, but I do not appreciate when your nonsensical statements hit close to the truth."

"Ha! They do get divorced. Or are they just taking a break? Tell me, are they going to counseling?"

"I will ignore your tendency to give human attributes to nonliving objects, Sara. Both salt and water are slightly lopsided. When atoms share electrons, the combined molecule is slightly imbalanced. In water, the hydrogen end carries a slight positive charge, and the oxygen end carries a slight negative charge. Likewise, in salt, the sodium end has a slight positive charge and chlorine end has a slight negative charge."

"Got it, but Dina, what is a charge?"

"A charge is an energy characteristic of matter that in simple terms, expresses a type of movement. Scientists arbitrarily call the charge in a proton a 'positive' charge, and the charge in an electron a 'negative' charge. Matter is stable when the positive and negative charges are balanced. So, if matter is in a position where its charges are imbalanced, it tries to rebalance itself. Since electrons are the smallest and lightest part of an atom, they move around until they are balanced with nearby protons. When a water molecule meets a salt molecule, the positive hydrogen end of water attracts the negative end of the chlorine atom in the salt and pulls it away from the sodium atom, breaking the bond between the sodium and chlorine in the process."

"So, the salt literally splits into its atoms. That's why we say salt dissolves in water."

"Yes. This concept is important: change occurs when an energy imbalance exists. In every environment, energy moves until it finds itself in a strong and stable arrangement that cannot be readily torn apart. If two imbalanced molecules meet, a struggle ensues. Each molecule attracts the oppositely charged end of the other molecule. The one with the stronger internal energy bond will retain its original form and will tear apart the molecule with the weaker internal bond."

"So, in our case the bonds between hydrogen and oxygen in the water were stronger than the ones in the salt?"

"Yes, but the environment plays a crucial role. Have you read about Stanley Miller's famous experiment in 1953?"

"Nope."

"In his experiment, Miller mixed water and other molecules believed to be present on the early Earth and charged them with electricity to simulate volcanic activity. After a short time, some of the matter inside changed into amino acids."

"Are you saying the power of Earth's early volcanoes turned simple molecules into more complex ones?"

"Indeed, but it happens at higher temperatures and forces. Some of the internal bonds of the gases coming out of a volcano were stronger than the bonds inside water. As a result, the water molecules split, and the now separate hydrogen and oxygen atoms then bonded with other nearby matter."

"Hmm. The strongest force wins."

"Yes, Sara. This is the same process that forms elements in stars."

"And the difference is in the amount of energy in the environment. To make elements, you need a huge amount of energy, like that found in a supernova. To make molecules, you need a lot of energy, but not as much as is in a supernova."

"Yes. Let me impart another important idea, Sara. As the energy from any explosion dissipates, whether it is from a supernova or a volcano, there is no way the newly formed matter can revert to its former construction. This is a universal characteristic of our universe."

"Hoo boy. I can see how the Earth environment got really complicated really fast."

"Indeed. Within the oceans, four critical types of matter formed in Earth's early years. The first are lipids, often called fats. These molecules cannot dissolve in water, so they float in water without losing their shape."

"Oh! So, that's why they say oil and water don't mix. That makes sense."

"The next type of matter is carbohydrates. Carbohydrates are molecules made of carbon, hydrogen, and oxygen. They bond with one another and form long strings. Sugars are a prime example. Carbohydrates can bond to many molecules and elements and store large amounts of energy in those bonds.

Next are amino acids. Amino acids are complex molecules that can join in long strings to form proteins. Proteins are quite impressive. Depending on their components and shape, they can perform a multitude of actions, from starting chemical reactions to forming different structures like bones and blood."

"Oh. Is that why we need to eat spinach? Because it contains proteins? Was my mom right, all those years ago?"

"Indeed, Constance was correct. And if she were here now, she would ask you to stop veering."

Sara laughed out loud. "Snap, Dina. You're right! I can hear her now. I bow to the mere thought of her warning. Go on."

"You veer often, Sara, and in doing so often touch on an important point. Human bodies do not store proteins, and some of the most essential proteins a body needs can only be obtained by consuming them from food."

"Sorry, Mom. I shouldn't have snuck the spinach to the dog under the table."

"Your mother was aware of your actions, Sara."

"Wait. Wha-"

"There is one last group of matter to discuss," Dina continued resolutely. "These are the nucleic acids. Nucleic acids are long strings of nitrogen-based molecules that are very powerful."

"Okay. We've got the Fab Four: Lipids. Carbohydrates. Proteins. Nucleic acids. So, why are they important?"

"About 4 billion years ago, when the Earth was 500 million years old, small particles of matter roamed the newly formed oceans, colliding into each other. Any combination that achieved stability in the aquatic environment remained. If the combination wasn't stable, it disintegrated. The combined stable forms also roamed the oceans and collided to create even more complex forms.

"About 3.8 billion years ago, when the Earth was 750 million years old, these four types of matter began to exist in large numbers. Some proteins developed the ability to pry open the outer shell of a lipid, allowing water inside before the shell closed again. This proved to be a significant capability."

"Why?"

"The water that entered the lipid carried tiny matter particles with it. Once inside, they were protected by the outer shell. The lipid became its own environment. What do we know about environments, Sara?"

"Aha! The things inside them contain energy and can change."

"Yes, and that is what happened. The particles interacted and combined into even more complex forms within these protected environments. There were likely billions and billions of different versions of such particle-filled lipid units. As with every other changing environment, some of the lipid forms did not last. They dissolved back into their components. Others did last. Those new units became the basis for even larger combinations.

"The nucleic acids played an instrumental role at this stage. In the early days, many types of nucleic acids formed, but only two survive today. Ribonucleic acid and deoxyribonucleic acid. Scientists shorten the names to the acronyms RNA and DNA. Both nucleic acids contain the same components, but they are different in structure and behavior."

"RNA and DNA? Okay, I've heard of DNA. Human cells are made of DNA. But RNA? I don't know much about RNA."

"RNA is quite important in Earth's history."

"More important than DNA?"

"At least as important. Ribose sugar bonds with a phosphate molecule on one end and a nitrogen base on the other to form an RNA molecule. The sugar-phosphate part of the molecule can hook together with the sugar-phosphate parts of other RNA molecules to make long RNA strands. They resemble a ladder cut in half down the middle of each rung. The middle rungs are highly reactive.

"Each RNA molecule performs a single function. When its middle rung, which is a nitrogen base, bonds with other molecules, it can create a protein or start a chemical reaction or perform multitudes of other actions. A strand of RNA performs an entire set of functions as specified by each individual RNA molecule in the strand. There are billions of RNA strands, and because RNA is so reactive, it is quite effective at creating change."

"I had no idea," Sara said. So, how is RNA different from DNA?"

"DNA contains the same basic molecules as RNA, but it contains a different sugar. The difference is significant. DNA uses deoxyribose sugar, which means the ribose sugar possesses one less oxygen molecule." Dina stopped. "What has made you chortle, Sara?"

"OMG, my ribose lost an oxygen molecule. What am I going to do?" Sara squealed in a high-pitched voice.

"I cannot begin to fathom the meaning of your statement."

She giggled. "Sorry, I keep picturing ribose losing an oxygen molecule. Eek. Help. I've been robbed."

"Should we retire for the day, Sara? Your statement makes no sense. Once a ribose loses its oxygen, it is no longer ribose, it is deoxyribose. There is no ribose left to say OMG."

"Ha! Good one, Dina!"

"I did not intend to make a humorous statement, Sara."

"Those are the very best kinds of jokes."

"Proceeding while ignoring. Deoxyribose sugar and phosphate bond like they do in RNA, but on the other end of the sugar molecule, the nitrogen base is very tightly bound to the sugar. When two nitrogen bases of DNA molecules attract each other, they stick together, rather than fly apart as they do in RNA. The deoxyribose sugar holds on too tightly to allow the nitrogen base in DNA to react like RNA. Compared to RNA, a DNA strand resembles an entire ladder. The middle rungs, that is, the nitrogen bases, are so securely bonded to the deoxyribose sugars on either side of the ladder they cannot easily come apart. If the strand is long enough, it twists around in what scientists call a double helix, making it even more stable."

"Aha! So that baby's not going anywhere."

"We are not yet to babies, Sara. Let us return to RNA."

"The ugly stepsister of early Earth? Okay."

"I recognize your reference to the fairy tale. Why do you label RNA this way?"

"Because RNA appears to be amazing and powerful but is unstable and overreacts a tad too much. RNA doesn't get any credit for her talents because super stable, always put together, Cinderella DNA, gets all the attention."

"Sara, I fail to see how this is an adequate comparison, but you are correct about RNA's power. Over time, RNA developed the ability to do something DNA never did."

"Oh? What's that?"

"Remember RNA is shaped like a half-ladder, and the hanging rung of the ladder is very reactive. Those rungs are quite particular. Each hanging molecule can only bond with a particular molecule. Scientists call that particular molecule its complement."

"Hmm. That's weird. That doesn't sound like how nature works. Doesn't nature try everything? Wouldn't RNA be more successful if its rungs could bond with lots of other molecules?"

"It depends."

"On what?"

"On what happens after RNA bonds with another molecule."

"I don't understand."

"Yet once again you have made an astute point that gets to the core of RNA's unique role. Other nucleic acids might have existed that did bond with many types of molecules, Sara. They would produce many different results depending on the different molecule with which they bonded. But because RNA could only bond with a single complementary molecule, it could only ever produce one result. That limitation turned out to be significant."

"How? I don't get it."

"Imagine a long RNA strand with many hanging rungs that bonds with its complementary RNA molecules. What does this strand look like?"

"Hmm. It would look like a full ladder. It would look like DNA."

"Now, let us say this doubled RNA strand splits apart. Now, we have two RNA strands, our original and its exact complement."

"Okay. I'm with you. Two RNA strands, one original and one complement."

"Now let us say each of those RNA strands bonded again. What do we have now?'

"Well, we have two identical strands."

"And when they split apart?"

Sara scrunched her eyes tight, trying to picture the two strands unzipping from each other. W-w-w-wait a malty mellifluous minute. We start with the original and a complement. And each of these has their complement. Hmm. . . A complement of a complement is . . . the original thing we started with. Okay, I've got it. We end up with four RNA strands: two originals and two complements."

"Exactly. What do you think about this, Sara?"

"It's sort of . . . circular. You start with one thing, mix it with its complement, then end up with the same thing again. You end up where you started."

"You are not where you started, Sara. More strands exist than when you began. Think, Sara. What is significant about these strands?"

"Well, from the one strand we started with, we now have two complements and one duplicate of the original."

"Yes." Dina stopped and waited. And waited. And waited.

"What is it, Dina? What?"

"Repeat what you just said, Sara."

"I said we have two complements and-"

"And?" Dina prompted.

"And one duplicate of the original."

"Stop. Right there. One duplicate. What does that mean, Sara?"

"RNA should have patented the Xerox machine? Har har."

"Sara. Think. What does it mean that RNA produced, albeit indirectly, a duplicate of itself?"

Sara stopped abruptly. "Oh my God," she whispered. "Oh. My. God. It copied itself. The RNA copied itself."

"Yes, Sara. RNA developed the ability to copy itself. Indirectly, yes, but still, it could create an exact copy."

"Holy cow! Without any intention at all."

"Of course, there was no intention. We are discussing minute particles."

"I know, but still, it's huge!"

"The molecules are actually quite small, Sara."

"No. I mean the idea."

"Why do you think so?"

"Well, because before this, matter particles rammed into each other willy-nilly, and whatever happened just happened. Now . . . it's a different ballgame altogether. The RNA still rams into other particles willy-nilly, that part doesn't change. But now, RNA can only ram into its complement, and when it does, only one specific result occurs. The complement can only make an exact copy of its original RNA strand. It takes a couple of steps, but it always happens the same way. So, not only can RNA copy itself, but it can do so repeatedly. As long as enough of the right material is hanging around, RNA can make lots of copies of itself. That is truly amazing."

"Indeed. This only happens because of the one-to-one relationship between the nitrogen bases. If each nitrogen-base-rung could bond to many types of bases, reactions would still occur, but the likelihood of producing an exact duplicate is much lower."

"Ohh. So that's why RNA was so important. I thought it was a limitation to only be able to bond to one molecule, but because of that limitation, it could reliably make copies. No wonder the multi-copiers lost out."

"Indeed. Such particles might have evolved on Earth, but they never produced copies in great numbers. RNA was far more prolific at this activity *because* of its bonding limitation. Given a lipid full of RNA and other nucleic acids, the RNA replicated more rapidly and more accurately. The others could not match RNA's capability, so they survived only in small numbers."

"Oh, I see. So, RNA was the best copier around?"

"RNA was the only successful copier around. And still is, I might add."

"W-w-w-wait a major minor minute. What about DNA? DNA was hanging around. Can't DNA copy things too?"

"DNA is unable to copy itself."

"Whoa Nelly in Nantucket. We are made with DNA. Our DNA is copied all over the place. If it can't copy itself, how do we exist?"

"DNA does not do the copying in your body, Sara. RNA does."

"Whaaaaaat? No way. I don't believe it."

"DNA is too stable to react. DNA's nitrogen-base-rungs react the same way as they do in RNA, however, they are too tightly bound in their double helix formation to come apart easily. So, DNA uses RNA's reactive ability to make a copy of itself."

"How the heckadoodle does that work?"

"If an RNA strand contains enough energy to collide into a DNA string and break the bond between the DNA's rungs, then the whole string of DNA will unravel down the middle."

"Oh, wow. Then we would have two halves of the DNA strand that look a lot like RNA. Hmm. Then they would be as reactive as any other RNA strand. . . each side would attract their complementary bases. "

"Correct."

"Oh my gosh," Sara said. "Then there would be two DNA strands, two IDENTICAL DNA strands all helixed up like they had never been split apart at all! That's how DNA gets copied. I get it! DNA uses RNA to split the strand, and then the strands collect their complements as if they were two RNA strings."

"Simplified, but basically true."

"So, that is why DNA is the true power, and why we hardly hear about RNA. RNA is too reactive to sit still. But DNA? Boy, DNA's a slug. Once formed, it doesn't move much. If DNA could use RNA to make a DNA copy and then revert to its original form, then DNA's very stability would allow it to stick around, and in big numbers."

"Yes."

"Wow, I think this is the first time I ever understood what DNA is all about."

"We have barely touched the surface of the significance of DNA, Sara. However, I think you have earned some rest before we continue. Let me provide one additional thought, before I leave you."

"Okay."

"Imagine the Earth at 750 million years of age. Earth's temperature ranges from 0 to 100 degrees Centigrade. The oceans are full of all kinds of matter, not just lipids and carbohydrates and proteins and nucleic acids, but combinations of these materials as well. One type of unit is of particular importance to us. Imagine a spherical lipid that can repel water. Some proteins in the ocean move with enough energy to bump into the lipid and wedge open its outer shell to let water inside before the shell closes again. The water entering the lipid is filled with a mix of molecules. Once the molecules are trapped inside, RNA reacts to create energy and other types of proteins. Occasionally, the RNA collides with its complement and copies itself. Or RNA collides with DNA and copies it. If this lipid unit expands to a size where the lipid shell cannot hold all the materials inside, it breaks apart, and re-forms into two new smaller units filled with the same materials. Are you following me, Sara?"

"Yep. Sounds reasonable so far."

"Many different versions of these units existed, all performing these same basic functions but in a variety of ways. Do you know what these units are called, Sara?"

"Hmm. Let me think. No. It's probably some scientific term I've never heard of. Something like, 'Lipid-RNA-DNA-Protein clump.'"

"No, Sara. Try again."

She sighed. "I guess I am tired. Maybe you should just tell me."

"All right. We have small, self-contained units that can ingest materials from outside without losing their shape. They convert these materials into other forms and copy the RNA or DNA inside. These little units are what humans call cells. And all the activities happening inside the cell? That is what you call 'life.'"

Sara closed her eyes, put her hands to her head, then brought them outward, fingers splayed wide in the universal human sign that her head was exploding.

Chapter 6

HELLUVA DREAM

"Just a few more, Mrs. Wallace," the technician said as he attached the electrodes to Sara's head. "Dr. Chen asked for the deluxe package today."

A second technician bent over the electroencephalogram machine. "Everything checks out. We're ready to go." He stood up. "Wasn't Mrs. Wallace in here a couple weeks ago?"

The first technician studied the chart hanging off the gurney. "Yeah, but we went lo-cal, then. We're doing full fat today. Dr. Chen wants to see results over a two-hour window."

"Hmm. What did the previous test show?" technician two asked.

"Er, average brain activity. Not typical for a coma."

"I wonder what Dr. Chen is looking for."

Technician One read further. "It says here 'a more comprehensive EEG is needed to identify the precise location of brain activity.' No skin off my nose, but it seems like overkill for a comatose patient. Okay. Last one. Ready?"

"Yup." Number Two flipped on the machine. "Whoa. Hold on, something's wrong. I'm gonna turn this baby off for a sec."

Number One walked around to check the monitor. "Holy shit. What's that? You better reboot before we try again."

Particles. collisions. Protons bashing into electrons and turning into what? Wild images swirled around in Sara's head. Volcanoes spewing golden lava and making blue oceans. She couldn't make sense of the thoughts galivanting through her head. Lipids and proteins and Cinderella dancing at the ball with glittery ladder rungs hanging from her dress. The little RNA engine roving around transforming proteins into coaches and footmen. RNA. DNA. . . Life.

Life. What is life? What was once an existential question had become a merely practical one. What had Dina said? To be life, a cell must do three things. Think, Sara. What are they? What makes a thing entirely made of non-living matter become . . . life?

Her mind raced. Back and forth. Energy. Change. Matter. Life. Oceans full of material that joined together or disintegrated into pieces. Energy. Bonds. Movement. Change. Settle down, Sara. Think!

Okay. First, a unit of life must be protected from the outside environment. Hmm. And protection from the outside creates a new environment inside. In an environment, the materials inside react. Matter with the strongest energy bonds wreaks havoc on matter with weaker energy bonds. Now, what protected the cell, again? Oh, that's right. Lipids. Little fat molecules that repelled water on the outside but contained liquid on the inside. And the liquid inside was loaded with all kinds of stuff.

"Okay. I've rebooted the machine," said Number Two. "Let's try again."

Number One rechecked the electrodes. "Okay, Mrs. Wallace. Sorry about that. The old machinerator decided to go into test mode to make sure all the frequencies were working. I think we're set, now." He nodded to Number Two." It's a good thing she's not awake right now. She would have had a cow if she'd seen our faces."

"Ha-ha. Probably. Here we go."

The machine whirred on and ran through its normal startup with no problems. "All right, we're on our way." The screen began displaying wave readings. "Holy shit!" Number Two stared at the display. "Hey, come look."

Number One stepped in front of the monitor. "What is that? What's going on?" Every wave on the screen careened up and down in stiff spiky shapes. It reminded him of disco lights flashing on a wall. "Are you kidding me? How can we have betas mixed with deltas and thetas. Shit, there are some alpha waves here, too."

"Whoa," said number Two. "This is not a brain in coma."

Okay, so a cell needs protection. What else? Well, the particles inside swim around willy-nilly. Based on what Dina told her about energy, those little guys would move any which way they could.

Collide. Break apart. And if RNA was hanging around, it would waste no time reacting to make new materials. Eventually. . . what would happen? Well, at some point, all the activity stop, wouldn't it? Sooner or later, all the material would be used up. Then what? The cell would sit there, stagnant and unchanging, until the lipid shell weakened, or the outside water broke in and destroyed it. Well, that probably happened a zillion times, Sara thought.

Dina said the cell created a workaround for this problem. It had something to do with proteins. Proteins do all kinds of things depending on their shape. Oh, that's right. Proteins found a way to pierce the outer edge of the cell and let water in without wrecking the shell's protective mechanism. That's how the cell got more material to make what it needed. If that happened, the cell could live for a long time. Hmm. As long as a cell obtained a steady stream of resources, it could theoretically live forever. Yes, that was the second thing. Life needs resources to keep the parts alive.

Dina mentioned one last thing a cell must do to be considered 'life.' Oh, yes. It needed to be able to copy itself. Now, why is that necessary? Couldn't the cell survive forever if it had a constant supply of resources? If so, why was duplication necessary? Sara tried to think. I don't understand. I'm missing something.

"It's not the machine. I checked the diagnostics. We've run the EEG protocol twice *and* the full two-hour test." Number Two shook his head. "Nothing makes sense here. Something is happening inside Mrs. Wallace's brain, but what it is, I don't know."

Number One stood up. "Let's send the data to Dr. Chen ASAP. Then wrap it up. We can't do anything else, right now." He began to remove the electrodes from Sara's head. "If I didn't know you were in a coma, Mrs. Wallace, I would say you were having one helluva dream."

Chapter 7

THE SECOND LEVEL

"Dina! Dina! Where are you! I need to talk to you. I have questions. I need answers."

The gold mist appeared. "Hello, Sara."

"Dina, I've been thinking about what you said. About cells and life. I understand why cells need protection and resources, but I don't understand why they need to copy themselves. If they have enough resources, why can't they live forever?"

"This is a mystery that humans have wondered about from the beginning of their existence. May I rephrase your question?"

"Of course."

"Why do we die?"

"Hmm. I guess that *is* what I'm asking."

"Let us return to the basics. What is the universe made of?"

"Energy. And energy moves, to answer your next question before you ask."

"Correct. How does energy move?"

"Any way it can."

"Let us consider a cell. The materials inside swim around and collide with each other. What happens during such collisions?"

"Well, the particles with the strongest energy bonds break those with lower energy bonds."

"Indeed. Sometimes the collision will benefit the cell and sometimes it will damage the cell. If the latter event occurs repeatedly, what happens to the cell?"

"Well, if enough parts are damaged, it will be destroyed."

"Exactly. And we can surmise this occurred often. Now, what happens if RNA and DNA also exist inside a cell?"

"Well, RNA can copy itself and DNA, so, my guess is some serious copying was going on."

"So, if a cell continues to ingest resources, and RNA uses the resources to make proteins and copy itself, what happens over time?"

"The cell will get bigger."

"Yes. And if the lipid shell stretches too far, what happens then?"

"It explodes. Why not? Everything else in the universe explodes when things get dicey."

"Well, Sara, I am sure some of the cells did in fact explode, but they are not the ones that survived. In some cases, when the surface tension stretched too far, the cell split into two smaller cells."

"Oh! That's sort of like an explosion."

"No, it is almost the opposite of an explosion. Now, we have two intact, albeit smaller cells, each with outer shells strong enough to provide protection from the environment. Can we say these new cells have twice the chance of survival as when they existed as one cell?"

"Oh. I see what you're saying. Copying is a way of increasing the chance of survival. One cell could get smacked by some high-energy, cell-killing protein, and the other could still carry on. So, copying is a way to counteract the risk of death."

"Over time, the cells that could make many copies survived in greater numbers. Remember, Sara, life is a form of energy. It developed as every other energy form before it. Energy tries all possible options to produce as many forms as possible. The environment determines which ones will remain. Forms that last are the ones that best fit the environment. Though life is an energy form, it is different than all previous ones that came before. In fact, we can look at life as the second level of energy."

"The second? What was the first?"

"The first level is all the energy in the universe before life. All forces and non-living matter make up the first level."

"Like particles and elements and molecules that bop around and join together and break apart, but never do anything else?"

"Yes, Sara. The second level of energy is life; protected forms that consume resources, convert them into nutrients, and copy themselves. They use resources to change themselves."

"Oh. Ohhhh. "

"Here we are again, Mrs. Wallace." Technician One attached the first electrode to Sara's head. "Dr. Chen was not happy about the results of yesterday's test."

Technician Two readied the machine. "I'm damn curious what's going to happen today."

They continued their startup protocols in silence.

Technician One stood up after securing the final electrode. "All right. All systems go. Ready?"

"Let 'er rip." Technician Two flipped the machine on. The waves held steady and smooth, bumping up and down as if the patient was chatting over coffee with a friend.

"Sh-i-i-i-t," Technician One said. "Normal brain activity. Boring. Ordinary. Average. What is going on with this woman?"

"The cell we have been talking about was the earliest living being, Sara. It survives even today."

"Today? Those little puppies are still around?"

"They are not small dogs, Sara. Scientists have a name for them. They are called prokaryotes."

"Pro carry oats? That's a funny name."

"They are your scientists, Sara, not mine. Prokaryotes are bacteria and other types of single celled organisms called archaea. They live all around you; in the soil, in ponds of water, and in remote areas that other living things find inhospitable, like hot vents at the bottom of the ocean. They even live inside your body."

"Wow, I had no idea." Sara paused. "W-w-w-wait a slippery salty second. Are you a prokaryote? Is that what you are, Dina? Taking over my brain and making me think about things I never thought of before?"

Dina remained quiet for a long moment. "You are wandering far outside the boundaries of your usual unusual comments, Sara."

That's not a no, Dina."

"Nor is it a yes. Now, let us continue. Prokaryotes contain all the components of life, but their internal structure is relatively disorganized."

"That makes sense. Everything starts simple and grows more complicated with time, right?"

"Indeed. Prokaryotes were the only type of life on earth for almost two billion years. During that time, they were constantly forming different shapes and structures. When a modified structure made a cell more efficient at using energy or producing nutrients, the cell was more likely to survive and flourish in numbers.

"About 1.8 billion years after prokaryotes formed, nearly two billion years ago, a one-celled model emerged that was more efficient than all the others nature had produced before. Your scientists call these cells eukaryotes."

"U carry oats?"

"Yes, Sara. Eukaryotes were one-celled organisms ten to a hundred times larger than prokaryotes."

"Wow! They must have stood out like the Incredible Hulk in the prokaryote world."

"I do not know of this . . . thing you call Incredible Hulk. A Large Object that is Not Believable?"

"Ha! That's it exactly. Plus, it's green."

"I do not believe eukaryotes had particular colors. Perhaps some were green."

"What? Really?"

"Sara, I am doing something you like to do, tugging your restraints."

"What?" she asked. "Oh! Yanking my chain. Dina, you tickle me silly!"

"You describe an impossible feat, Sara. Now, let us continue without further nonsensical interruption. Eukaryotes are large for a logical reason. They are more organized. As prokaryotes evolved, they grew larger and more structured, primarily by consuming smaller prokaryotes that retained their structure inside their host. Over time the cell used these undigested units to perform specific functions. Even the DNA inside the cell, which ran free in a prokaryote, became encased in its own protected membrane, something scientists call the nucleus."

"The nucleus! Why is that important?"

"If DNA is present without protection, RNA could make copies or produce proteins at any time. But when DNA was protected inside the nucleus, it was able to regulate when RNA was allowed inside to do its work. That design proved to be more efficient."

"I have a question, Dina. If eukaryotes are just fancy prokaryotes, why do we give them a different name?"

"That is a human attempt to organize and classify the life forms they see in the world. Nature needs no classification. It just . . . is. But these classifications do make sense to humans. The distinction between prokaryotes and eukaryotes marks a significant change in how life existed on Earth. The disorganized structure of a prokaryote limited its ability to grow. A eukaryote, on the contrary, held its components inside special compartments, or organelles, which could only perform specific functions at specific times. This increased efficiency enabled a eukaryote to grow much larger. In fact, eukaryotes were so successful they became the basis for an entirely new set of life models."

"What happened to the poor prokaryotes?"

"They could not compete with their more efficient relatives. They responded as life forms have responded ever since. They moved to locations where eukaryotes were unable to compete. Though eukaryotes were more efficient, they needed far more resources to survive. In areas where resources were scarce, prokaryotes were not bothered by eukaryotes who could not exist in the harsher environment."

"Clever."

"No, Sara, not clever. These cells did not possess brains. They just did what life does. Entities move around and survive in places where they fit and die in places where they do not fit. This is how life works."

"W-w-w-whoa negative Nelly. You just blew my mind."

"No, Sara. Your head is still intact."

She giggled. "No, silly. I mean you made me think of something I never thought of before. Still, I kind of sympathize with the prokaryotes being beat up by the bully eukaryotes."

"Sara, you do this often, give human attributes to non-human entities. It is quite illogical. Prokaryotes still exist, today, but from that point on, eukaryotes also existed. And each one proceeded along different trajectories. Prokaryotes, with their relatively disorganized construction, still live in favorable niches but have never grown in size. Their disorganized structure allows them to exist only as one or two-celled organisms. However, far more prokaryotes live on Earth than eukaryotes. There was an advantage to their two-billion-year head start."

"Oh, I feel better. What about the eukaryotes? What happened to them?

"Eukaryotes, with their more efficient construction, fanned out more widely. They have become the basis of every plant and animal on earth today."

"What? Every plant and animal? Every single one?"

"Yes, Sara. Including humans. Change is an ever-present activity, and eukaryotes have never stopped changing. Like the prokaryotes before them, they continued to try every option available to them in the earth environment. At first, they retained their one-celled structure but grew larger. With their ability to compartmentalize functions, they began to join with other cells to act as a single unit. They also developed the ability to copy themselves without fully dividing into separate units."

"What do you mean?"

"Let me give you an example. Imagine a cell copies itself, giving each new 'sub-cell' the full set of components it needs to survive. If each new cell changed over time to perform half of the activities needed to survive, then the unit would use less energy overall. In other words, it would be more efficient. Over time, two-celled entities that shared functions began to dominate. With more time, those two-celled organisms evolved into multi-celled entities and settled into every part of this Earth."

Sara was quiet for a long moment. "So, what you're really saying, when you get right down to it, is that God is a prokaryote."

"No, I have not discussed the concept of a deity at all. You have sprung to a terminating point I do not understand."

"Ha. Jumped to a conclusion."

"I am curious why you think that the prokaryote is a deity."

"Simple. Prokaryotes were the first living thing on Earth. Eukaryotes evolved from prokaryotes. Humans are eukaryotes. Humans are made in God's image. Ergo, God is a prokaryote."

"It surprises me to say this, Sara, but I am not prepared to respond to your statement."

"Aha! Probably because I'm right."

Chapter 8

BUILDNG BLOCKS

"I can't explain it, Mr. Wallace. Your wife's condition is highly unusual. Typically, with comatose patients, we see low neural activity. Your wife is presenting the opposite. She is showing variable high-level activity in the areas associated with memory production. More uncommon, she presents with elevated heart rate and blood pressure during active brain occurrences. It's as if she is fine in every way, except the mechanisms that should wake her up are turned off. I've never seen a situation like this before."

"Is there anything you can do, doctor?"

"Well, I called the experts at the University Brain Center. They would like to examine her."

"When?"

"As soon as we can arrange transport. I just need your permission."

"Of course."

"Thank you, Mr. Wallace." Dr. Chen made some notes in the chart. "The nurse will be back with the forms in a few minutes. Once you sign, we can prepare for the trip."

Randy nodded. "All right." He picked up Sara's hand as the doctor's steps receded down the hallway. "Sara, I know you don't like to do anything the normal way, so I don't care how you do it, but please wake up. Come back to me, sweetheart."

"Sara, you have an affinity for oceans. Where are we now?"

"Hawaii. The Big Island. The rainy side."

"Indeed."

She was sitting on a log wedged into the sand of a pointy outcrop. The sky was gunmetal gray, and the waves crashed against the rocks and shattered into a wild spray. The air was wet with salty mist. There seemed to be no boundary between the water and sky.

Sara breathed in the tangy air. On the shore were luxurious trees and bushes. Pink jasmine, yellow orchids, and a host of other flowers her mind remembered but whose names she could not recall. "It's so lovely."

"There are many such places on Earth."

"I'm sure you're right, but this is the only one I've ever seen. It's funny, Dina. Our talks have made me realize how little I really understand about this earth."

"There is too much information for any one person to understand everything. However, you humans are quite impressive in your efforts to learn about the world you live in."

"Which brings me to the question that's been nagging me since we talked last."

"What is that?"

"You explained how prokaryotes are formed from inorganic parts. Essentially, a bunch of nonliving stuff combines to form a living thing. That's amazing, Dina. Like, amazingly amazing. And it makes all kinds of sense and is sort of . . . obvious. So, how come I've never read about this? This idea should stand front and center in every seventh-grade biology class."

"That is due to how human science works, Sara."

"What do you mean?"

"Humans have derived a sound way to discern what is and is not a fact. For a fact to be accepted as true, it must be replicated. If there is even one instance when a fact cannot be repeated, then it is deemed to be not true; it cannot be considered a fact."

"Oh. I never thought about it like that."

"It is an arduous test. Over time, humans have built a collection of facts that serve as the base for further knowledge."

"W-w-w-wait a mildly misunderstood minute. I recognize that pattern. That's how the whole universe operates. Complex things form from stable, simpler things that managed to survive in their environment. You're saying complex ideas form from simpler ones?"

"Correct. This process occurs over and over. It is the way of the universe. We will return to this pattern often, Sara, but first we must talk about life in more detail."

"Okay, but finish what you were saying about facts."

"Very well. The set of accumulated facts people agree on is what you call science. This set of facts grows every day as scientists learn more and more. As to your question, Sara, your scientists have not yet been able to definitively prove the link between inorganic materials and life."

"What? It's kind of obvious, isn't it? Well, once you explained it to me."

"Obvious is not a scientific standard, Sara. Your scientists are getting closer to proving this exact thing, but the work is not easy. You see, they do not know the precise conditions of the early Earth, so it is difficult to replicate them in a laboratory. Remember when we discussed how we can look backward and discover the one path life has taken? Humans can do this quite well. They have worked backwards to identify the cell's components and how they operate together to keep the cell alive. It is much more difficult to go back to the beginning and chart all the available paths and work through them to arrive at the specific outcome we call life. Did meteors seed the planet with the molecules to create early life? Did volcanoes create a combination of compounds that formed life? Did the constant wetting and drying of molecules in tide pools produce proteins that blew into a warm ocean to form life? Or all? Or in a particular sequence? How many millions of possibilities occurred and in what order? Recently, using existing knowledge, scientists have discovered it was possible for life to form from inert materials. They mapped out the temperatures, atmospheric conditions and matter they believe existed at certain times in Earth's history. They weaved together several possible paths along which one-celled organisms could emerge, while fitting those conditions."

"Cool."

"Your scientists are close to proving the link between inorganic matter and organic life, but they have not done so yet. In doing so, they may create a new type of living entity in their laboratory. It remains to be seen how these efforts will transpire."

"Isn't creating a new life form kind of risky, Dina?"

"Your species undertakes many risky activities, Sara."

"True. We do tend to run headlong into unknown territory before considering the consequences."

"You describe a common human trait. I answered your question. Now, shall we continue with our discussion of life?"

"Of course."

"These eight topics are quite important. We will discuss them now as life relates to non-humans. Later, we will revisit them when we discuss humans."

"Oh. I get it. Compare and contrast as we redundantly said in school."

"Are you prepared and ready for this discussion?"

"Ha!"

"Topic one. Nucleic acids. DNA and RNA are the building blocks of all life on earth. Do you know why?"

"No."

"This is one time I would welcome some expansion of your verbal proclivities, Sara. You are being unusually unresponsive."

"That's because I don't know anything about RNA and DNA."

"You understand more than you realize. Tell me what you think you know."

Sara took a deep breath. "All right. RNA is a clump like thingy-dingy that makes proteins and can copy itself. It can connect its molecules together in long strands to reel off a slew of reactions one after another."

"I did not use those words, but yes, this is true. What is RNA's relation to DNA?"

"They're cousins."

"I do not understand. Explain without conferring human status on non-human life forms."

"Well, they're not close enough to be siblings, but they're still a lot alike. They both contain the same parts, but their structures are different. RNA is reactive, DNA is stable. RNA has a single strand with the bases hanging off the middle rungs, and DNA has a double strand with its bases connected and protected. But they get along. They work together to get things done, though personally I think DNA takes complete advantage of RNA."

"No human characteristics, Sara."

She giggled. "Not exactly sorry."

"Let us explore RNA, Sara. Assume an RNA strand contains three molecules. The first molecule creates a protein that produces energy. The second creates a protein that can travel to the edge of the cell, wedge it open, and allow resources inside before closing again. And the third stops any additional reactions from occurring."

"That's cool. It's like a little computer program. Start, do something, stop."

"And as long as you give the program the same instruction, that is, have an RNA strand with the same molecules, it will perform the same activity?"

"Heck, yeah! That's why computer programs are so useful. They can do the same job over and over when you run them."

"So, would you say RNA is like a computer program? Like a set of instructions?"

"Why, yes, yes you can."

"Now let us talk about DNA."

"Cinderella? Of course. Throw the ugly RNA stepsister aside. She gets tossed away as soon as princess DNA shows up."

"Sara, did you hear what I said about applying human characteristics to non-human entities?"

"Well, I'm a human and it's something I do."

"Try not to. Tell me about DNA."

"Well, DNA is like RNA except she sits alone in her ivory castle and does her nails all day."

"Sara…"

She giggled. "Okay, okay. DNA is made with the same stuff as RNA, except for the different sugar it uses. Deoxy something. It is a lot more stable than RNA because it has two rails of the ladder and is held tight in its double twisty shape."

"The double helix. Correct. So, if the strand of RNA is essentially a computer program that makes certain things happen, how would you describe a strand of DNA?"

"Well, it could be a set of instructions, except each side is already matched with its complement and bundled up in its own protected cocoon." Sara paused. "Spoiled little coddled princess," she whispered under her breath.

"Ignoring your childish outburst. What if an RNA molecule was able to collide with a DNA strand with a strong burst of energy? What would happen?"

"Well, RNA would get in trouble and be sent to her room."

"Ignoring."

"Ha! Okay, okay. Let me think." She stopped. "W-w-w-wait a madly maddening minute. The RNA would be able to break the DNA strand down the middle. Hmm, then the DNA halves would not only look like RNA, but they would act like RNA. Hmm. They must be computer programs too. Wow, that's fascinating!"

"This is how RNA and DNA work together in a cell. DNA holds all the instructions needed to make the components the cell needs to survive, but it stays protected in the nucleus. The DNA lets certain RNA strands into the nucleus at specific times to enable proteins to be made or its entire strand to be copied."

"Hmm. Teamwork makes the dream work. Dina, this is all so intricate and complicated. How could RNA and DNA figure out how to do this?"

"Sara, this did not happen overnight. It took millions of years and trial of countless scenarios for this model to emerge. DNA molecules were added to strands, and as they were, the instructions enabling a cell to survive were altered. Far more versions perished than survived along the way. If a cell with a specific strand of DNA fit the conditions in which it found itself, it would remain."

"So, are you saying that cells with different sets of DNA were constantly forming?"

"Yes. Even to this day. Remember our timeline? It took 1.8 billion years for prokaryotes to change themselves into a much more efficient eukaryotic form able to dominate the environment. But cells did not stop changing, even then. In another billion years, eukaryotic cells tried enough scenarios that efficient multi-celled entities emerged. In multi-celled creatures, every cell contains the same DNA, but each cell uses the DNA differently."

"And humans are merely an extension of this process, aren't we?"

"A late extension, at that, Sara. Did you know every creature carries a unique set of DNA? None are the same."

"I think I understand. My DNA is different than that of a grain of yeast, because if it were the same, I would be a grain of yeast."

"Correct. And I will point out that grains of yeast do not talk."

"Ha! Dina, I believed you just cracked a joke."

"I was pointing out a true statement."

"You don't fool me. I'm not rising to the bait. Ha-ha. Get it? Rising? Yeast?"

"I should not try to relate to you as a human, Sara. We both end up veering down a cottontail abyss."

"A rabbit-hole. Ha! You crack me up, Dina."

"Let us continue. Indeed, you are not a grain of yeast, because your DNA is different than that of yeast. Yet, about 32% of your DNA is identical to yeast DNA."

"32%? Are you serious?"

"Yes, I am. You share about 70% of your DNA with fish. And with primates? You share about 99%. As time went on, DNA strands changed. Occasionally, a particular DNA structure became so successful that entities containing that model would dominate. They produced many offspring, all sharing a similar DNA structure, yet differing slightly from each other. Their descendants, in turn, produced more offspring also with slightly different DNA structures. Over time, a variety of lifeforms emerged."

"Oh, you're describing the tree of life. I read about this! The trunk is a single celled organism containing the original DNA creature, probably a prokaryote. Over time, DNA evolved, and the creatures grew more complicated. When an entity with a particularly successful set of characteristics emerged, a new branch of the tree was created. The branch then produces a bunch of little branches all sharing the same basic structure but also changing it. So on and so on, until you have a flourishing tree full of different branches."

"A useful analogy, Sara. Different characteristics means they have different sets of DNA. What is important to remember is that your DNA extends back all the way to the beginning of life. Some of your DNA is so successful it has existed for billions of years and is present in every living creature existing today."

"So, we are all really just one big happy family."

"Big, yes. Family, yes. Happy? That is a human quality people have not always been generous in sharing with others." Sara started to speak, but Dina interrupted. "I want to explain a few more things about DNA before you initiate more linguistic acrobatic events."

"But-"

"The DNA in your cells is grouped into what you call 'genes.' Genes are a subset of a DNA strand that start, create, and stop the instructions to build a protein. These three parts of the gene are all molecules of DNA, but they are grouped in this order, so a protein can be built only under specific circumstances. All the genes linked together compose an entire DNA strand, which is called the genetic code."

"So, every living thing has its own unique genetic code."

"Correct. Your genetic code is like that of other humans, but they are all unique. Your son Josh shares half of your genes, so his genetic code is different from yours."

"How do you know Josh? Have you visited him too? What does he say about all this?"

"Sara, please focus. Let me continue. DNA holds all the instructions needed to ensure an entity's survival. With the help of many types of RNA and enzymes, which are special proteins that can start chemical reactions, DNA instructs RNA to make proteins. Proteins then perform all the functions the entity needs to survive.

"Every cell in a body contains a complete set of DNA, but DNA does not operate the same way in all cells. Each cell has evolved so the RNA-DNA interactions are tailored to produce particular structures or perform specific functions at specific times. You, Sara, have identical DNA in your liver and your heart, but when you were in utero, the RNA-DNA collaboration orchestrated a sequence of protein production to form your liver and your heart. You can think of this process like one of your symphonies. All the instruments produce their particular type of music at particular moments to produce a pleasing audible episode."

"Yes! And if the instruments do not play at the right times, or they're out of tune, they produce an unpleasing audible episode."

"True."

"Then no one will listen to the unpleasing symphony, and it will die out."

"And the pleasing symphony might gain more listeners."

"Wow. Life is pretty darn amazing."

"Sara, you do not often understate the reality, but in this case you have done so. Now, I must leave you with one more thought before we end our conversation."

"Uh-oh. That usually means my head is going to explode."

"No. This is a logical extension of our discussion."

"Uh-oh. Double head explosion coming."

"DNA is the most successful life form on Earth, Sara. It has changed itself into a myriad of forms which have survived for billions of years and burrowed into every corner of this planet. Yet DNA is indifferent to the forms it makes. It does not matter which entities survive, as long as some of them do. To DNA, a particle of yeast is just as valuable as a fish or a human. Do you understand, Sara? It is DNA that rules the world, not humans."

Sara closed her eyes. I knew it, she thought to herself. Double head explosion.

Chapter 9

FIT OR DIE

"I can't explain it, Dr. Chen. We performed extensive testing on the patient, and there is nothing in our case database similar to this presentation of symptoms. The brain function is high, as if the patient is dreaming or actively processing memories, but her base level maintenance functions are operating at a bare minimum." Dr. Ritan, University Brain Center's lead physician on Sara's team, plunged his hands into his lab coat pockets as he faced Dr. Chen.

"So, you've never seen a case like this?" Dr. Chen asked.

"Nothing even close. However, we found an aberration that might provide some answers."

"What kind of aberration?"

"The patient's bacteria levels are highly erratic. They shift into and out of normal range without any apparent consistency."

"What bacteria? Are they in any specific location?"

"That is the puzzling part. They appear in every part of her body. Every time we test, we obtain different results. So far, we haven't been able to glean any meaningful pattern."

"Do you think there might be a single cause to explain these shifts?"

"We don't know yet. We are still analyzing the data. What is puzzling is that even though these bacteria levels are all over the map, her system remains remarkably stable."

Dr. Chen sighed deeply. "What is happening with this woman?"

"This place appears different from the last time we met, Sara."

"Yes. We're on the Big Island again. It looks like a different world when the sun is shining, doesn't it? The ocean is smooth as glass today. Look how deep the shade is beneath the trees. And the ginger's in bloom. I never knew ginger had so many colors." She glanced around. "I guess I'm getting used to our chats, Dina. Even though I can't see you, I know you're here when I see the gold mist."

"What do you mean, Sara?"

"When you're here, everything glows with a gold haze."

"I did not realize this."

"I like it. It helps me recognize you're here. So, what are we going to talk about today?"

"Fitness."

"What? Are we working out?"

"No, Sara. I mean fitness to the environment. Life forms survive by fitting the conditions of their environment. Let us start by defining the term 'survival.' Survival is the process of transforming resources into the specific nutrients an entity needs to live. If a creature consumes resources and lives long enough to produce and rear offspring, it will have successfully converted those resources into a new generation of entities."

"I'm with you so far."

"To understand fitness, let us consider a particular scenario. Assume many living creatures exist in one place, and they all consume the same resources to survive. What happens if the environment changes and the food these creatures eat is reduced significantly?"

"Yikes! Not everyone is going to make it."

"Can you guess which ones will survive?"

"Well, those that get to the food first."

"True, but there are other ways an entity can survive in this situation. Perhaps a creature can eke out an existence by eating less. Or perhaps it can consume a different resource. Or move to a location where food is abundant."

"Hmm, so the critters that can find some way to eat when everything goes belly up will be the ones that survive."

"Yes, although I will point out that the environments are not defined by the existence of vertical abdomens, Sara. An entity only survives if it can live according to the conditions of the environment."

"But you said environments constantly change. So, really, no one is safe if everything is always changing. Any creature could die at any minute."

"Indeed. Now let us alter our scenario. Only two creatures now exist in our environment. Creature A has produced a dozen offspring. Creature B has produced none. Now, once again our environment undergoes a major change, and a significant portion of the resources are destroyed. Who will survive?"

"It's the same answer as before. The ones that can get to the food first or live on less or move."

"Indeed. But the numbers affect the outcome, do they not? There are 13 of creature A and only one of creature B."

"Hmm. I see. So, the sheer numbers of A give it a survival advantage. Even if eleven A's die, A would still have twice the presence in the environment as B. If each one then produced a dozen offspring, there would be 26 A's and 13 B's."

"Yes. Can you understand now why entities produce offspring? On the face of it, it seems illogical for a creature to spend some of its precious resources to produce babies when it could use them for its own survival."

"I'm on to your tactics, Dina. Early creatures tried both ways to survive. It's just that the ones producing babies were more successful. Your example shows it. More numbers result in a better chance of survival. Think about it. If a creature didn't produce offspring, sure it would keep all its food for itself, but sooner or later, something would happen, and it would die."

"Please explain, Sara."

"You just explained. Environments change. Sooner or later, some disaster is going to come along. Floods. Heatwaves. Volcanic explosions. The creatures with the most offspring have the best chance of survival. Wait. That's not exactly true. The creatures may die, but some of their offspring might survive. It's their genetic line that survives, even if they die."

"True. So, what determines which offspring will survive and which will die?"

"Well, besides sheer numbers? Let me think." She thought of her sweet children, so alike and yet so different. Now, only Josh would carry on their family line. Would his children have his dark eyes, or would they carry the bright blue of Maggy's? She caught her breath. Of course. "Each baby is different, Dina. The offspring are all unique. Didn't you say every entity's DNA is different, even in members of the same family?"

"Yes, Sara. Each entity carries slight variations in how they look and act."

"So, the creatures with the variations that enable them to get to the food first or eat less or move will be the survivors?"

"Exactly."

"But Dina, how do the offspring get their different variations?"

"Remember the basic characteristics of the universe."

"Energy. Everything contains energy, and energy moves, so everything changes."

"Correct. Every time DNA is copied, it is a change event. It is also an opportunity for an unexpected outcome."

"You mean a mistake? Are you saying Cinderella isn't so perfect after all?"

"This is not a fairy tale, Sara. Mistake is a pejorative term not applicable to nature. Unexpected outcomes are merely logical results of a changing environment or an unexpected energy event. There is no right or wrong in nature. Things happen because of existing conditions. The outcome just. . . is. And every outcome redirects how life proceeds into the future. Without exception. This is why you cannot predict what might happen given a particular starting point. Any number of outcomes can result, which makes accurate prediction nearly impossible."

"That thought makes my head explode. Let me go back to variation. What you're saying is that all offspring contain slightly different DNA. So, by definition, they must all be slightly different from each other?"

"Correct. Because DNA directs the production of proteins, a different set of DNA means a different set of proteins will be made. The different proteins will make each creature different in how it looks and acts. The differences may be slight, but if an environment undergoes dramatic change, those small variations might be the difference between life and death."

"So, the ones with the most suitable DNA survive to produce more offspring."

"Yes."

"Well damn, Dina. You're telling me life is a crapshoot. How can anyone expect to make any sense out of this?"

"This is how life works, Sara. Humans have given this process a name."

"Really? The crapshoot theory?"

"No. Your description did not make it into the lexicon. You are familiar with the work of Charles Darwin. He described this process as 'descent by modification.' He first proposed it in The Origin of Species, in 1859."

"The phrase is familiar, but I confess, I don't really know what it means."

"Mr. Darwin proposed that each generation changes, or evolves, based on the impact of the environment on existing life forms. He used the term 'natural selection' to explain how entities with characteristics most suitable to the environment would survive and those with the least suitable characteristics would die. The accumulation of many small changes over millions and billions of years gave rise to the different creatures existing today. Your scientists estimate four billion different species have evolved since life began."

"I read somewhere that almost all of them are extinct, and the species living today represent only 1% of all the ones that ever lived."

"That is true. It interests me how you perceive this idea. Humans tend to label it as negative."

"Well, it's kind of sad that 99% of all living species have died."

"This is my point. You think of them as dead."

"Fact check. Dinosaurs. They lived for, like, 165 million years, and now they are gone from this planet."

"Turn your telescope around, Sara. View life from the other direction."

"What do you mean?"

"You humans gaze from the perspective of the entity. You study all the forms of life and organize them into tidy classifications and then lament that many forerunners no longer exist. Look from the other direction. Tell me the story of the cells composing every life form on Earth."

"The cells? Hmm. Let me think. The first DNA cell formed about 3.8 billion years ago." She took a sharp breath. "W-w-w-wait a marvelously manic minute. I see! Life containing DNA has never died! Lots of DNA versions have died; those are the extinct species. But DNA itself is still here. True, only one percent of all the DNA forms that ever existed are still plugging along, but those numbers obscure the real point. *DNA* is still plugging along. Given your figures, at least 40 million species exist today. To put it another way, 40 million different DNA versions are running around the current Earth environment. If we consider it this way, DNA is remarkably successful!"

"Indeed. All creatures are unique containers of DNA. Some containers survive and others do not. But DNA has always survived and still survives today. You are right, Sara, the 99% of extinct species are, in some way, an illusion. Life can be defined as one long flow of DNA, continuously repackaged into different forms. The forms don't die as much as they change to match their environment. If you view it from the perspective of DNA, nothing has ever gone extinct. Evolution is DNA's survival strategy. Constant modification enhances the chances of survival given the certainty of an uncertain environment. Humans and all the plants and animals alive today are the current set of DNA containers that best fit the present planetary conditions."

"Well, that's all well and good, but I'm a human first, not DNA."

"Are you sure that is true, Sara?"

She paused. "Wait a minute. Okay, okay, I have DNA inside of me, but-"

"No, Sara, you are entirely composed of cells containing DNA. Your body is made up of more than 35 trillion cells and nothing else. And every one of your cells contains DNA. You are not a human with DNA, you are a collaboration of DNA laden cells made into human form."

"Maybe that's true, but I'm still packaged as a human, and as a human, I want my package to survive as its own thing."

"Spoken like a true human."

"Am I wrong? Containers are expendable, and DNA created enough of them so that some will survive regardless of what happens. If my human package dies, other DNA packages will take up the slack. I'm going back to my earlier statement. Life is a crapshoot. A DNA crapshoot."

"I am not sure how excrement bombardment relates to our discussion."

"Don't you see, Dina? To DNA, any container will do. It's just chance as to which ones make it."

"Not chance, Sara. Entities suited to their conditions survive. Every living creature today is the descendent of a long line of DNA containers that fit their environment. Yes, the mix of DNA containers has changed over time, and from many combinations attempted, only a fraction ever survived. But humans are in that group."

"Hmm. I guess I should be grateful nature selected us."

"I never agreed with the term 'natural selection.' It implies nature possesses the ability to pick one surviving entity over another. Nature does no such thing, Sara. Nature is a collection of energy and matter that moves as it can. Outcomes occur with given conditions, without any intention. I much prefer the term, 'natural fitness.'

"Hmm, that is clearer, Dina. It puts the responsibility on the entity to fit into nature, not nature to choose the entities that fit. Plus, I know you don't like when we give non-human entities human traits."

"True."

"But Dina, if what you say is true, we're not special . . . none of us. Not humans or fish or yeast. To DNA, we're just useful idiots."

"I will not pass judgment on DNA's methods, Sara. Nor should you. They are what they are. Energy produced matter. Matter produced life. Life produced different forms to ensure its survival."

"You're blowing my mind again, Dina. So, what's next? After life?"

"We are not ready for that discussion yet, Sara, but keep your question in mind; we will arrive soon enough. But first, we must address one particular subject humans take pains to avoid."

"Only one? I can think of at least ten."

Chapter 10

SURVIVAL STRATEGY

"Let's talk about death, Sara."

"So that's the topic we don't like to discuss. I was betting on religion. Or sex."

"Why do humans dislike discussing death, Sara?"

"Come on, Dina. I don't think we're much different than any other critter. No one wants to die. No one wants to think about dying, so no one wants to talk about dying."

"Yet every entity dies. Every single one."

"Wa-wah. Are you feeling okay, Dina? You're being a real Debbie Downer."

"You are exhibiting a clear example of avoidance."

"Maybe people think it's better to not think about death and live as if they'll never die. Then we can get the most out of each day without living in fear."

"Do you fear death, Sara?"

"Hmm. Less so than I used to. Since Maggy died, I fear a lot of things less than I used to."

"Explain, please."

"Well, the worst has already happened. Nothing will ever be as bad as the loss of Maggy. It changed me; changed my perspective. I guess I'm not afraid bad things will happen because they do. They have. They will again. Maybe I'm just getting old. I know death will come sooner or later. Maybe I'm more at peace with the idea."

"Why do you think entities die?"

"Is this our third topic, Dina?"

"Indeed."

"DNA gets tired of us and throws us out like an old pop can."

"Once again you unintentionally collide with the truth. How do you think this works, Sara? DNA possesses no intentional thought process. It does not suddenly decide to dispose of a creature. It does not 'decide' anything. So, something must happen to make a creature die."

"Hmm. Maybe over time we run out of the stuff we need to keep living. We are born, grow, and if we're lucky enough not to be hit by a bus, we die of old age."

"But how, Sara?"

"I haven't the first clue, but I bet you know, or you wouldn't have brought up the subject. And I bet it's important and will make my head explode."

"I hope that is not true. It has not happened yet. I am beginning to think it is an expression that means something else."

"Ha! Yes, it means you blow my mind with new ideas."

"That is not a helpful description."

Sara giggled. "Somehow, just chatting with you makes me feel better, Dina. Okay I'm ready. Let's attack topic three. Tell me how death works."

"The basic reality is that with life comes death. The two are inseparable."

"Why? Why can't living things just ingest food and live forever?"

"What is the nature of the universe, Sara?"

"Energy. Movement. Change," Sara recited.

"Exactly. Energy is embedded in every single particle of matter on this earth. The particles inside an atom move; the atoms inside a molecule move; the molecules inside a cell move. Energy in the form of nutrients and proteins move to nourish a body. If the energy within a body stops moving for a long enough time, the cells will not be able to function, and the body will die."

"Wow, don't sugarcoat the truth, Dina. Just give it to me straight."

"My explanation was succinct, Sara, Also, I do not understand how one can paste sucrose onto words." Dina paused. "I understand now. You are attempting to avoid this topic."

"Yep."

"Then we shall continue. Death can occur by external forces. For example, a large rock can fall on your head and crush your skull."

"Ouch!"

"If your brain cannot direct the functions of your body, you will die, or more precisely, all your cells will die. Or you can succumb to operational failure. Your heart may clog up and stop pumping, thus keeping the blood from nourishing the body as it flows through your circulatory system, once again killing all the cells. . . and you. Humans call this type of external cell death necrosis. A cell dying by necrosis releases its contents into the body in a disorderly manner. This can endanger nearby cells, resulting in inflammation, which can kill more cells and even your entire body."

"Oh, so that's what inflammation is. So, when I sprain my ankle and it swells up, necrosis is at work?"

"Yes. You are describing a mild non-lethal event. The damage to your ligament caused the death of some cells whose contents spread throughout the area, causing swelling and bruising. Sara, necrosis is not the only way a cell can die. In fact, individual cells are constantly dying in a living body."

"Really? In mine too?"

"In every living body. In certain circumstances, if a cell encounters negative stimuli, it will commit suicide. Humans call this process apoptosis. Events such as exposure to hazardous chemicals, immune reactions, infectious agents, even high temperature or radiation, may result in damage or, in a worst-case scenario, cell-suicide. In this case, the cell shrinks and is then eaten by phagocytes, which are small cells whose job is to absorb a dying cell's material before its contents can be released into the body."

"Wow. That is the ultimate form of recycling."

"You are correct. Apoptosis is a natural process. It starts early and continues throughout the lifetime of all living creatures. It happens in every single living being. Do you know why, Sara?"

"Obviously, it must have been better than all the other survival methods the early cells tried. And as much as I hate to admit it, it seems efficient. If an entity can remove injured cells and use those parts as resources for the remaining cells, it would have a much better shot at surviving than those without that capability. As long as most cells survive at any one time, the body won't die from the loss of a few cells."

"Indeed. The cells of a body work together to keep the entity alive. The ability for a cell to kill itself is a survival strategy designed to keep most cells and the body alive. To be more precise, the DNA within the cell instructs itself to die if the condition of the cell will endanger the survival of the body. As long as some cells survive, DNA survives; that is what is important. In fact, once an entity produces offspring containing DNA, there is no actual need for an older model to exist."

"Oh, I get it. So, DNA places its efforts into shiny new baby objects with shiny new DNA versions rather than keep an old gray creature alive. So, that's what happens, doesn't it, Dina? The DNA offs its own old geezers. That's why things die of old age!"

"Sara, human oriented as your logic is, it has collided with the right answer. Indeed, just as DNA follows strict timing and protocols to enable new offspring to be born and develop into adults, it also developed a mechanism to enable them to die if they do not succumb to accident or illness."

"What giveth can also taketh away. Rotten Cinderella."

"Your words make no sense, Sara. This Cinderella you refer to is the basis of your life."

"I don't always need to make sense, do I?"

"It would be helpful. Do you want to learn how this happens? How things die?"

"No, but I suppose it's always good to know when Cinderella's going to stab you in the back."

"Enough. No more talk of nonsensical fairy tales."

"But-"

"No, Sara. No more. DNA is what it is. Your judgement and opinion only prevent you from understanding its nature. You must focus."

"I'm sorry, Dina. I didn't mean to make you angry."

"Anger is not relevant to this discussion, Sara. Facts are. Our discussions grow only more complex from here. I need to ensure you understand the nature of life before we can get to the core of our discussion: humans."

"Fair enough. No more talk of Cinderella. Go on."

"Very well. Inside the nucleus, the DNA is organized into genes and wrapped with proteins in tight bundles called chromosomes. At the end of each chromosome is a protein called a telomere. The telomere is a long string with many sections, like beads on a necklace. Its job is to keep the chromosome geometrically stable. Throughout your lifetime, your cells are regularly copied to replace your dying cells. Every time a cell is copied, every bit of the chromosome is copied except the last bead of the telomere string. This process repeats and with every repetition, another bead of the chain is lopped off.

"When no telomeres are left, the cell can no longer replicate. If a negative stimulus then hits the cell, damage may result. It may die, or it may commit suicide. If enough cells in the body die this way, the body dies. This is how a body dies from old age. Sooner or later, energy in the form of deadly stimuli catches up to enough cells, and the body dies."

"Wow. I see what you're saying, but why does it work that way? Why didn't DNA develop a more efficient process that copied the entire string and allowed the cell to live forever?"

"Perhaps DNA tried that alternative, but the arrangement did not survive. Telomeres serve as the stop sign for replication. Once the enzyme gets to the last telomere, it stops copying and unhooks from the gene. The telomeres evolved to make sure the entire chromosome remained stable while the process occurred. However, telomeres are like RNA, highly reactive. The longer the string, the more likely it will react uncontrollably. This can result in cancer or other life-threatening illnesses. If the string is too short, the beads may run out before the entity can produce offspring. So, a balance evolved; short enough to prevent uncontrollable copying, but long enough to enable procreation. When I say, 'a balance evolved,' do you understand what I mean?"

"Hmm, the entities with either longer or shorter telomere strings tended to die before they could procreate."

"Correct. Only the lengths enabling the occurrence of both events, long enough to procreate but short enough to prevent unwanted copying, survived."

"I see now. DNA did try to live forever. It just realized it couldn't."

"No realization was involved, Sara. The cells with very long strings did not survive. The cells with very short strings did not survive. The telomere strings that exist in each creature today are balanced not because they are perfect but because they survived. Part of the legacy that each entity inherits is an imperfect, yet adequate, telomere length."

"So, how long are human telomeres? How long do our cells replicate before becoming vulnerable to the energy forces around them?"

"Would you like to guess?"

"Heck yeah! Well, they probably last beyond age 40, when women can still bear children. Wait a second. Women can have babies until they reach menopause, which happens around age 50. Men can father children up to any age. Hmm. Clearly, I don't have a clue. I'll say. . . at age 60."

"Close, Sara. On average, human cells stop replicating at the age of 55."

"There is no difference between men and women?"

"Women possess longer telomeres then men, though your observations about male procreative ability are accurate."

"Hmm, I thought it would be the opposite. Either way, it all goes downhill after 55."

"It does not need to be this way. However, from that age on, humans would be wise to follow much more stringent maintenance habits if they want to live a longer healthier life."

"Ooh boy, no more sour cream potato chips for me," Sara muttered under her breath.

"The human body, like all living bodies, is a mechanism for DNA to store itself. From the DNA perspective, as long as a human can pass their DNA to a new generation, DNA's survival is ensured. Over time, the entities that created new generations of entities with fresh new telomere strings survived more easily than the entities with longer telomere strings. DNA's ability to pass a version of itself on to the next generation of bodies ended up being the most efficient method for DNA to survive. The death of the older entity is not important if a new entity with new DNA now exists."

"So, what you're telling me is death is a survival strategy."

"Yes, but it is the DNA's survival strategy, not the entity's. It is a matter of perspective. Every creature dies, but DNA does not. At least it has not died yet."

Sara sighed. "Death is a survival strategy. That's a paradox if I've ever heard one."

Dr. Ritan, took off his glasses and rubbed his eyes tiredly. "Dr. Chen. We think we found some answers but not a complete picture of what is happening to Mrs. Wallace."

"Well, that is more than I know. Anything helps at this point. Her family is beside themselves with worry."

"We think we pinpointed the main issue to the RAS, the reticular activating system. As you're aware, the RAS covers a broad portion of the brain stem."

"Yes, and it's responsible for the wake-sleep cycle."

"Among many other functions. It appears the RAS is not functioning properly. The neurons causing arousal from sleep are not working. Her body is producing the neurons, but the brain isn't carrying them to the right locations to awaken her. There are no lesions or blockages along the pathways, so we are not sure what is causing this."

"So, she's . . . sleeping? She has a form of narcolepsy?"

"Well, if it is, it's an extreme case the likes of which I've never seen."

"What are her hypocretin levels?"

"They're low, but not low enough to cause such extensive symptoms. We're still investigating. You can take her home, now. We have run all the tests we need. We need to analyze them in detail and do additional research into past cases."

"What about the bacteria levels?" Dr. Chen asked.

"Inconclusive. We continue to see varying levels in different parts of her body at different times, but we can find no cause. We are looking to see if there is a link between the changing bacteria levels and the RAS."

Dr. Chen nodded. "One more thing, Dr. Ritan. I went over our case files last night. I noted a few mentions of an odd yellow cast on the patient's skin, not all over her body, but in several areas. Have you or your staff seen this phenomenon?"

"Oh?" he scrolled through the notes on his computer. "I see no mention here of anything like that. I wonder if the variable skin changes are linked to the different bacteria levels?"

"Oh, that's an idea."

"If you observe the yellow coloring again, run some tests on the areas nearby."

"I will. Maybe it will lead us to some clue about what is happening here."

"Right. Let us know if you find any link. Once we analyze our findings, I will be in touch."

The doctors shook hands.

"Why did you come back here, Sara?"

Sara looked around. They were back in the forest, the fake forest to be exact, where they first met. "I wanted a place where I felt safe."

"Are you feeling unsafe?"

She sighed. "Sort of. Our last discussion unsettled me a bit."

"In what way? Does the concept of death bother you?"

"No, it's . . . I thought humans were . . . unique."

"You are unique."

"Not when we are 99% primate and 32% yeast. We're just one of a bunch of trials in a big experiment."

"Sara, every single life form on Earth is related. Yet every single life form is unique. Why does this idea unsettle you?"

"Ooh, maybe that's not the right word. I thought humans were. . . 'special.'"

"Define 'special.'"

"Unusual. Or . . . exceptional. Or . . . superior to all the other life forms. Now, I find out we are a tiny leaf on the edge of the tree of life, so complex we could be snuffed out of existence with the next energy burst that comes along, poof, like the one that killed the dinosaurs."

"All life forms are unusual and exceptional and superior in their own way, Sara. Each animal existing today emerged from a long line of survivors. Each developed their own way of surviving. Humans, too, as we shall discuss soon. But before we can address humans, let us complete our discussion about the nature of life."

"Okay. I'm glad I'm here in this place. I know it now, Dina."

"I wondered if you would recognize it."

"It was buried deep. This is the forest I pictured in every fairy tale I ever read when I was a kid."

"Please do not tell me we will visit Cinderella again."

"Ha! You tickle me, Dina. No, now that I think of it, there wasn't a forest in the Cinderella story, but Snow White and Hansel and Gretel and Sleepi-"

"No need to continue, Sara. I understand this memory provides you with a sense of security."

"It does. I always knew living was a risky business, Dina, but I guess I didn't realize life itself is a risky business."

"Once again, you have segued into our next topic. Every moment of living is accompanied by the risk of death."

"Topic four is as big a buzzkill as topic three."

"Only from the human perspective; not from DNA's perspective. All the instructions that make up DNA exist to keep the cells within a body alive. As long as the cells survive, the DNA survives. The fact that the container holding all the cells survives too, is a secondary outcome. The primary goal is the survival of the community of cells. If a body dies, it's DNA dies too. And with death, the DNA's chance to carry forward to a new generation. DNA no more wants a creature to die than it wants its cells to die."

"That's comforting. Still, no matter how you look at it, risk of death is a basic characteristic of life. Risk of dying is the cost of living, Dina. End. Stop."

"Indeed. It is a harsh truth to accept, but it is true nonetheless."

Randy kissed his wife as she lay in her bed. She was finally home, here in the sunny bedroom she loved so much. The IV pumping nutrients into her body hummed methodically. No answers, he thought. Still no answers. "Sara, sweetheart, look what Josh and Nina brought over. Lilacs; your favorites. It's been a good season for them." He picked up the vase full of lavender stalks and wafted them in front of her nose. "You always told me these reminded you of your childhood. I hope you are dreaming happy dreams."

She lay silently, almost peacefully. Randy looked around at the pale walls reflecting the bright sunshine. He stared sadly at Sara, so quiet and far away. "Rest, sweetheart. I know you'll wake up when you're ready." In the brightness of the room, he did not notice the golden tinge creeping down her face and onto her neck.

Chapter 11

RESOURCEFUL

"You changed our location, Sara."

"No. It's the same forest."

"But it appears different. There are large bushes with purple flowers everywhere."

"Those are lilacs. Aren't they lovely? They've always been my favorites. These look like the ones from my yard when I was a child. A row of lilacs lined our back fence. For the few weeks in spring they bloomed, they formed a wall of purple perfume."

"Lilacs were not part of your fairy tale memory."

"No, but I was thinking of them, and . . . here they are."

"This is interesting. You blended two memories together while in this state. Quite unexpected."

Sara inhaled deeply. "I know this is a dream, but I can almost smell their heavenly scent."

"Interesting indeed. Are you ready for our fifth topic?"

"Yep. What are we going to cover?"

"Let us talk about resources. Resources are the raw materials a living entity consumes to survive."

"Oh, food!"

"That is a limiting word. Organisms can consume a variety of materials. Air. Minerals in rocks. Plants-"

"Cheetos!"

Dina remained quiet for a moment. "Sara, we could veer off to have a long conversation about food and food-like substances, but I choose not to do so. We must continue our discussion."

"Okay, okay. I guess I'll be adding Cheetos to the taboo list. At least the sour cream potato chips will have company."

"As I was saying, resources are found in an entity's environment. All creatures need two types of resources to survive. They consume proteins and minerals, which are the raw materials needed to build and maintain their cells. They also consume carbohydrates and convert them into energy to fuel the cell building processes. All bodies must obtain a balance between the two. If an entity eats only carbohydrates, it does not have proteins and minerals on which to build the structures the cell needs. The carbohydrates provide the fuel, but since the cell has no proteins with which to work, it stores the potential energy in the body for a time when they are available."

"Ooh. I bet they like to store things in hips. At least in this body."

"In hips. In stomachs, In arteries. The body stores excess energy wherever it can."

"So, if I then eat protein and minerals, the body uses its energy to process them into materials that are useful to the cell."

"Exactly. This is how your skin regenerates after a cut, or how bones are kept strong or how your heart continues to beat even though cells are continually dying."

"So, Mom was right. I did need to eat my spinach."

"Yes, as it is a source of proteins and minerals."

"I guess I regret wrapping it up in my napkin and storing it in my drawer."

"In addition to feeding it to your canine companion? I daresay, some small creatures benefitted from your action."

"Ha! Caught. Again."

"Let me ask you this question, Sara. We now know creatures must consume resources to survive. Does this same requirement exist for matter?"

"Hmm. I don't think so, but to be honest, I never thought about the two being linked."

"Imagine you eat a bowl of spinach. Your body breaks the spinach down into parts and gives your cells energy and proteins to make your cells grow stronger. In a way, your body converted the spinach into a slightly different human body. . . and some waste products it didn't need. Does inert matter work this way too?"

"Hmm. I don't think so. Matter is stored energy, but when two non-living clumps collide, they can only do two things. They can bounce apart or stay joined."

"Indeed. Let us consider a collision between hydrogen atoms. If they join together, they become a helium atom and release excess energy into the environment. Together they make a different element with different characteristics. But the components themselves do not change. If we bombard the helium atom with enough energy, it will split back into two hydrogen atoms."

"Right. But if you split me apart, you could not pull the spinach out of me intact, no matter how hard you tried. So, that's the difference between life and matter. Life consumes materials and changes them into body parts in a process that can't be reversed. Matter can't do that."

"Indeed, this is a significant difference between the two. Given this difference, can you surmise how living beings might affect their environment?"

"Oh, man. I need to think about this. Can you give me a hint?"

"What is the basis of the universe, Sara?"

"That's a question, not a hint. Energy. Movement. Change. I need more, Dina."

"Think of a rabbit."

"A rabbit? Like Bugs Bunny?"

"No, an actual live rabbit. A rabbit consumes resources and converts them into a larger version of itself and most likely, many offspring rabbits."

"I'll say. Josh had a pet bunny when he was nine. It got out once, and lo and behold, we ended up with a whole slew of-"

Dina proceeded resolutely. "Each time the rabbit goes through this conversion process, it shifts the balance of resources in the environment."

Sara stopped. "Oh. More bunnies, less lettuce."

"Which, by definition, changes the environment. The environment will never be the same again. It is irrevocably changed by the bunny's actions."

"By just one rabbit? Really, Dina that's hard to bel-"

"Think, Sara. If all creatures, not merely one rabbit, but all creatures consume resources, make larger versions of themselves, and then produce offspring, do they not alter the mix of resources in the environment?"

"Hmm. So, that's why environments change."

"Environments change because they are composed of energy, Sara. However, you make an important point. Environments containing life, what you call ecosystems, change faster than inert environments made only of energy and matter. Both systems move from simple to complex, but life changes much more quickly."

"How can you tell?"

"Let us compare the complexity of matter in space to life on Earth."

"Okay. Hit me with it."

"This must be an expression. I find it impossible to strike you with words possessing no physical form."

"Oh, Dina, you're adorkable. I mean, please explain."

"Which forms dominate in space, the simple or the complex?"

"Hmm. You said hydrogen is the most abundant element in the universe, so, simple forms dominate. Only 2% of the matter we see are elements heavier than helium."

"Indeed. Based on mass, 74% of mass in the universe is contained in hydrogen, and 26% is in heavier elements. Based on this information, what life forms do you think dominate on Earth, simple or complex?"

"Hmmm. Well, if it works like the rest of the universe, I'd guess simple organisms dominate. Since hydrogen is the big winner in the matter world, I'm going to guess that, based on the numbers, 75% of all life on Earth is made up of one-celled organisms. But based on mass, let me think. If 74% of the mass in the universe is still hydrogen, and life moves faster than matter, I'd say 50% of all life mass is made of simple organisms."

"A logical conclusion, Sara. You are quite close when it comes to the numbers of entities. Scientists estimate 60 to 70% of all living things are single-celled organisms."

"Woot woot! I'm a physics genius."

"No, you are not."

"Buzz kill alert. So, does this mean I'm way off on my mass guess?"

"You are right that you are wrong."

"What?"

"If you compare the carbon mass, which is how scientists measure relative mass, single-celled organisms only make up 17% of all organic matter."

"Only 17%? That's all? But they make up 60-70% of all life forms? I'm shocked. Are humans the heavyweights? I bet we are. We have a problem with obesity, you know."

"I refuse to provide commentary on the relative carbon mass of humans, but I will say no, humans do not win the heavyweight award. Terrestrial plants make up the bulk of biomass on Earth. They represent 83% of all carbon mass, which is about 450 billion tons."

"Plants? Hmm. So how much is human biomass?"

"Animals as a group represent only 0.4% of all carbon mass. Ants are the most successful of this group. There are about 10 trillion ants weighing approximately 3 billion tons. In comparison, the 8 billion humans on the planet weigh about 350 million tons, which is only 12% of ants by weight."

"Well, if some of us lost a few pounds, we could drop that 350-million-ton number!"

"It will have little effect, Sara. Humans make up only .025% of all animals."

"Wow! So, we are .025% of 0.4%. That's, let me see, that's one ten-thousandth of a percent. We are mere pipsqueaks in the biomass world."

"True. But in terms of impact on the earth, you strike over your heft grouping."

"What? You mean we punch above our weight class?"

"That is the phrase, but it is not a compliment, Sara."

"No, I suppose it isn't."

"Resources are essential to living entities. Not only are resources required for survival, but their use also irrevocably changes the environment."

"Say that again, Dina."

"Not only are resources required for survival, but their use also irrevocably changes the environment."

"Thank you. That is my summary statement for topic number five."

"Very well, Sara, but to be precise, it was my statement."

"Dad?"

"Hey son."

Josh sat down next to his father and watched as he held his mother's hand. "How is she doing?"

"No change," he said quietly.

"Let me sit with her for a while. Maybe you can take a walk or something."

Randy closed his eyes tiredly. "I don't want to leave her for even a minute. What if she wakes up and I'm not here?"

Josh smiled. "You're not going to be much help if you become a walking vegetable, Dad. You're already a pretty good asparagus stalk. Especially with that haircut."

He stood up, hugged his son. "Thank you, son. It will be good for me to get out." He walked toward the door. "Josh?"

"Yeah, Dad?"

"I love you, kid. I just want you to know."

He nodded. "I love you too, Dad."

"I won't be long."

"Take whatever time you need." Josh looked down at his mother as he listened to the hum of the feeding tubes. "How are you doing, Mom? I decided I'm not going to nag you to wake up. You never did nag Mags and me. But . . . it sure would be nice if you did. Wake up, that is. Besides, Nina and I have some news. We're going to have a baby. A baby girl to be exact. It would be great if you woke up, at least before our new little Maggy heads off to kindergarten." He squeezed her hand. "I love you, Mom."

Chapter 12

AS THE WHEEL TURNS

"We have returned to your blended memory, Sara."

"Yes. I can't get enough of the lilacs. They bloom for so short a time. I can remember holding bunches of them and inhaling the aroma when I was a child." And there they were. A bundle of lilacs, their stems bound together with aluminum foil, filled her arms. "Oh my," She gasped. They're lovely."

"I do not want to concern you, but it might not be helpful to blend your memories like this."

At once, the fragrant bouquet disappeared. "Why?"

"I'm afraid my explanation might be confusing. I ask you to trust me in this regard."

"All right, Dina." The wall of lilacs disappeared. "I trust you."

"Thank you. Are you ready to discuss our next topic?"

"Of course. What aspect of life are we going to visit now?"

"I want to talk about complex life on Earth. The Earth is 4.5 billion years old. Life formed relatively quickly, but it was nearly 3 billion years before the first multi-celled entity evolved. Why do you think complex life forms took longer to emerge than the first simple ones?"

"Hmm. It took 380,000 years for hydrogen to show up, and that's like the simplest thing ever. But it took billions and billions of years to make elements larger than helium, so I guess life works the same way. Everything takes time, Dina. Change is slow."

"Indeed. It takes time for energy to interact with an ever-increasing number of forms to produce a complex form that will last."

"Hmm. You can take a lot of time and energy making new versions that don't make it. Is that what you're saying?"

"Yes, Sara. The universe does not require change to be efficient. History is laden with combinations that did not work. Only a few attempts are ever successful. Complex forms result from those few successful changes that fit with the environment. Each change is made step by step. It is a simple process. As complexity increases, more combinations must be tried before success occurs."

"Dina, the universe makes new elements using energy bursts from star explosions. How does evolution on earth happen?"

"This is a good question, Sara. To consider how life evolves, we must think small. Tell me, what is every complex creature made of?"

"Cells, and each cell is loaded with DNA."

"Are cells big or little?"

"Don't make fun of me, Dina."

"It is a logical question."

"Okay. Small. Tiny. Eensy weensy, super itty bitty, not even noticeable to the naked eye-"

"Do not be difficult, Sara. Yes, cells are small. So, how large an energy dose might be needed to cause a single cell to be modified?"

"Are you saying cells are susceptible to change because it doesn't take much energy to affect them?"

"Yes. An unusual fluctuation in temperature or contact with natural radiation, may be enough to alter the DNA of a one-celled creature. If the alteration causes the organism to survive more easily than its unchanged neighbors, it will likely proliferate and crowd others out of existence."

"What if a modification makes it more difficult to survive?"

"Then the cell will die, and the unchanged survivors will proliferate instead. Change is slow. A modification to a gene here. A mutation caused by radiation there. It takes many, many years for small changes to accumulate and produce creatures with different shapes and structures."

"So, one-celled organisms kept evolving, but so did complex organisms. Wait. I'm confused. If complex entities are bigger and well . . . more complex, they would need to consume more resources, right? Wouldn't they be at a disadvantage in an environment where everyone needs resources? I understand how they might form, but how could they survive?"

"Yet they did survive, Sara. Complex creatures thrive on Earth. They are far more successful than the heavy elements that formed from stars. Remember matter heavier than hydrogen makes up about 26% of all the matter in the universe. But on Earth, complex entities represent over 80% of all life. How can this be if they had an evolutionary disadvantage?"

Sara paused. "Well, obviously it wasn't."

"Wasn't what?"

"Wasn't a disadvantage. Complex forms must have been way better at surviving than their single-celled cousins. Otherwise, they wouldn't dominate."

"Correct, Sara. Think about the prokaryotes and eukaryotes. What is the fundamental difference between the two?"

"Hmm. Both are one-celled organisms, and both contain the same components, but the eukaryotes are more organized."

"Right. The prokaryotic structure is less organized and as a result, less efficient. It can expend much energy without earning a beneficial outcome. In eukaryotes, the internal functions are arranged and separated in organelles. The organelles are only permitted to perform their specific functions. By limiting their functions, the cells overall use less energy."

"I see. That's the definition of efficiency, isn't it? Getting more done with less. So, that's why the eukaryotes stomped all over the prokaryotes. Prokaryotes were not organized enough to create efficient multi-celled organisms."

"Prokaryotes and eukaryotes were not enemies, Sara. The prokaryotes developed first, and gradually evolved into the more efficient eukaryotic form."

"Oh, so prokaryotes are the parents of eukaryotes."

"In a way. Whenever a new form of life significantly diverges from its previous form, humans give it a new designation. Prokaryotes didn't die out when eukaryotes emerged; they merely moved over and began to share the Earth with them. But they needed to adjust to the changed environment where eukaryotes were more efficient. They retreated to small niches and harsh areas with scarce resources. In these areas, eukaryotes had a difficult time consuming enough food to maintain their larger structures. In those places, complex organization carried no advantage."

"And eukaryotes dominated everywhere else?"

"Yes. This is how evolution works. Every time a DNA-RNA interaction created a more complex variation with increased efficiency, those individuals carrying that modification would be more likely to survive. Those DNA changes would then be incorporated into the next generation of the entity. This cycle was repeated many times. If efficiency leads to better organization, and better organization leads to more complexity, then it follows that entities will grow more complex over time. After millions of years, the creatures resulting from this long line of modifications might look very different than the ones from which it descended.

"Over time, individual single-celled eukaryotes transitioned into multi-celled organisms where groups of cells performed specific functions. Those groups developed into tissues and organs and circulatory systems. Consequently, the organisms evolved into larger and larger beings. These changes still go on today. You just don't notice them because they work on such a slow time scale. Evolution is still going on right here, right now."

"So that is why things grow more complex with time. They're more organized."

"Indeed. I want to emphasize a few points about life with our final topics, Sara. These might seem obvious, but they will become important later."

"When we talk about humans?"

"Yes. Our sixth topic concerns the cost of organization."

"Oh, you mentioned this before. Organized entities use more energy."

"They do, but not as much as if they were haphazardly joined together. Organization requires each cell to perform a specific function. Though every cell contains the machinery to do every single job in an entity's body, they are not permitted to do so. Individual cells in a eukaryote use less energy because they perform a limited function, but in exchange, the cell gives up its freedom to roam inside the body."

"So, what you're saying is a eukaryote is a socialist commie."

"I did not say such a thing. We are not talking about humans at this moment."

"Sorry. I was just parroting the freedom-loving, right-wing faction of the world. FREEDOM!!!"

"Veering, in a most unhelpful way."

"No, Dina. Prokaryotes are free to do as they please. Never mind that they can't grow large enough to become a large body, they are FREEEEEEEEE. Eukaryotes work differently. They make a trade. They exchange freedom for security. They become part of a complex body by giving up some of their freedom."

"Well, it is a bit more complicated than that, Sara."

"Of course, it is, but let me use a simple example to explain what I mean. Let's say I'm a heart cell in a human body. All I need to do every day is pump blood cells from the heart cell behind me to the heart cell in front of me. Borrrrrring. But in exchange, what do I get? I'm part of a huge team made of a bunch of cells that perform all the other functions I don't have to do. They consume resources and see objects and hear sounds and, well, do all the other stuff a body does. I am fed and protected. If a virus comes along, the body's army cells protect me. When my body eats, I receive my fair share of nutrients. Every cell in the body, including little ol' me, is doing their one little job and, together, the whole body . . . works."

"You oversimplify the actual mechanisms, Sara, but you grasp the idea. Every cell in an entity, whether yeast or bee or human, must be nourished for the entire body to survive. If a life form is to become complex, mechanisms to bring resources to every cell must exist. Every multi-celled being possesses such a system, though they vary. Humans have circulatory systems that pump nutrients into the blood to arteries and capillaries and individual cells, and then back through veins to the heart for refueling. Bees operate differently. Their internal organs exist in a mixture of blood and fluid from which they draw nourishment. No matter the method, every entity nourishes every cell in its body. Without cell nourishment, the survival of the entity itself is threatened."

"Hmm, so the cells and the body work together to make a win-win."

"More precisely, the cells work together to form a mutually beneficial relationship. The body is a convenient side effect of the process."

"Ooh, I never thought of it that way. Dina, how many cells are in a human body?"

"Your scientists estimate the number at about 35 trillion."

"35 trillion? Trillion? Holy caboly, that is some serious cell machinery."

"Indeed. And every time an entity changes at the cell level, the change must allow the entire system to work, or the entity will not survive."

"So, as an entity gets more complex, successful changes are more difficult to achieve."

"Yes, Sara, which implies fewer will result. The actual numbers reflect this reality. Far fewer complex forms exist, whether we are speaking of elements or bodies. Generally speaking, life changes faster than matter. Relatively more complex life forms exist compared to complex matter."

"Hmm, but when a living entity is successful, it can dominate others and alter the entire panorama of life."

"Well said."

"Dad, what do you think is going to happen?" Josh sat across from Randy at the kitchen table.

Randy shook his head. "I don't know. Everyone is mystified. Leave it to Sara to become a new medical mystery phenomenon."

Josh smiled. "I wish she could see how she's confused all these doctors. She'd get a kick out of that."

Randy sighed. "I feel like I'm in suspended animation, son. I am afraid to walk out of the house in case she wakes up or needs help. But I'm going crazy just sitting here, waiting for . . . what I don't know."

"You can take a break, Dad. The caregivers are keeping regular hours. Mom wouldn't want you to grind yourself down."

"No, but she would also understand how I feel. That's the problem, she's the one who knows me best. I just . . . miss her so damned much."

Josh covered his father's hand with his. "Let's do our best to get through this."

Randy choked back a tear. "I don't want to lose her, Josh."

"Where is this place, Sara?"

Sara sat on the steps of an old country church perched on the crest of a hill. A road meandered up the hill, then wandered down again, twisting into the trees, bright green in the full summer light. Across the road, a wooden fence penned in a gray horse. How many times had she petted that horse? Two bikes rested against the outside wall of the church, her small one dwarfed by Randy's large one.

"Sara?"

"Oh, hi Dina. This is one of my favorite places. I forgot about it until now. It was one of my favorite bike rides. It took a ton of effort to ride to the top of the hill. We always stopped here to rest and pat the horse. Randy used to take me on so many adventures in the day."

"Why are we here?"

"I was thinking about Randy. It feels like I haven't seen him in years. I know I'm in some strange dream where time is passing in weird ways, but he seems so far away all of a sudden." She smiled. "I know I'm being silly, but I miss him. He's my completer."

"I do not understand this term."

"Ha! That's because I made it up. That's what I call Randy. He fills in my gaps. He makes me more whole . . . more complete. He is my opposite in so many ways. Through him I see the world from a different perspective. He opens my mind."

"Would you like to forego our discussion so you can think more about him?"

"No, that's okay. He's never far from me anyway."

"All right. We must discuss a few more facts of life."

"Ha! That's funny. Are we going to talk about the birds and the bees?"

"Yes, but I fail to perceive why this should elicit laughter."

"Facts of life is a human code word for talking about sex."

"Interesting. We will talk about sex and procreation soon, but we are not at that stage yet."

She giggled. "Not sex as a natural thing, but sex as a . . . you know . . . human thing."

"Sex is a natural 'thing' for many creatures, Sara. It is not unique to humans."

"Tell that to the porn industry."

"I do not understand."

"Oh, Dina. I'm not sure I could even explain it to you."

"Then let us move on."

"Great. What's next on our facts of life tour?"

"Two key ideas remain. Let us turn to the concept of self-interest."

"Now you are delving into an area I understand, but, Dina, I must admit my bias right off the top. I do not like this term."

"At least you are honest about this, Sara. As you are familiar with the term, will you define it?"

"Okay. Well, in nature, all living beings strive to survive, so they'll do whatever they must do, including kill and eat others. This is the natural state of the world; kill or be killed, eat or be eaten."

"Your statement is true, as far as it goes."

"There's more to it?"

"Of course, living beings strive to survive. It is also true they need to consume resources, which may include eating other entities. But living beings can survive in many ways besides killing and consuming another entity. Think of the bee that draws nectar from a flower, and in so doing carries pollen to other flowers. This is a mutually beneficial relationship. The bees obtain resources, the flower procreates, neither one dies."

"A mutually beneficial relationship? Like the cells in a body? Tell me more. Can creatures survive in other ways?"

"Of course, but this is not the point of our discussion at the moment."

"What is?"

"Life forms, at least non-human ones, act without intention. They do whatever they can to survive given the environment in which they live. Remember how the DNA-RNA collaboration works, Sara?"

"Yes, they team up to make proteins the cell needs."

"Right. RNA and DNA can only do what they have already evolved to do. If they create a protein and the environment suddenly shifts so that protein confers a disadvantage, the entity may die. And a similar entity that did not produce the protein may now gain an advantage. All non-human life is reactive, Sara. It can only do what it is programmed to do and no more. There is no intent, and, consequently, no self-interest."

"So, are you saying self-interest is not an accurate description?"

"Yes. The term implies entities can choose among different options to prefer themselves over others or prefer one action over another. Nothing could be further from the truth. Non-humans have no choice or have a limited number of choices at best. They are provided their instructions for living by their DNA, which evolved over billions of years of survival and fit within the environment."

"Hmm. Those who fit best, survived best, and those that didn't fit are not around to tell their sad tale."

"Exactly. Entities perform as instructed because they must. If the environment changes, they will still perform as instructed, even if that behavior results in their death."

"So, let's not call it self-interest. What should we call it?"

"Adam Smith used the term 'self-love' in his book The Wealth of Nations."

"Yikes. No way. I don't think it's fair to load that fraught term onto nature's creatures. They don't deserve that."

"Perhaps we can use the term self-benefit. It implies the same result without any suggestion of choice."

"Self-benefit. I like it. I think I can agree that all living entities strive for self-benefit."

"Very well. We have now reached the last characteristic of life before we begin the next phase of our discussion. I want to talk about order."

"Order? Whoa natural Nelly, I didn't expect that. From everything we discussed, this universe is more Stravinsky than Beethoven, a chaotic cacophony rather than a beautiful symphony."

"That is an interesting analogy. Yet, symphonies do exist inside this vast chaos. You are a symphony of order and structure inside an unpredictable and untidy Earth inside an unpredictable and untidy universe."

"Hmm, you're right. So, what do you mean by order?"

"Picture the Earth as a wheel. Inside the wheel, every square inch of Earth is covered with a host of beings that have survived through multiple generations, and which all fit within the environment. Each entity survives the same way: they consume resources, convert them to energy and proteins, procreate and die. When they die, their bodies break down and are consumed by other organisms. Every single life form is both prey and predator. This process occurs over and over and over.

"With each turn of Earth's wheel, a different mix of life forms is produced. If, by chance, the environment goes through a massive energy event, many entities may die. Those dead entities then become resources for the survivors. The mix changes, but the wheel of life continues as if nothing happened.

"If you arrived on this Earth, as humans did 200,000 years ago, it would look as if all living beings were designed to fit on the Earth. Which in fact, they were. They have been designed by the environment and eons and eons of trial and fit.

"But that's only on the surface. It just appears orderly from a distance. Order is not harmony. On close examination, we find each entity inside the wheel is trying to survive as its instructions allow. Non-humans do not possess any external tools to help them; their bodies are their only tool. They are always on the move to procure resources without becoming a resource. Though the wheel of life appears stable from the outside, from the inside, it is a constantly moving, ever-changing, network of resource-seeking beings, striving to survive."

"So, you're saying an overall order emerges from a chaotic base."

"Well stated, Sara."

"Thank you. Are we're done with the facts of life, now? Yippee! Where do we go next?"

"We will now consider the core ideas that enable us to address the predicament humans find themselves in."

"I don't understand."

"It is now time to open the black box named survival. I leave you with this question, Sara. How do entities survive?"

Chapter 13

GOTTA DO WHAT YOU GOTTA DO

Sara's mind raced. The ideas Dina planted in her head chased each other around. Dina made it clear she should remember them because they would discuss them later when they talked about humans. And now a new question. What was the secret sauce that enabled entities to survive?

She should start with what she knew. That's what Dina would do. Start with the eight 'whats' of life and determine if they led to the 'how' of life.

First: RNA and DNA. They are the building blocks of life.

Second: An entity must fit into its environment to survive.

Third: Death is a survival strategy. For DNA.

Fourth: The risk of death is a characteristic of life. The risk of dying is the cost of living.

Fifth: Resources are required for survival, and using resources changes the environment.

Sixth: Complex life forms result from efficient organization.

Seventh: Self-benefit is programmed into DNA and is not intentional.

Eight: Order is an illusion. Chaos rules!

Okay, start at the beginning. Energy. Movement. Change. Matter evolved into life, and life did something matter couldn't do. Aha! Creatures consume resources, process them into energy and nutrients, and procreate. She laughed to herself. I'm trying too hard. It's obvious. If an entity does these three things, it survives. If it doesn't, it dies. But, if it's so easy, why did Dina make such a big deal of it? Hmm, because living beings die. They can perform all these actions and still die.

So, the question is. . . what causes death? How does an entity minimize its chance of death in a dangerous world? She chased the thoughts around before she gave up. Never mind. I'm tired. I want to rest. She took a deep breath, and in her arms appeared a giant bouquet of lilacs, freshly cut from the blooming bushes in the back yard of her childhood. She inhaled deeply. "Ahhhh, lovely."

"We are back in Hawaii. Is there a particular reason, Sara?"

"I like it here, Dina. It's noisy with life. The beautiful flowers, the birds. It's . . . full."

"It is an apt backdrop for our discussion. Have you thought about my question? How do entities survive?"

"Well, you told me the answer before. An entity that consumes resources, processes them into energy and nutrients, and procreates, survives. They pass their DNA to a new generation."

"This is true, but we must now expand the concept."

"How, Dina?"

'Remember our wheel of life, Sara. There is not merely one entity on Earth; millions and billions and trillions of life forms exist. Why does one survive, and another perish?"

"That's what I missed. I didn't think about the Earth as a system where everyone must compete-"

"Stop. You said the very word I want to discuss."

"System?"

"No."

"Oh. Competition? Ohhhhh. *Competition*."

"Yes." It seemed to Sara the golden glow grew more intense as Dina continued. "Competition is the default survival strategy for every living creature on Earth."

"You sound like an Economics book."

"You don't agree?"

"Not entirely. I always thought competitive theory was . . . incomplete."

"Indeed. Your assessment is valid. As we delve further into our discussion you will understand why."

"Now, you've piqued my interest."

"The Earth is a finite system, so, by definition, resources on Earth are also finite. We can conclude, therefore, that Earth can sustain only a finite number of entities."

"Hmm. Makes sense."

"Every creature born on Earth must eat to survive. When beings vie for the same resources, they compete. Competition was the first survival mechanism to exist, and it still exists today. It occurs at every level of life, from the one celled organism competing for nutrients in a pond to elephants eating only plants, to cheetahs eating other animals. Every single creature vies for the resources it needs to survive."

"Of course," Sara said. "I'm embarrassed I didn't make the connection. Me, with my business background, where everyone worships competition as the be all and end all of human behavior."

"This is your bias, Sara. You do not trust competition."

"W-w-w-wait a secret sanctimonious second. I never said that. I said it's not entirely true-"

"This is your secret, is it not, Sara? You have long believed competition is not to be trusted as the focal point of human behavior. It is why you are so . . . what is the word I want to use . . . uneasy. . . about the very business environment in which you operate."

"That's true, but it's not something I share. How did you know?"

"Your skepticism is why we chose you. Nevertheless, it is essential you understand how competition works. It is a fundamental mechanism of survival-"

"Maybe, but sometimes competition causes a lot of problems."

"Indeed. It is critical to discern when competition is useful and when other mechanisms are preferable."

"There are other survival mechanisms?"

"Yes. But competition came first. And like all forerunners, it has staying power. Competition is beneficial, but it has drawbacks as well. I want you to understand when it should be used and when another alternative is more effective."

"Hmmm. Clearly, I need to open my mind. Go ahead, Dina, explain away."

"The hallmark of competition is pressure. Every waking moment in the life of a competitor is stressful. Most competition occurs among individual members of a species, where individuals are physically close and consume the same resources.

"Four types of pressure affect competitors. First, competition is determined by the availability of resources. If food is abundant, all creatures can eat their fill and procreate with little trouble. In this situation, there is little pressure to survive. But if food is scarce, competition intensifies, and those who are more capable of obtaining resources have a better chance of surviving than those who don't."

"So, a bigger animal may have an advantage. Or one that's faster. Or one with sharper eyesight. Or-"

"Correct. Any characteristic enabling one competitor to obtain resources more effectively than another confers a survival advantage on that competitor. When an organism consumes more resources, the less capable creatures struggle to eat and are more likely to die. Also, the successful eaters procreate more readily and produce more offspring that grow to outnumber less capable organisms. The scarcer the resources, the higher the pressure, and the faster this process occurs."

"Ooh, that's why life changes faster than matter. Competition accelerates the winner's chances of survival not only by enabling them to produce more offspring but also by getting rid of the weaker players."

"Yes. The second cause of pressure is the number of competitors within a particular environment. Ironically, procreative success can breed the seeds of destruction. If a living entity is efficient at ingesting resources, it procreates more successfully. What happens when many offspring of the same species exist?"

"Well, it's not going to take long for competition to intensify among them. All the outsiders are gone, and now the environment is full of the same type of critters that are all proficient at eating the same resources. Some of those little guys aren't going to make it."

"Precisely. Which ones will survive, Sara?"

She thought for a moment. "Well, the ones best at obtaining resources. Though the whole group is efficient, if resources are still scarce, the less capable of the group won't make it."

"Yes. Whenever there are more competitors than food, the stronger individuals will weed out the weaker ones. This cycle repeats as long as resources are scarce."

"So, competitive pressure rewards even tiny differences that improve survivability."

"Indeed. And those differences occur only at the individual level of each species. Individuals drive all change, Sara."

"So, are you saying scarce resources and the number of competitors will result in different types of creatures over time?"

"Yes. Sara, still more pressures exist impacting how entities compete."

"Right. What is our third culprit?"

"As with all of life's characteristics, no judgment is involved, Sara. Choose a more appropriate word."

"Okay, okay. Hmm, how about cause? What is the third cause of pressure?"

"Better. The environment is our next cause. All beings must comply with their surrounding conditions. For example, if the temperature gets too hot for an entity to survive, it dies. However, if one or more individuals in a group possesses a genetic characteristic enabling it to withstand higher temperatures, it may survive. Let us say one entity has a light-colored external coat that enables it to better reflect sunlight. If this slight difference allows the entity to survive at higher temperatures, it may produce offspring with lighter coats that can also withstand higher temperatures."

"Oh, and then the offspring hand down the favorable characteristic to their offspring. I see. Pretty soon, there will be a whole bunch of blonds running around!"

"Let me return you to the non-veering zone, Sara. As we discussed, an environment is a relatively stable area in which things can last long enough to change form or procreate, but it is not unchangeable. And environments are massive compared to the individuals living inside them, so it is not possible for a single entity to alter the conditions. It must fit or die."

"Hmm. So, it doesn't matter to the environment which organism survives, does it? Some survive, others don't. Nature doesn't care, does she?"

"Nor do the organisms, Sara. Intention does not exist in the world of non-humans. Creatures ingest resources, convert them to energy, discharge their excrement, and procreate. If resources are available, they continue to exist; if the resources are not available, they die."

"That's a stark reality, Dina. So, what you are saying is there is no meaning in life, except to eat, metabolize, poop, and have sex."

"You are missing the key idea, Sara. This is the physical reality of non-human life. And it is the foundation of your human existence as well. Meaning is not involved. Meaning is not a physical consideration."

She sighed. "I don't understand."

"We are not ready for that conversation yet, Sara, but we are making progress. Let us instead discuss our last cause. Pressure is also caused by the number of predators existing within the environment. The existence of predators does not affect competition among entities sharing the same resources, but it does affect how many competitors exist in a given area. As we know, each predator may also be prey. So, not only do entities need to compete for their preferred resources, but they also need to be alert so as to evade those who would treat them as a resource."

"Eat or be eaten; now that's the definition of competition I know and do not love."

"Did I ever tell you the story of how your mom and I became friends?" Randy asked Josh as he cupped his hands around his hot cup of coffee.

"Yeah, Dad, about a thousand times. You and Mom were at prom but with different dates. And your dates got together, so you ended up dancing all night. The rest, as they say, was history."

"That's technically true."

"What does that mean? Technically?"

"There's a little more to the story, son."

"Spill, Dad. You can't start a convo like that and then let it hang."

"Actually, I can."

"Dad, don't pull that precise language shit on me. Spill."

Randy laughed. "Okay. Okay. It was the junior prom, so it was springtime, May, I guess. Your mom was dating this first-class A-hole, football jock and dumb as a post senior. She was, in my opinion, the prettiest and nicest person in school, but she kept dating idiots. All of them were good looking, but my god, it would take a whole drawer of them to make one sharp knife."

"I take it you didn't make the cut, Debate Club president."

"Obviously not, smart ass. I wasn't in her league. But we were friendly. We saw each other in class, worked on a project or two together over the years."

"Did you have a crush on her?"

"I would say I kept my eye out for her. But she was always dating someone else."

"So, you both went to the junior prom with different dates."

"Yeah."

"Who did you take?"

"I don't remember her name. But she was a friend of a friend, a senior whose date got sick at the last minute. So, she asked me to step in, and I told her I would take her."

"Gallant. What happened?"

"It didn't take long for me to understand why my date's first choice dropped out. From the minute we arrived, she was hanging all over Mom's boyfriend. Now, what was his name? Rick, Rich, Rob? Rob. That was it." he smiled, remembering. "Your Mom was horrified and embarrassed, so I asked her to dance."

"That was nice of you."

"I didn't know what else to do, especially since Rob wasn't exactly pushing my date away. After a little while, we didn't see them on the dance floor. Mom said she needed to go to the ladies' room, but I could tell she was about ready to burst into tears."

"Poor, Mom. That must have been rough."

"It got worse. She walked into the bathroom and found the two of them, er, going at it, shall I say, in the bathroom foyer."

"You are shitting me."

"Nope. Suddenly, she comes bolting out and runs straight into me. I asked her what was wrong, but she pushed me aside and headed for the door at about 80 miles per hour."

"That sounds like Mom."

"At first, I thought she was sad or hurt, and she was. But what she really was, was mad. M A D. I haven't seen that look too many times over the years, but when I do-"

"You run for cover. I know. Same," Josh said.

"Ha! I can remember a time or two when you two squared up. You're a lot like her in that way."

"Yup. That's why it was hard for me to get away with anything. So, what happened?"

"Well, I followed her out. This was the era before Ubers or Lyfts. She stood outside, alone, trying to figure out what to do. I came out and asked her what happened. She wouldn't tell me. She just shook her head and pinched her lips and put on her super disappointed face."

"Yep. Seen that look a few times."

"So, I told her I had an idea. I drove her to her house and asked her to change into some regular clothes and grab a flashlight. I'll never forget her look. Curious, but suspicious. I remember telling her it would be okay; that we were going to do something to make her forget about what's his face. That's what I called him, and she laughed. She returned in about 90 seconds. Then we drove to my house. She came in and met my family. They were more than surprised I left with one girl and came back with a different one. You could just feel all the questions, and your mom was great. She told them how I rescued her from, what did she call it, 'a bathroom disaster,' and explained that Rob was too 'incapacitated' to drive her home. I came out looking like a hero, and off we went to our fictional post-prom party."

"So where did you go?"

"We went to the local disc golf course. A new one had opened that year. One of the first in the country. That's where my friends and I hung out when we weren't-"

"Debating or making 8mm videos in the parking lot."

"Low blow, son. She had never played disc golf before, but your mom was coordinated. She picked up how to throw a disc pretty quick. It was just what she needed. She didn't say much, but she hucked those discs as if she were aiming at Rob's head."

"Ha, I can picture that."

"Yeah, there we were, with the flashlights pointed at the baskets, hurling those things as hard as we could. They flew all over the place, and then we would grab the flashlights and run around until we gathered them all up. By the third hole, she was laughing and chatting. That was the moment we became friends. But it was on the last hole, as we fished out our discs from the basket, she said something I'll never forget. She said, 'I'm never going to date a person who can't be honest with me. Never again.' She looked up at me with a frown and said, "I don't need to be treated like a queen. I just need to be an equal.' I never forgot her words."

"Wow, I never knew this story. And you started dating after that?"

"Hell no. She needed a lot more time to recover. But we did become best friends."

"Whoa. Better than your debating buddies?"

"She became another buddy."

"Wow. I didn't know she was on the debate team."

"She wasn't, but she used to help me practice. She was better at debate than me, but she never wanted to join the team. She pursued her own interests. That's how we rolled. We flowed. We did things together. We did things in groups. We did our own things. She needed her space. During senior year, she grew grounded and happy."

"Did she date anyone else?

"No, she didn't, which surprised me. Years later, she told me she dated those guys so she could reassure herself of her own worth. By dating a popular, good-looking guy, she had 'value'. After Rob, she realized she needed to value herself first. That's what she did during that year; she learned how to value herself."

"Good for Mom. Now I understand why she is such a stickler that I say what I do and do what I say."

"That's your mom. She doesn't want you to be a Rob."

"I get it. But face it, Dad, she kinda went overboard on the lesson at times."

"I can't argue with you, son. This is your mom we're talking about. She is who she is."

"Let's continue our conversation about competition, Sara."

"Okay, but I must say, Dina, I'm still not sure how beneficial competition is. It seems to me a lot more things die than survive on this Earth."

"True."

"Then, how is competition beneficial? Life is special. It should be valued."

"Which life, Sara? Which life should be valued? Who decides which entities should live and which should die?"

"Well . . ."

"This is what the Earth's environment does, Sara. It determines value without bias or favoritism. Those fitting its conditions survive, and those that do not, die. At every point in time, only the most suitable life forms exist."

"I see what you're saying, but it's so brutal."

"It is neither brutal nor gentle. It just is. It is the way things work. It is a system of total accountability."

"Accountability?"

"Non-humans directly bear the consequences of their actions. If they perform better and procreate, they still die, but their descendants benefit from an improved chance of survival. If they get eaten by a predator or starve to death before they produce offspring, they not only die, but they are further penalized by having no descendants. Their DNA is gone forever. This is the definition of accountability."

"Hmm. Competition is a win-lose arrangement."

"Precisely. The pressures of competition combine to separate the fit from the unfit. With each generational change, entities emerge with variations in their DNA. Every single modification is tested by competition. An organism evolving a minor variation still must compete in an environment in which all its competitors are the latest winners in life's contest. Only the most fit emerge as victors. Over time, the victors grow less and less similar. With even more time they evolve into different species."

"And each surviving creature is born with a structure that fits the environment, which means it's 'designed to survive.'"

"Yes. If the environment changes drastically, the fit of these creatures may also change drastically. Recall when Chicxulub hit the Earth and plunged the planet into a multi-year darkness. Many entities, including the dinosaurs, who needed sunlight, died. But a few survived. And those survivors found themselves in a world of abundant resources with few predators. Natural fitness continued to work on the survivors, and in time, the Earth was once again full of diverse creatures, but all of them were descended from the few who survived the meteor strike."

"So, the variety of life forms is another way DNA can withstand an unexpected change."

"Indeed. Diversity is arguably the most noticeable outcome of competition. But competition provides other advantages as well."

"Hmm. Isn't diversity enough? I feel like you might be selling the whole competition idea just a little too hard."

"I have no need to make available for purchase an idea when I have facts, Sara. Diversification is another way of saying the DNA of a creature has changed. Those changes alter their structure or behaviors in such a way to increase their chance of survival. Each change is an innovation."

"I'm not sure what you mean."

"Let me use an example provided by Mr. Darwin when he studied the birds on the Galapagos Islands. A few finches flew from mainland South America or floated on debris and reached the Galapagos Islands. Over time, the finches evolved unique structures enabling them to eat the plants growing on the particular island they inhabited. Each alteration in the birds' structure was an innovation that improved their ability to obtain resources, and thus improved their chances of surviving. This is how innovation works. Over time, if an environment is stable enough and resources are scarce enough, entities produce different structures. If the new structure is innovative, that is, more efficient, the difference is incorporated into future generations. Since all life forms work this same way, over time the Earth becomes full of diverse and efficient entities that seem to be designed to live in their environment."

"They are designed, Dina. By the environment itself."

"True, Sara. The environment is the silent unrelenting force shaping all life forms. Without exception. We can extend this idea further. If compounding changes increase efficiency among all living things, then the ability of the Earth to carry more creatures increases along with their efficiency."

"So, the Earth is full of super-efficient life forms that have been shaped over countless generations."

"Yes. Remember, not all changes result in innovation. Nature tries all kinds of changes, but only some result in improved chances of survival."

"But those few become the big winners in the survival game. Hmm."

"Dad, do you remember when I stole the candy bar from the store?"

Randy started laughing. "Oh God, yes. That is one of your mom's favorite stories. I've lost count of how many times she's told the story."

"I only have a hazy memory of it. I'm not sure if what I know is her version or what I remember."

"Let me see. You were three. Maybe four. It was a Saturday morning in the summer. In those days, you and I would do the grocery shopping on Saturday mornings and give your mom a little time with Maggy. While you were sitting in the grocery cart in the checkout line, you took a candy bar."

"I remember. It was a Snicker's Bar."

"Yep. That's it. A full-sized one. And I discovered it when I put you in your car seat. I think I asked you something like, 'What's this?' And you said . . ."

"A Snickers!" they both said in unison. The two of them laughed.

"That's right. Back we went to the store, where we, I mean, you, explained you took the candy bar. The clerk was a sixteen-year-old kid, and he was like, 'Wtf?' He leaned down close and said: 'Good thing you're honest, kid, otherwise the man would have thought I stole the candy bar and fired my ass." Then he stood up and gave me this smug smile. "I remember telling him, 'Thanks, buddy. May you have at least a dozen children.' He shrugged and said to you, 'Got it, kid?' You nodded, but I could tell you had no clue what was going on. On the way home, you asked, 'Dad, what's an ass?'"

"Wait, what? I have never heard this part of the story. I don't remember that."

"Yeah, I never told Mom. I knew she would over-react about the whole thing anyway."

"What did you tell me? When I asked the ass question?"

"I explained the term was a crude way of saying bottom. And you said," Randy laughed at the memory, "'Why would the man want to light that guy's bottom on fire?'"

Josh and Randy broke into laughter. "I don't remember this at all."

"When we got home, I gave your mom an abbreviated version of the story: you took a candy bar, we returned it, everything was fine."

"And Mom went into hyperdrive."

"Well, you know your mom. She was focused on you becoming the best person you could be. She didn't waste any time. I can still picture her, bent down in front of you, eye-to-eye, earnestly explaining that shopping was an economic exchange where each side needed to give something of value to the other to make the trade work."

"Yeah. I give her credit. She was a good explainer."

"You were three."

"But I understood what she was saying."

"Yeah, I remember you looking very serious and nodding at every word. When Mom finished, you were quiet for a few seconds before saying, "But, Mom, I didn't want to trade. I just wanted the candy bar.""

They both laughed. "You could have knocked your mom over with a feather, she was so shocked, as if it was the first time anyone had ever rejected the classical economic argument. She wrapped you in a hug, mostly because she didn't want you to see her reaction."

"Did she say anything? I don't remember."

"Yeah, she said 'Well, honey, let's think about that.' But that night, she said to me, almost wistfully, "Oh, to have the openness of a new mind. I would like to be more like Josh. Wide open and flexible in my thinking. It's amazing what that little guy keeps teaching me.""

"Sara, you said competition was brutal. Do you remember?"

"Yes, and I still think it is, no matter how natural it may be."

"Would you be surprised to learn other creatures shared your sentiment?"

"W-w-w-wait a massive maleficent minute. What?"

"Well, in a manner of speaking. As they do not possess intentions, they display no sentiments, but their actions are quite telling."

"Whoa. You're going to have to explain, Dina."

"Given the constant pressure they faced, entities developed different strategies to enhance their chances of survival by avoiding competition."

"Now, that is interesting. What do you mean?"

"What if a change enabled an entity to eat a different resource than its competitors? What do you think would happen?"

"Hmm. Well, then they wouldn't compete anymore."

"Correct. The pressure to compete for the same resource no longer exists if an entity can eat a different resource."

"Ooh, thus, avoiding competition."

"Now, what if an entity does not change, but instead moves to an area in which there are many resources but fewer competitors?"

"Hmm. Different tactic, same result. Competition disappears."

"Indeed. Entities survive in the most efficient way possible. Not by intention, but by responding to the pressures surrounding them. If survival is threatened and an entity can move, it will move. If it moves to a new environment in which it does not face as many competitive pressures, its chances of survival increase."

"Hmm. Competition is a survival strategy, but avoiding competition is a survival strategy too."

"Yes. If a variation enables an entity to avoid competition, it is as valuable as a variation enabling an entity to compete more efficiently. But this is only part of the story. Creatures that avoid competition have a greater tendency to diversify."

"What do you mean?"

"If an entity avoids competition by moving or consuming different resources, then it may produce many offspring, which leads to many variations. Variations leads to diversification. As time goes on, many different types of entities will emerge."

"Oh. So, that's why creatures are stuffed into every nook and cranny of this Earth. They're trying to get out of Dodge and find a nice quiet place of their own."

"No, I did not mention-"

"Ha! What I mean is they try to find their own environmental niche with as few competitors as possible."

"Yes. I was worried you had misunderstood my words."

"Oh no, I understand. I knew it. I knew competition wasn't the be all and end all economists insist on as they bore us to death with their multitude of books."

"Competition is a powerful force, Sara."

"Yeah, it's so great and powerful even the creatures who invented it try to get away from it."

"Entities did not invent competition, Sara. They were born into it. They compete or die. It is the default strategy for survival."

"That's only because they haven't figured out how to avoid it, so they're stuck with it."

"True."

"W-w-w-wait a slippery serendipitous second. Are you telling me competition is a last resort? That creatures would do anything else rather than compete?"

"In a way. Some entities can do nothing else. They cannot avoid competition and since they must survive, they must compete. But those who can avoid competition, tend to do so."

"Whoa Nelly in a Nutshell! This turns everything I ever learned on its head. Tell me more. Tell me the other ways entities avoid competition?"

"As we just discussed, they can consume different resources or move. Some creatures also do something called self-separation, that is, they consume the same resources at different times, or consume different parts of the same resource. For example, birds fly in the daytime and bats at nighttime in the same environment, thus not getting in each other's way. Some bees developed different lengths of proboscides so they could access nectar from different depths of the same flower."

"That's brilliant!"

"No, Sara. That is evolution. The bees with the long proboscides evolved from the short proboscis group. The bees with short proboscides were so numerous many died out. The bees with long proboscides survived because they could reach nectar their fellow short-proboscis bees could not. Over time, that variation grew into a different species. That is not brilliant, Sara. That is life."

"Then life is brilliant."

"That is a statement I cannot contest. Another way entities avoid competition is by attacking or even killing their rivals. They tend to perform this action more often when competing for mates, but it also happens to protect a group's territory or take over a competitor's territory. Infanticide is the most common form of intraspecies murder."

"No. Killing babies? Hmm. I suppose it makes sense. If you want to survive, destroy the rival's offspring so their DNA can't continue. Brutal, but effective."

"Indeed. It is an action programmed into the DNA of some creatures. Animals of the same species rarely fight over resources such as food or space. It is too perilous. The risk of dying when attacking a rival of the same approximate size and strength is quite high, so most animals use other less risky techniques to avoid competition."

"That makes sense."

"Even plants can take actions to reduce competition, Sara."

"Plants? How?"

"Some trees emit poison so their own seeds, or seeds from other trees, cannot grow too close to their trunk so as to crowd out the parent."

"That's kind of brilliant, too." Sara stopped. "But Dina, I don't understand something. I get competition is a basic survival method. But I can also see how entities can avoid it. So why does competition still exist? If it's easier to live without it and entities try to avoid it, why does competition exist at all?"

"Ease has nothing to do with it. Sometimes competition is the only way to survive. Competition exists because the need to survive exists in an environment where resources are limited. This is the nature of the Earth. It is ironic, but no matter how hard an organism tries to avoid competition, their reprieve is ever only temporary. Competition always comes back."

"What do you mean?"

"Reproductive success plants the seeds for future competitive pressure, Sara. Producing many offspring forces the numerous members of the new generation to compete. The more offspring, the higher the pressure. The pressure creates competition, and the competition creates not only diversity, but also renewed efforts to find new resources or new environments. Competition drives life outward to more and more unpopulated areas. Over time, Earth became loaded with efficient organisms."

"Oh man, this is disappointing."

"Why?"

"Don't you see? If competition always wins, the Earth is doomed to be a brutal and heartless place, just like all my business and economist friends say."

"Do not despair, Sara. Competition is the original and very successful survival mechanism, but it is not the only one. Some entities did figure out how to permanently avoid competition."

"Really? Sara smiled. "And it worked? Tell me."

Chapter 14

THERE'S GOT TO BE AN EASIER WAY

"Dina, Dina. Are you there?"

"I'm here, Sara. Is everything all right? Are you in pain?"

"Oh no, I'm fine. I'm just anxious to hear how living creatures beat competition."

"Why are you so opposed to competition, Sara? It is one of the most powerful forces on Earth. Why do you bear such animosity toward a force that was essential in shaping you?"

"Well, it seems to me that humans use competition as an excuse to do rotten things."

"You make a valid point. How competition operates in the non-human world is different from how humans use it."

"Oh? In what way?"

"This is a broader subject, which we are not yet ready to address. We will do so soon, but we are making progress. Will you be patient for a little while longer?"

Sara sighed. "You're driving this bus."

"Sara, that is a nonsensical statement."

She giggled. "It's an invisible bus, okay?

"Let us proceed before you veer our invisible bus off an invisible cliff."

"Ha! I declare, Dina, you're getting the hang of human-speak."

"Your human-speak, Sara. I am not convinced you are a fair representative of your species."

"Is that a good thing or a bad thing?"

"A factual thing. Now, shall we continue?"

"All right, Dina. How do entities avoid competition?"

"Let us start with this question. What is the opposite of competition?"

"Hmm. Competition sets entities up against each other and results in winners and losers. So, the opposite would be . . . joining with others to make a win-win."

"That is indeed what happened. To outsmart competition, so to speak, cells joined together and worked toward a common goal, so they were not forced to compete."

"Wow! Those smart little cookies. How did they do it?"

"I fail to understand how combinations of flour, sugar and egg relate to this conversation, Sara."

"Ha, and here I thought you were progressing in human communication. It's just an expression, Dina."

"Then I will ignore it. Let us start at the beginning and examine this activity that is the opposite of competition. Let us call it collaboration. Collaboration is the interaction of two or more units that results in an outcome neither unit could achieve by itself."

"Hmm, that's not what I think of when I use that word. I think of collaboration as an activity combined with conscious intent."

"Which is why we need to be clear in our definition. Non-humans are incapable of intent. When we broaden our meaning to include unintentional activities, we see collaboration has been going on as long as this universe has existed."

"What? You are kidding."

"Not at all. In the early universe, force particles combined with matter particles to create hydrogen atoms. This is a type of collaboration. Hydrogen atoms combined with other hydrogen atoms in the depths of a star to create helium and energy. This is also collaboration. And exploding stars forced elements together to create heavy elements like gold and platinum and uranium. These elements represent a type of collaboration, too. Each of these smaller units combined to produce a form that could not act in the same way as one unit acting alone."

"Interesting."

"Now, let us proceed to early Earth. Early in the history of Earth, inert material floated in the oceans. Units of matter joined, broke apart, and rejoined with other units. Whenever a complex unit was able to retain its structure, it became the basis for more complex forms to emerge. This combinative process happened repeatedly. Around 3.8 billion years ago, cells emerged that ingested material from their environment, converted it into nutrients and energy, and made copies of themselves."

"Aha! Our prokaryote friends."

"Yes. Prokaryotes are the earliest living organisms containing DNA. If we opened one of these cells, we would see a number of diverse materials inside."

"Like RNA and DNA?"

"And many more, Sara. There is an outer shell that gives the cell its shape, a membrane that allows resources to enter and exit the cell, and a gel-like fluid inside called cytoplasm. The cell has external growths, pili, and flagella, that enable it to move, and it contains small vacuoles that store energy. Within the cytoplasm are amino acids, proteins, lipids, carbohydrates, RNA, and DNA."

"Holy cow! That's a lot of stuff packed into a little package."

"All these components work together to make the cell alive. Yet, none of the components themselves are alive. In studies, scientists separated organelles of a cell, and discovered their specific functions. They found that though the organelle performed its function outside the cell, it was no longer alive. It was merely matter causing something to happen. If organelles are broken down further, the parts exist only as inert organic matter."

"Dina, you're exploding my head again. What you're saying is all these tiny pieces of matter joined, or collaborated, to make life."

"Indeed, this is exactly what I am saying. Because inert matter collaborated in this very particular way… we can say it changed the course of how energy exists on Earth."

"How do you feel today, Ms. Sara?" Shelley was just starting her shift as Sara Wallace's caregiver. She drew the shades halfway down the window. "The sun is getting stronger every day. We don't want you to be too warm." She turned toward her patient. "How about I brush your hair out? It's such a pretty color, and so long."

"She pulled the brush out of the drawer and pushed the button on Sara's bed to place her in a sitting position. "Ah, that's better." She began brushing Sara's tawny hair, speckled with gray strands. "You have thick hair, Miss Sara." She scooped it into a pile. "Let's make a ponytail, for you."

Holding the ponytail in one hand, Shelley reached down to the table to grab a hairband. She put down the brush to use both hands to secure the ponytail. "Oh my God!" she screamed and jumped back. "What is that?"

"Dina? Dina? Where are you?" No answer. "Dina?"

Only silence. "Dina, I'm sorry. Maybe you didn't want me to learn about . . . what you said about life being made from matter. I won't tell a soul, I promise."

Nothing.

It seemed like a long time before she heard Dina's whispery voice. "Sara, are you awake?"

"Yes. What happened? Where did you go? Are you okay?"

"I had to leave suddenly, but I am here now. Where were we in our discussion? It is crucial we continue without delay."

"Is everything okay? Dina?"

"All is as well as can be expected, but I need to ask you a question not related to our topic.

"Sure."

"You mentioned you see something when we talk."

"The gold mist. Yes."

"Describe this to me."

"Well, it's more like a haze. It looks like there's a golden filter in the air. It pulsates when we talk, from light yellow to deep gold. It's lovely. When I see it, I know you are here, though I can't actually see you."

Dina paused. "I must consider this, but for now, let us continue."

"You were telling me about prokaryotes being the first link between matter and life."

"Yes. I would also say prokaryotes are Earth's first example of successful collaboration by a life form using DNA. It was different from all past collaborations between inert materials in an important way."

"Oh? How?"

"Inert matter can collaborate to produce a more complex structure, without changing the basic composition of its components. For example, when hydrogen forms into helium, the hydrogen nucleus still exists, but it is embedded in a new structure, the helium atom. If you applied enough energy to split the helium atom apart, two hydrogen atoms would result. With non-living matter, the fundamental components can be recovered.

"Living entities act differently. They break down the food they consume into energy and nutrients, and they can make copies of themselves. They use the matter they consume to change themselves. In cells, resources are transformed into different materials by the introduction of energy. It is the collaboration between inert matter and energy that enabled life to emerge."

"And prokaryotes are the bridge."

"I do not understand."

"Prokaryotes are the bridge between matter and life. Everything inside the prokaryote is inert, just proteins and carbohydrates and RNA and DNA. None of these things are alive, but they collaborate to produce a living thing. They are the basis of all life. All other life forms just extend from prokaryotes, right?"

"Indeed, but this did not happen overnight, Sara. Collaboration is a time-consuming process. Unlike competition, which is the immediate result of existing in a high-pressure environment, collaboration is a response to competition that took a very long time to occur. When the conditions became hospitable for life to emerge on Earth, life did not immediately result. It took 750 million years for many small changes in matter to develop into the prokaryote. In nature, many variations occur, but not all of them produce operable forms. Most variations fail. Each altered structure must fit its environment to survive."

"Ooh. So, that's why it took so long for the prokaryote to develop. Millions, perhaps billions of variation cycles occurred before a workable cell emerged."

"Correct, Sara. Prokaryotes are small. As you humans learned from the coronavirus epidemic, single-celled organisms can replicate and mutate quickly. Within mere months, several new strains of coronavirus developed, each with slight variations affecting their ability to infect and sicken humans. Any organism with access to an abundance of resources takes this path."

"And we had eight billion resources for the coronavirus to work on. Hmm, let me summarize. Take small cells that can rapidly mutate, give them millions and millions of years to lie around on a relatively stable, resource-abundant Earth and . . . a lot of different models can result. One of them was the prokaryote."

"Correct. Now, let us proceed a little further forward in Earth's history. Do you remember when we talked about complexity?

"I do. Organization allows entities to become more complex."

"And what was the difference between prokaryotes and eukaryotes?"

"W-w-w-whoa notorious Nelly. The eukaryote contains all the same components as the prokaryote but is more organized. But . . . what does this have to do with collaboration?"

"You are almost there, Sara. Remember when we talked about cell freedom? How cells trade their relative independence for security?"

"Hmm. Are you saying organization is just another word for collaboration?"

"Yes."

"So, eukaryotes are more collaborative than prokaryotes."

"Now, you have made the connection. Eukaryotes were more organized than prokaryotes because each organelle performed a specific function. Each organelle, performing their unique function, joined together with the other organelles performing their unique functions to produce a cell that could survive and replicate itself. Humans call the act of dividing responsibilities 'specialization.'"

"Specialization?"

"Correct. With specialization, organelles expend only the energy needed to perform their single function."

"Which makes them more efficient than prokaryotes."

"Yes. Over time, the protected organelles were able to evolve more intricate structures to perform their function even more efficiently. What happens when one entity is more efficient than another?"

"Hmm. The less efficient one dies out."

"Or, in some cases, the less efficient entity confines itself to locations where the more efficient one cannot compete. That is what happened in the early days of Earth. The complex eukaryotes, with their specialized architecture, were so much more efficient they dominated great swaths of Earth's environment and pushed the prokaryotes into tight niches where only they could survive. Prokaryotes still exist, Sara. In fact, they dominate in pure numbers, even today, but they have hardly advanced past the one-celled stage."

"So, collaboration can produce large, complex entities, but disorganization keeps things small. So, how did this happen?"

"Your scientists do not possess a good fossil record of early multicellular organisms, so they do not know for sure. They have developed two major theories to explain how this change occurred.

"The first theory is that single-celled eukaryotes of the same species joined together into clusters. Over time, variations in the clusters produced specialized functions. With enough time, different areas of the cluster could specialize until a completely self-contained, yet interdependent, entity formed. Scientists have shown that environments with scarce resources or many predators enabled these types of multicellular clusters to survive more successfully than individual cells."

"Interesting. What is the other way multi-celled entities might have formed?"

"Scientists have surmised that single-celled organisms transformed into multi-celled entities over time. As we know, the eukaryote evolved separate functional areas. In doing so, the eukaryote's DNA evolved to instruct RNA to create and nourish specialized organelles under different conditions. Over time, the evolutionary process produced a feedback loop between the cell components and DNA that would allow them to keep beneficial changes in sync with one another. If a cell replicated but did not fully separate, the same feedback loop that worked within the cell now could operate on the merged group, with the most fit arrangements surviving. Over time, they grew into a diverse array of multicellular creatures."

"Hmm. Both theories make sense. So, which was it? How did cells become groupies? Were they joiners or growers?"

"What do you think, Sara?"

"I knew you weren't going to answer my question. I'm not a scientist, so I have no clue, but I'm going to go out on a limb and say they're both right."

"Yes. Perhaps there were other ways, too. The point is eukaryotes developed a process by which the full DNA structure was incorporated into each cell, yet the cells only performed limited functions to keep their part of the complex entity alive. This is the first case of collaboration between living entities, Sara. It is quite significant."

"Right. Now, we're over the bridge. Prokaryotes bridged matter and life. Multi-celled eukaryotes involve life only."

"This is true."

"So now, we have multi-celled critters. What's next?"

"Around the same time eukaryotes evolved, sexual reproduction emerged as a survival strategy."

"Ha! That's funny."

"I do not understand the humor in this factual statement, Sara."

"That's because you're not human and particularly not a teenage boy discovering sex for the first time."

"I may not understand humans, but I have surmised by watching human behavior that one important issue they cannot cope with, besides death, is sexual reproduction."

"Oh no, we can cope with sexual reproduction, it's just the idea of sex we can't handle.'

"Why do you think this is, Sara? I am interested to learn what a human thinks about this subject?"

"Besides the fact we are immature, and we cover up our ignorance with stupid jokes?"

"That is a telling statement."

"I'm not an expert, and God knows, I fumbled through my own two attempts at trying to explain sex to my kids. Maybe if you tell me how sexual reproduction began, I can develop a better strategy?"

"Somehow, this does not seem to be a plausible outcome, but let us make the attempt anyway. On the surface, sexual reproduction would appear to be an unlikely survival strategy. It takes more energy and time to produce offspring than the previous mechanism of asexual reproduction. Asexual reproduction involves an organism creating offspring identical to the parent. It can occur in a variety of ways, but all involve the process of mitosis, in which the DNA of a cell is meticulously copied and separated, after which the cell divides into two identical units."

"Well, if we humans operated that way, we would still figure out a way to be immature about it."

"You are more familiar with human behavior, Sara, so I accept your conclusion."

"Ha!"

"Sexual reproduction is a complicated approach."

"You can say that again."

"Sexual reproduction is a complicated approach." Sara snickered. "A cell undergoes the process of meiosis in which a full DNA strand is divided into two identical strands."

"Oh, right. RNA bumps into DNA and splits it into two parts. Then the two parts attach to their complements."

"It is a bit more complicated, Sara. In sexual reproduction, half the genetic code comes from each parent. The genes are stored in sections called chromosomes. These chromosomes perform an unusual activity once joined during sexual intercourse. The genes mix themselves up before settling into a chromosome slot."

"Mix themselves up? What the heck does that mean?"

"It is akin to the frivolity of melodious seats."

"Melod . . . you mean musical chairs? The chromosomes play musical chairs?"

"Yes, except there is no music, and no chair is removed."

"Ha! Other than that, it's exactly the same."

"You are missing the key point here, Sara. This re-organization of the genes forces the joined cells to be different than those of their parents, even though all offspring share the same DNA. The offspring exhibit different characteristics depending on how the genes in the chromosomes are arranged. This process also repairs any unworkable combinations resulting from the blending activity. The cells then divide again to create four cells. Two of them become egg cells in females and the other two become sperm cells in males. When a male and female engage in sexual activity, these different eggs and sperm produce offspring with an even wider variety of characteristics."

"Nope. This is not going to help in the sex education department."

"Perhaps not. It should not make sense regarding energy use either. This process is much more energy-expensive than asexual reproduction."

"So why did it evolve?" Sara paused. "I know the answer, even if I don't actually *know* the answer. It must have improved the entity's chance of survival."

"That is indeed the answer. You are beginning to understand the patterns of the universe, Sara. If an organism finds itself in an environment of scarce resources or many predators, sexual reproduction allows it to produce offspring with slightly different characteristics. Even if only a few of them survive, they and their method of reproduction, remain to procreate another day. On the other hand, if the asexual organism cannot survive in a harsh environment, its identical offspring will likely fare no better; the organism may die, leaving one less organism practicing asexual reproduction."

"So, sexual reproduction is another example of collaboration. Hmm. It also speeds up the ability to diversify."

"Yes. Instead of waiting for variations to appear as mutations, the entity produces variations as part of its procreative process. As each new form is born into the competitive environment, variation and change accelerate. This acceleration is evident in the fossil record. Eukaryotes began to sexually reproduce about 2 billion years ago. Multicellular organisms emerged 1 billion years ago. And multicellular animals appeared about 600 million years ago, a mere 400 million years later."

"Hello Ms. Sara. Look what I brought for you, this beautiful morning." Shelley strode into the bedroom clutching a large bouquet of summer flowers. "Mr. Wallace told me how much you love flowers. Aren't these beautiful?" She laid the bouquet down on the bed next to Sara. "Now, let me find a vase."

She returned a few minutes later with a vase and her phone in her hand. The phone was encased in a wallet that folded into a triangle as if it were a tiny stand. She placed the phone on the bed, just beside Sara's head. Then she picked up the bouquet. "Let's see if this vase works for our colorful bundle of joy."

She hummed as she fussed with the flowers, occasionally glancing over at the phone. The video was rolling now. Perhaps it would finally capture proof that she was not imagining things. How many times had she seen that bright gold patch on her patient's neck? Every time she looked, she would glimpse it for an instant, and then it would disappear. Sometimes she wasn't sure what she saw was real. It was almost as if the patch sensed her presence or heard her voice. Why else would it disappear whenever she came near Sara's upper body?

She felt like an idiot sneaking around to film Sara, but she was frustrated. And curious. What if this color-changing occurrence was making her patient sick? Was it a virus? If it moved so fast that she couldn't be sure it was real, maybe the doctors had missed it too.

She finished arranging the flowers. "Perfect. I hope you enjoy these, Ms. Sara." She straightened the blanket as she unobtrusively picked up the phone. She played back the video. "So, I'm not crazy." There on the screen was a gold patch. It spread from the base of Sara's neck up to her scalp. It pulsed, changing from a pale yellow to a bright goldenrod to a deep bronze. "Oh, Ms. Sara, what is happening to you?"

"Sara, sexual reproduction caused the rate of change in entities to accelerate. With sexual reproduction, each new generation now inherited different characteristics. And because the fight for survival continued relentlessly, those best fit to the environment were more likely to survive to produce offspring with even more varied characteristics."

"Oh, so once sexual reproduction started, it turbocharged the evolution of all kinds of different critters."

"Correct. Around 600 million years ago, a particular model of multi-celled entity emerged. We can surmise this model was more efficient than all other previous models, because all of today's eukaryotic creatures share its format.'

"All? All of . . . us?"

"Indeed. Fish and bees and humans are remarkably similar, far more similar than different. They each possess brains and hearts and appendages and circulatory systems. They appear quite different on the outside, but they operate similarly on the inside. This is because they all emerged from a common ancestor embodying this basic structure.

"From that basic structure, a multitude of multicellular entities has emerged. Cells did not stop collaborating because they existed in multicellular units. They continued to produce more efficient functions and more specialized roles. Complex entities produced cell groups called tissues. Tissue cells then collaborated to make larger units called organs. Each new development resulted in more efficient functions: hearts to pump blood; kidneys to process waste, stomachs to convert raw resources into nutrients and energy. With every functional change, the bodily structure changed as well. The cells within the tissues and organs collaborated to create appendages to ensure the creature better fit into its environment. Arms and legs for living on land, or fins and gills for living in water, tails for balance, fur for warmth."

"Ooh, I am starting to put the whole picture together."

"Yes, Sara. Competition and collaboration worked together to produce the diverse ecosystem that exists on Earth."

"Hmm. And both work simultaneously."

"Indeed. Did you know there is one additional level of collaboration more complex than a multi-celled entity?"

"Hold on. I'm still marveling about how lions and tigers and bears formed. More complex than individuals? I have no idea."

"Think, Sara. A group of entities acting as a larger unit."

"Hmm. A lot of animals stick together. A pride of lions. A gaggle of geese. A murder of crows. That's my favorite. A colony of ants."

"Sara, once again you have stumbled on the truth without any awareness of doing so. You have a talent for this."

"Ha! I know, right? What did I get right this time?"

"Tell me, is there a difference between a pride of lions and a colony of ants?"

"Yeah, lions are bigger. Duh."

"Let us consider the smaller of the two."

"Ants? Ants live in colonies."

"Indeed. Ants form colonies. As do bees. A colony is an unusual formation in nature, one in which the individual entities collaborate to ensure the colony survives. Feeding baby bees or ants different amounts of food after hatching produces individuals that perform different functions: foragers; guards, caregivers, or queens. In a sense, the colony acts like a living entity, and the individuals act more like cells within a body."

"That's amazing. And those individual bees do what collaborative cells do; they give up some of their freedom to specialize in one function, and in turn receive protection and nourishment from all the other bees performing their functions."

"It is the same pattern. This is how nature works. Over and over and over."

"But why did bees and ants develop colonies, but not humans?"

"Bees and ants branched off into their own groups around 120 million years ago. They have had a long time to evolve this survival technique. Do you recall when humans branched off into their own species, Sara?"

"Oh. About 55 million years ago. So, with another 65 million years, we might form colonies too?"

"No. I do not believe this will happen."

"Why not? If bees, can do it, we should be able to do it, too."

"You humans diverged down a path too different to perform like bees or ants."

"How? In what way?"

"Mr. Wallace, I'm about ready to leave for the day. May I speak to you for a moment?"

"Of course, Shelley. Come and sit down. Would you like a cup of tea? The water's still warm."

"That would be nice. Thank you."

Randy pulled a mug out of the cupboard. "Oolong, okay?"

"Fine." Shelley said.

He plopped in a teabag and filled the mug with steaming water from the electric kettle. "There you go. So, tell me, what's on your mind?"

"I – I – I'm worried about Ms. Sara," Shelley stammered.

"What? Is she getting worse?" Randy stood up and moved toward Sara's bedroom.

"No, Mr. Wallace, her condition hasn't changed. But. . ."

Randy slowly sat down again. "But . . .what?"

Shelley took out her phone. "I made this video today. I noticed something on Ms. Sara's neck a few days ago, but it disappeared whenever I touched her or came too close to her. So, today, I filmed the spot while I was putting flowers in the vase by her bed."

"I don't understand."

"Please look at the video, Mr. Wallace. It sounds crazy, but please, just look."

She held the phone out and pushed the play button. Randy watched in silence. "What the hell is that?"

"I don't know. I wish I did."

He stood up and headed toward the bedroom. "Well, let's go see."

Randy walked to Sara's side, gently turned her over and lifted her hair.

There was not a mark on her. No discoloration. No shifting of colors. Nothing. Shelley sighed. Randy looked at her. "It would be about the stupidest thing you could do to make this up."

"I'm not making it up, Mr. Wallace."

He took a deep breath. "I know you're not." He turned toward his wife, lying quiet and still. "So, what the hell is going on?"

Chapter 15

300 TRILLION MUSKETEERS

"Dina? Are you there?" It wasn't often Dina didn't answer when Sara called. Sara sighed. She was worried about her new friend, if it were possible to be worried about a thing she couldn't actually see. She had noted an urgency in Dina the last few times they spoke. As if Dina was in a hurry. Or was anxious about something.

Sara couldn't fathom a reason to be in any hurry for any reason. She was merely having a crazy nonsensical dream. Except Dina said things that made sense. She had to admit, this wasn't like any of her other dreams where stupid, unbelievable events happened that made perfect sense while she was dreaming but were even more stupid and unbelievable when she woke up.

Randy sat by his wife's bed. He piled her hair to one side, so he could see her neck and waited quietly. Waited to see the golden color Shelly showed him on the video. He sat for over an hour. Nothing. He leaned over to kiss her cheek before leaving. "Sweet dreams, Sara Mia."

"Dina?" No answer. Well, perhaps she could think about these ideas on her own. Now, what had they been talking about? Oh yeah, collaboration. Collaboration that started within a prokaryote and grew to work in complex entities acting as a unit. There were even colonies of entities that worked as a team. But humans did not operate in colonies, and never would. Dina was adamant about that. Hmm. I wonder why?

Did anything more collaborative than a colony exist? Can different species ever work together for mutual survival? Of course. The evidence was all around her. She needn't travel any further than her own garden. How many times did she watch bees pollinating her flowers? Dipping into one blossom, then another, then another, before zipping back to a hive hidden in some safe spot. She read about how honeybees and flowers developed a. . . what was the term? Oh yes, a 'mutually beneficial' relationship. The bees ingest nectar from flowers to consume the nutrients they need to survive. While feeding, they collect pollen on their furry legs which they deliver to other flowers as they continue to feed. The cross-pollination made possible by bees enabled the flowers to procreate.

Hmm, she was pretty sure the bees and flowers didn't sit down and convene a meeting to decide to collaborate. She giggled to herself; but she would bet Randy a million dollars a Far Side cartoon existed somewhere depicting that very thing. Nope, bees and flowers collaborate the way all living things do. They evolve into it.

In the early days, lots of early insects and plants existed, she thought. Each one developed different survival characteristics. Plants evolved all kinds of structures to optimize their chances of procreation. Insects tried all kinds of ways to obtain energy and procreate. Over time, the plants and the bees co-evolved a mutual survival strategy. The bees found a high source of energy in flower nectar, and flowers created a sexual organ covered in pollen the bees needed to wade through to get to the nectar. Did the flower need nectar to survive? Maybe not, but the bee did. So, the flowers that produced nectar attracted the bees and obtained greater pollination, thus guaranteeing more flower offspring. Do bees need furry legs to survive? Maybe to help regulate temperature, but generally not. But the bees carried more pollen when they had fur on their legs, so they did a better job of pollinizing plants, thus ensuring the availability of lots of nectar. Together each co-evolved to ensure the survivability of both.

"Quite well thought out, Sara. I am impressed."

"Dina! Where have you been?"

"I was . . . unavailable for a time. I am here, now."

"Are you all right?"

"You continue to ask me this, Sara. Why?"

"Well, because I care about you."

"I am familiar with this word care, but I do not understand how it applies to our discussion."

"I've gotten to know you, Dina, even though I can't see you. I like you. You're interesting and funny and super smart. Those are qualities I admire."

"Our discussion is important, Sara, not my qualities."

"You definitely must not be human, because people mix those things up all the time."

"You are quite correct. I should have been able to discern that. We will discuss this quality later when we talk about-"

"Humans," they said in unison.

"We will get there," Dina said, "but we must work at a faster pace now."

"Why? These things we are talking about are getting more complex. I might need to go slower if I want to be sure to understand them."

"That may be true, Sara. Yet the need for acceleration is becoming clearer to me each time we speak."

"Is something wrong?"

"Some matters concern me."

"What?"

"A problem I must correct."

"Oh. Can I help?"

"You can assist best by increasing our pace, if possible."

"Okay, will do. We were talking about collaboration."

"Yes. Let me start with this important point. When cells collaborate, it becomes impossible for them to compete. When they join together to form an entity, they work together for the benefit of each other and the entity. Competing against one another would provide no benefit, only harm."

"Hmm. That makes sense. Imagine what would happen if heart cells went to war with the stomach cells in a body? If the heart could start a war and win it, it would be a short-term victory at best. The entity won't survive without a stomach to process food into resources. It will be deader than a doornail with two hearts and no stomach."

"Sara, I do not believe pegs fastened to an access unit are alive to begin with."

"My point exactly."

"Your statement is nonsensical, but you make an inadvertent yet valid point. Collaboration pushes competition out of the way."

"Ha! Competition can't compete with collaboration! Irony alert!"

"Your bias is showing again, Sara. Collaboration does not destroy competition. It merely moves it up to a higher level of existence."

"Huh? Wait. What do you mean?"

"The entity made of collaborative cells still exists in a world of scarce resources, Sara. Competition does not disappear; it rises to the level at which the complex entity exists as an independent unit. The heart of a tiger beats, and its stomach processes food, but the tiger itself now must compete with other entities that consume the same resources the tiger does. If the tiger does not eat, it does not survive, and the whole collaborative structure, heart, stomach, and every other organ making up the tiger, dies."

"Hmm. Bees in the colony work this way too. They collaborate to keep the hive alive; they don't compete with each other. When the individual bee eats its fill of nectar, it doesn't decide to go on vacation in the Bahamas with its extra nectar in tow. It carries the nectar to the hive so the caregivers can feed the babies. No one is on the take in a colony. Everyone cooperates."

"That is basically true, Sara. Let us continue. Do you remember the effects of competition?"

"I do. Diversity. Innovation. Productivity. Those are the big three."

"You missed accountability. That is not a surprise. You humans often skip over this concept."

"It's not my fault!" Dina paused for a moment. "Joke alert."

"Non-humorous, Sara. Let us talk about what collaboration offers to living entities."

"Well, there must be something, otherwise it wouldn't exist."

"Indeed. Collaboration is a complement to competition. It produces quite different results."

Sara considered for a moment. "Oh, I think I know. Hip hip hooray! 'All for one and one for all. The 300 trillion musketeers!'"

"I do not understand."

"It's from a book by Alexandre Dumas. He wrote about the Three Musketeers. Because a tiger has 300 trillion cells, I took the idea of the musketeer and transferred it into the idea of cells, and then-"

"I believe this is the largest degree of veering you have done since our discussion began, Sara."

"It was a joke."

"Another joke? Then, please progress to the forceful slap mark."

Ha! The punch line. There is my sweet Dina. I was afraid you were losing your sense of humor!"

"Sara . . ."

"Okay, okay. I mean the cells are interdependent. Each of the tissues and organs in a body are alive; they are composed of living cells, but they can't function without the rest of the cells in the body. The survival of the cells in a tiger is not just linked, it is utterly dependent on every cell doing its job. Cut out the heart, a living organ, made of living cells, the whole entity dies."

"Correct. This is the most glaring difference between competitive and collaborative cells. Can you think of any others?"

"Let me think. W-w-w-whoa nosy nasturtiums. How about this? If competition is win-lose, then collaboration is win-win. I mean, they operate like the bees in the hive, they are mutually beneficial."

"Yes, Sara. Collaboration works because the cells participate in exchanges that benefit each participant as well as the entire unit. Each cell performs its dedicated function to help the body live. In exchange, it gets fed and is protected from competition. It is as you say, a win-win for those cells, unlike competition, which is a win-lose scenario. Mutual benefit is symbiotic. Not only does each cell function and survive, but together they create a body, which has an existence of its own that would not be possible but for the collaboration of the cells."

"Hmm. We see a tiger, not a collection of organs and tissues and cells. It has a value of its own outside its components."

"Yes. Fundamental to the process of collaboration is exchange. For collaboration to work, it *must* be a win-win. Each participant gives something to get something. The thing it gives must be something of value to the body, and what it receives must be of value to its own survival and, by extension, the body's survival. A heart pumps blood in exchange for nourishment and protection from competition. A kidney excretes waste in exchange for nourishment and protection from competition. Without the exchange, no collaboration can exist."

"So, mutual benefit and exchange must occur in every single cell for the body to survive. The fact the body also survives is kind of a bonus if you view it from the cell's perspective. The cells only care the entity survives because that's what assures the cells survive. Ooh, that is humbling!"

"No, Sara. It is life. This is how life works. Given collaboration as a survival mechanism, do you think entities grow more or less complex with time?"

"More complex."

"Why?"

"I'm not sure. I just know we started with one celled organisms and today the world is filled with every sort of critter made up of trillions of intricately related cells that all look alike but do different things."

"Indeed. This is caused by specialization. Collaboration drives specialization. Once started, it could not be stopped. Though a cell within a body is protected from competition, variations still occur at the cell level. Like all variations, the more efficient ones help the body survive, and the less efficient ones reduce the chance of survival. Let us consider the heart as an example. In small multicellular organisms 600 million years ago, the heart was a single tube. When the first fish evolved 420 million years ago, the heart became a two-chambered pump. In modern humans, the heart is a sophisticated four-chambered system. Over time, tissues and organs have evolved into more efficient members of the body. Once a stable complex form evolves, it becomes the basis for further changes. Life produces efficiency through complexity, rather than through simplicity.

"With these changes, the actual structure of a body changes too. Entities developed different bodily structures depending on their different environments. Fish and bees and spiders and tigers look different, though their internal structures are similar, because they are specialized to fit into their particular environments."

"So, specialization causes complexity."

"Indeed. But specialization cannot occur without interdependence. Let's talk about cell freedom."

"Freedom! Are cells tiny idealogues with tiny political agendas?"

"Veering, Sara. Freedom of a cell to move is different in competition and collaboration. In competition, a cell can avoid competitive pressure by moving to a new location or changing what it eats. Cells that are part of a collaborative unit do not have such freedom. The heart cells cannot abandon their body and lodge themselves in another body if the current one is failing in some way. No, they are forced to remain in the body they inhabit. To survive, they must play their role in the collaborative unit. In exchange, the heart cells are nourished and protected from competition from other cells or tissue or organs. The cells are still free to create mutations, but if a mutation harms the body, it might endanger the survival of the entire entity.

"Hmm. I know what we call cells that try to go it alone in a body. A super successful cell that grows uncontrollably? Yeah, that's cancer."

"Indeed. Cancer is an example of cells that stop being collaborative. The body does its best to kill a cancer cell before it spreads. What do you think happens if it does spread?"

"Hmm. Let's say a super successful Lone Ranger cell in the pancreas spreads to the body's other organs. Yes, this rogue pancreas cell is a winner. But now, every infected organ is being instructed to produce cells and act like a rogue pancreas. So, the stomach and the lungs and the liver can't do their actual jobs because they are being instructed to become one giant pancreas. What happens? The body dies. If a cell within a complex entity stops collaborating, it risks not only its survival, but the survival of the whole body."

"It is not that clear cut, Sara. Survival is a delicate dance between competition and collaboration. At every level of life is a symphony of the two. The inner workings of a cell, the tissues and organs of complex plants and animals, the members of a hive. All are collaborative units spurred on by the pressures of outside competition. Competition and collaboration work together to make a body function."

"So, which one is winning?"

"Both."

"Let me ask the question this way, Dina. Are there more collaborators or competitors on Earth?"

"You ask an interesting question. If we consider all life forms on Earth, we find more cells collaborate than compete."

"No way."

"Yes way. There are more single-celled organisms than multi-celled ones. But multicellular entities are large and contain many more collaborative cells than the total of all single-celled competitive cells. Remember the tiger? 300 trillion cells compose a single tiger."

"And those 300 trillion cells jumped out of the competition game and into the collaboration game."

"Exactly. With competition, each organism is completely accountable for its actions and alone bears the responsibility of its survival. That is a high-risk situation. By collaborating, organisms reduce their risk. Single-celled organisms that join together when dealing with harsh environmental conditions increase their chances of survival as a group. When a cell possesses the opportunity to specialize and become more efficient without having to continually fight off predators, it improves the chance the entity will survive as well. Even though the entity still must compete, the load on each individual cell is much lower."

"Oh. So, does this mean the component cells of a body are less accountable than those equivalent organisms who go it alone? I can just imagine some lazy laggard stomach cell sitting around, taking in resources, and then not sharing with the rest of the cells. We all know that cell. We've all had to pick up after that cell-"

"Sara, sometimes you make valid points but express them in childish human ways."

"Ha! I am so busted."

"I did not say you were broken, but you say things in a way-"

"It's an expression that means you hit the nail on the head."

"Sara, we are not discussing pegs on-"

"Forget it, I'm just leading you down the garden path. Go on."

"We are not on a walkway in an arboreal-"

"I never thought I would say this, Dina, but you're veering."

Dina paused. "Yes. Let me re-direct. How do complex entities keep their cells working in concert?"

"That's what I want to know."

"Would you be surprised to learn there is a form of collaboration devoted specifically for this purpose?"

"Well, knock me down with a feather. Explain."

Chapter 16

SYMPHONY

Dr. Chen examined the video for the third time. The reports she had read identifying the gold coloring on Sara Wallace were all anecdotal. The University doctors had not reported any instances of the odd occurrence. She nodded to the two nurses. "Let's take a look." She started with Sara's head and examined her face and scalp. "I don't see anything resembling what is on the video." She pulled Sara's gown down to her waist. Nothing. She then lifted Sara forward to examine her shoulders and back. Nothing. In the same manner, she examined the rest of her body. Nothing. "I don't see a thing."

"Didn't the caregiver say the gold patch seemed to sense her presence?" Tina, the nurse on her left asked. "Maybe we should try to repeat the video. Let's set a phone near her neck and record what happens. If that shows something, then we can look further."

Dr. Chen nodded. "That's a good idea." She set the video option on her phone and propped it next to Sara's head. "All right, let's wait five minutes." The three filed out of the room.

Those five-minutes felt like a hundred years to Dr. Chen. What was happening with Sara Wallace? How had she missed what was so clearly evident on that video? How had the University doctors missed it? And why could she not detect it now? What was happening inside her patient's body? Did the gold coloring cause the coma? Or did the coma cause the roving discoloration? The alarm on her phone rang out from Sara's room. She took a deep breath. "Let's go take a look."

The three medical professionals filed into Sara's room. Tina picked up the phone and stopped the recording. They huddled around while the video played.

Five minutes of nothing. Five minutes of clear unblotched non-gold colored skin. All they saw was the soft rise and fall of the carotid artery on Sara's neck. "You continue to perplex me, Mrs. Wallace," Dr. Chen whispered under her breath.

Sara had never waited so long for Dina since she started this crazy dream. Something was going on. She had noticed it before. The slight acceleration in Dina's speech. The frequent urges to hurry. If Dina were human, Sara would swear these were signs of worry. But her new acquaintance was not human. Hmm. So, what is Dina, anyway? She sighed. A magical dream fantasy, that's what Dina is. Okay, an endearing magical dream fantasy. Sara didn't really mind the break from her mysterious friend. It gave her some time to put together the ideas Dina threw at her in a constant stream. Their chats were going somewhere, and it had something to do with humans. But dang, it was slow going. Secretly, she just wanted to get to the part where people evolved and screwed up the world. At least that's where she thought Dina was going with all this.

"Hello, Sara. Have you figured out life's special collaborative mechanism that keeps cells from being lazy?"

She laughed. "I have. Strict Mama cells?"

"There you go, giving non-human life forms human traits again."

"That's only because I don't know the answer to your question."

"Let me begin by asking you some questions."

"That's how we always start when you throw a new idea at me."

"Indeed. Learning is best done when a question is internalized to produce an answer."

"Or at least a guess."

"In this case, a guess is as beneficial. It is the thought process that is important. Do you have a guess, Sara?"

"Besides strict Mama cells? Hmm. Well, if I were a body and I wanted to keep all my cells doing their jobs, I would need to communicate my expectations to them. So, boss cells. That's my guess. Each body contains boss cells that run the show."

"Yes, Sara."

"Really?"

"And no."

"What? How can I be right and wrong at the same time?"

"You are a human. This is a basic characteristic of all humans."

"Ouch. How come truth hurts worse than a bald-faced lie?"

"Because it is true."

"Smarty. So, explain this bossy but not bossy collaboration mechanism?"

"Do you know the meaning of the word 'coordination?'"

"Absolutely. Hmm. You mean, cells coordinate with each other? Duh, of course. They would have to."

"Let us define coordination, so we are clear. Coordination is the organization of different parts of a complex unit to enable them to work together to achieve a goal. Coordination is a type of collaboration. However, coordination behaves quite differently than collaboration, so it is better to think of it as a separate survival mechanism."

"So now we have three 'C's.' Competition. Collaboration. Coordination."

"Yes. We can think of competition and collaboration as actions an entity takes to survive. Coordination, on the other hand, is the mechanism that keeps all the parts of the entity working together. No living organism can exist without coordination. It has been part of life ever since the first life forms evolved."

"So, it sits underneath competition and collaboration; like a foundation."

"That is a useful way to look at it."

"How does coordination work?"

"Let us start by looking at prokaryotes."

"Our disorganized single-celled friends?"

"Yes. As chaotic as prokaryotes are, they are still somewhat organized. Early on, they developed mechanisms to detect conditions in the outside environment and communicate that information to the cell interior so the cell could respond to those conditions."

"How the heck did they do that?"

"In a variety of ways. Outside conditions come in many forms: temperature, pressure, light, presence of food or dangerous substances. These conditions express themselves in chemical forms. Cells developed receptors, specific string-like proteins with one end perched outside the cell membrane and the other inside the cell membrane. When the receptor bumped into the chemical signal in the environment, the receptor changed its shape. In response, the receptors inside the cell membrane emitted their own chemical signals causing the cell to react in some way. When some receptors received a signal about a change in conditions, they would deliver the message to other proteins and create a chain reaction that sent the message to DNA which then created a response. At first proteins used chemical bonding to deliver the message along the path. Over time, some receptors developed the ability to produce an electrical signal as the messenger mechanism. That was a significant development. Chemical signals are fast, but electrical signals move at the speed of light."

"Wow! This is fascinating. I guess I never thought about how a cell works, but it makes all kinds of sense. I can picture how these little pieces all started, and slowly by trial and error-"

"Not error, Sara. Trial and fit. Nature does not make errors. Nature gives every form of life an opportunity to survive. The environment, the conditions in which creatures find themselves, determines which ones survive. Over time, some cells created a web of these receptors covering the outside of the cell. They detected external stimuli on a continuous basis and sent messages to the interior cell for response. If the DNA responded so as to improve the cell's ability to survive, the DNA and its responses would be copied into the next generation. The cells with DNA that did not perform this function well would lose the survival race. With millions and millions of years to work with, prokaryotes developed an efficient and fine-tuned capability to sense the environment and respond in such a way as to enhance their chance of survival."

"So, are you saying coordination is the way the components of a cell work together to enable the entire cell to survive? Every component has a job, and the job must be done just right AND at the right time, or the cell goes kablooey. It's a bit like dominoes falling in a line. If just one of the dominoes is tilted slightly in the wrong direction, the whole line stops falling."

"Your analogy is close enough, though creating a living cell is a little more intricate than building a row of children's blocks."

"Isn't that the point, Dina? Each little component only fills a tiny role, right? One receptor guy checks out the environment, passes the info on to another little guy, who passes it to another who passes it to another and another and so on until the RNA and DNA respond and the cell survives or not. That's the genius of the whole system. Tiny little guys each carrying tiny little loads can make tiny little changes, and the cell eventually evolves into a complex being. And coordination is the key to making the parts work as a unit. As long as the little guys evolved to communicate with each other, they could transform into all kinds of creatures with enough time. Hmm."

"Correct. So, given a eukaryote is more organized than a prokaryote, tell me about its coordination capabilities."

"Ooh, they would have to be stronger."

"Yes. Single-celled eukaryotes converted the rudimentary receptors of the prokaryotes into more structured forms. They used a wider variety of chemicals to improve connections between signaling cells and even changed the length of some proteins to better connect."

"So, they improved how to transmit messages to the DNA."

"Correct. As eukaryotes evolved into multicellular organisms, the messaging systems became more sophisticated. Over time, multicellular entities gradually transformed signaling proteins into actual cells that transmitted messages. Today, scientists call these cells neurons and the system of neurons that resulted as bodies became more complex are called the nervous system, or the neural system."

"Neur way."

Yes, that was the way it worked, Sara. The receptors of the prokaryotes evolved into sense organs in multicellular creatures. And near these organs, a large cluster of neurons developed to process the continuous flow of signals coming from the sense organs. Do you know what that cluster is called, Sara?"

Sara shook her head. "Neur, I don't."

"I heard your non humorous joke the first time, Sara."

"Ha! Dina, you crack me up!"

"Do you know the name of the message processing organ, Sara? Think."

"You better tell me, Dina. I'm still laughing at my own joke."

"It is the brain."

"The brain! Oh, man. I should have guessed. Of course."

"Yes, the brain is the cluster of neurons. In every multi-celled eukaryote, the neural system takes signals from the outside world, like images you see, sounds you hear, aromas you smell, and objects you touch, and converts them into electric messages that are then passed from one neuron cell to another-"

"W-w-w-wait a snippy snappy second. That's how proteins and DNA work in the prokaryote."

"Yes. The mechanisms are no different, Sara, but the scale is much larger. The brain is a collection of cells, all of which contain DNA. The incoming message is dispersed throughout the brain where each cell, or more precisely, the DNA within each cell then processes a part of the message. The parts are then collected and formed into a response. The response is then returned through the neural system and instructs the appropriate parts of the body to respond.

"Each coordinating system, or neural system, developed to meet the needs of the organism it supported. Though every brain is different, they all function in a similar way. And they share the same goal; to enhance the body's chance of survival by coordinating its cell functions in its specific environment.

"Let me give you some examples. A bee has a brain volume of 2 cubic millimeters but contains 1 million neurons. Bees use their brains to process environmental conditions from their sense organs, mainly optical images and odors, but they can also communicate locations of food resources to other hive members. Bees have both short- and long-term memory."

"And they are bee-yootiful!"

"Fish have about 10 million neurons in their brains. Their brain size varies with their body size. Even though they live in water and ingest oxygen via water entering through gills, something other animals cannot do, their brains are quite similar to other eukaryotic brains."

"Sounds fishy to me!"

"Humans have about 86 billion neurons."

"86 billion! Wow!"

"Yes, if you compare brain size to body size, humans have by far the largest brain mass to body mass ratio."

"Really? We're freaks! Who's in second place?"

"Can you hazard a guess, Sara?"

"Well, since we came from the ape family, I would guess some type of gorilla or chimpanzee."

"That is a logical deduction, but incorrect. It is true primates as a group have quite high ratios, but dolphins have the next largest brain to body size ratio."

"Hmm, I think I read somewhere dolphins are highly intelligent."

"What is important to understand, is that with a greater number of neurons, an entity is better able to coordinate the actions of its cells."

"Right, because they can process more signals at once."

"Correct. Coordination efforts have even evolved outside of bodies. Bee colonies are an example of higher-level coordination. The formation of a colony is an extension of the way all other life forms evolved. With bees, the colony is the 'complex body' to be nurtured for survival. The queen acts as the coordinating system and ensures enough bees exist in every role so the hive survives. The bees are like cells inside the 'body,' each performing a specific function to make sure the hive survives. If the hive survives, they survive. A colony is a collaborative structure orchestrated by the queen coordinator."

"Ooh, I see the pattern. It is the same as how a cell works and how a body works. It really is . . . elegantly simple."

"In concept, yes. But in reality?"

"Hmm. In reality, living creatures are beautiful symphonies of the most complicated instruments ever to exist."

"Mr. Wallace, I am not sure what to tell you. We do believe something has invaded Mrs. Wallace's system. What it is, we don't know yet."

"That . . . that . . . thing, that gold colored patch? That is the 'something?'"

"Well, we are unsure. The discoloration could be a symptom, or it could be . . . "Dr. Chen's voiced trailed off.

"Could be what? A virus, or a creature, like a tapeworm?"

"We checked for a variety of parasites and other viral sources. We find no evidence of anything like that."

"You saw the video. Something is there."

"Unfortunately, we cannot replicate what was on the video. Whatever it was, maybe she has shaken it off."

"Or perhaps it's causing her coma."

"We don't know. I am hesitant to treat her when we are unsure of the actual cause of her symptoms. We have treatments for some parasites, but some of those treatments may result in damage if the parasite we are targeting is not present."

"What do you propose to do? Let her sleep forever? You said she's asleep. Can't you just give her something to wake up?"

"I need to tread carefully before I attempt a treatment. Yes, she is asleep, but I can find no cause for her condition. I can't take a blind stab at a treatment without data, especially when she is otherwise stable. I want to keep her at the hospital for more extensive tests. I've ordered another blood panel, and I plan to draw samples of different body fluids. I am going to run a more extensive EEG. Perhaps the coloration is related to brain activity."

"What do you mean?"

"Her brain activity levels are quite high for a comatose patient. It's not unheard of, but it is rare. What I want to understand is if the brain activity and coloration are somehow linked."

"And then what?"

"Let's see what we find. I need to have better information before I act, Mr. Wallace. She has been stable since she fell into the coma. I do not believe she is in any immediate danger. Can you wait a bit longer as I try to figure out the best way to treat her?"

"All right, Dr. Chen. When will you know something?"

"Soon, I hope. I put a rush on the tests."

"Hello, Sara. We are in your childhood lilac garden. You do have an affinity for this place."

She breathed in deeply. "I love it here. It's so realistic I can actually smell the lilacs." She inhaled again. "I can't get enough of them."

"You can smell the flowers?"

"Mmm, yes. They're lovely."

Dina was quiet for a long moment. "Let us resume our discussion about coordination."

"Of course."

"Do you remember the benefits of competition?"

"I do. Diversity. Productivity, Accountability, Innovation."

"Good. You included accountability. And do you remember the benefits of collaboration?"

"Mm-hmm. Specialization. Interdependence. Mutual Benefit. Complexity."

"Correct. Now, we shall discuss the benefits of coordination. As I mentioned coordination works quite differently than competition and collaboration. Recall a complex body is made up of cells grouped into organs and tissues that perform specific functions."

"Yep, the 300 trillion musketeers. All the cells work together to enable the body to survive."

"Correct. And what is the role of the neural system?"

"To send messages from the cells to the brain and back."

"Yes. Why?"

"Why what?"

"Why do they send messages?"

"To coordinate."

"You worked your way around the circle without answering the question, Sara."

"Oops."

"Even though the neural system is a component of a body, its role is to ensure the body itself survives."

"Ooh, that is different. Coordination keeps the whole unit alive. The body. But, why, Dina? if every cell in a complex body works in a self-beneficial collaborative mode, and the cells' survival ensures the body's survival, why does the body need any coordination help at all?"

"Collaboration cannot exist without coordination. The actions of the neural system ensure the cells work together the way they need to, so both the cells and the entity survive. If something happens to the body, an injury, or an accident, for example, the neural system works to protect the body, and most of the cells in it."

"Most? Not all?"

"There are cases, Sara, where the neural system kills some cells to protect the many."

"Oh yes. We talked about that. Ap- apt- what was it?"

"Aptosis. In these cases, the neural system instructs certain damaged cells to die. As a result, the majority of cells and the body, survive."

"The neural system makes decisions about life and death?"

"It appears so. Tell me, Sara, do you think that is actually what happens?"

"No."

"Why not?"

"Because you wouldn't ask if it did."

"Your logic is backwards, Sara."

"Yes, because I've been backwards into a corner. Obviously, I don't know, but I know this much; nothing in nature makes a decision. Living things follow instructions from their DNA and are stuck with the results, whatever they may be. Odds are in my favor with a No vote."

"Your thinking process is quite inventive, Sara. Let us talk about how a neural system works. Maybe this will help answer the question."

"Fair enough."

"Regardless of how brains evolved and how different they are in each creature; they all generally work the same way. Coordination requires the neural system to set a goal, obtain information, establish rules, and enforce those rules."

"Careful, Dina, I think you're giving brains some human qualities. Set a goal? Establish rules?"

"You are correct, Sara, but in this case, it is the best way to explain how a brain works to a human."

"Now, I'm curious. Go on."

"For coordination to be successful, the complex organism must have a goal."

"Dina, that goes against every single thing you have said about the universe until now. A goal? Nature doesn't have goals. Everything just exists in its environment, end stop. If an entity fits with its environment, it survives to procreate. If it doesn't, it loses out to its competitors and dies. Live or die. On or off. No goal anywhere in sight."

"Stop, Sara. Go back and think about what you said."

"You know what I said. You have been saying it all along too. Entities fit in the environment. They survive or die." Sara stopped. "Ohhhhhhh."

"Go ahead, Sara. Continue."

"Of, course. That's the goal, isn't it? Survival?"

"Yes."

"But. . . How does it work? Entities can't make decisions."

"The goals of entities do not work the same ways as goals of humans. Humans think of goals first, then act on them, but non-humans inherit their goals. You are correct, non-humans do not make decisions; they follow the instructions of their DNA. What happens to an entity that follows its instructions?"

"They survive or don't survive based on the conditions of the environment."

"Yes. If an entity survives, it goes on. If it does not survive, it is out. Permanently. It is gone. Its DNA is gone. It is forever lost to the Earth. So, tell me Sara, at any one point in time, what types of entities exist on the Earth?"

"Hmm. The survivors."

"Yes. So, given entities cannot make decisions and can only follow instructions in their DNA, can we say that DNA instructs them to survive?"

No, Dina. Creatures follow their instructions and if they survive, they survive. If the environment changes drastically tomorrow, they could die by following those very same instructions."

"True, but is not the result identical? DNA instructions have been honed over eons and eons to give the entity its best chance of survival, not by overt decision, but by attrition of the non-survivors. All creatures that followed instructions but did not survive have died. Only survivors are left. We can think of it like this: all survivors carry DNA that in the past promoted survival. So, we can say the goal of any non-human, not by overt decision, but by the fact of their existence, is to survive."

"Oh, what you're saying is survival is the only goal, because survival is the only game. Nothing else matters except that an entity lives. Hmm. So, DNA instructs the cell to do everything it can to live, because the only other option is to die. In that case, the DNA can't do squat. The cell's coordinating function didn't set the goal. It inherited the goal from billions of generations of survivors. It's the transitive property again. If all organisms strive to obtain resources, and obtaining resources enhances survival, then the transitive property would say all organisms strive to survive. Voila - we have a purpose, as unstated as it may be."

"Right, Sara. This is a difficult concept for humans to understand, but I believe you have grasped it."

"W-w-w-wait a merry monthlong minute. Are humans ironic and backwards from nature? Or is nature ironic and backwards from humans?"

"Who was here first?"

"I knew you would side with nature."

"There is no siding to be done, Sara. The point is true regardless of the direction from which it is viewed. Shall we continue?"

"Of course."

"It is important to understand that a neural system does nothing to directly aid the operational existence of an organism."

"What do you mean?"

"Sara, the neural system does not participate in the direct functions enabling an entity to survive, like eating or converting food into energy or pumping blood. So, given what living entities need to do to survive, a brain would appear to be a nonessential element of an organism."

"Unless it made itself useful by making the body more efficient. Isn't that how all life changes are incorporated?"

"Well said. A neural system enables the cells in a complex entity to focus their essential activities to be more efficient than if it did not exist."

"So, how does it work?"

"Though the coordinating system is composed of the same cells as the other cells in an entity, it is physically separate from them. Brains and nervous systems are covered by bones and tissue layers that serve to protect them from injury. These coverings also enable the neural system to efficiently perform its function. Some responses may not benefit a particular organ or cell, even though the overall body benefits. The coverings make sure any negative outcomes resulting from neural actions do not affect its ability to operate. For instance, the neural system may direct white cells to kill off cells infected with a virus. The protective coverings prevent any detritus from hitting the brain and affecting its ability to do its job."

"Hmm. Are you sure the brain is not just being a selfish prick, and doesn't want to be held accountable for its actions?"

"That is a human response based on your knowledge of humans, Sara. It is not an accurate representation of non-human behavior. I assure you, we will discuss this topic later. For now, it is essential you focus on how all other living creatures operate."

"So, non-human brains aren't selfish pricks."

"I am not sure why you label brains as egotistical perforations, but the answer is no. The neural system is fully accountable for its actions. If its responses enable an entity to survive, the entity, along with the neural system, survives. If its responses result in injury or death, the neural system suffers as well. They are intricately connected. Neurons are nourished as all other cells are; it gets no preference in nourishment over any other organ or cell in the body. Yet brains perform a vast amount of work to keep a body alive. Millions of signals come into receptor organs of a body every second. Each receptor cell transcribes the signals into a form that can be handed off to the neural system. Each neural cell in turn transfers the signals up the chain and through the nervous system until they get to the brain.

"The brain collects responses coming in from all the body's receptors. It processes a response for the entire body based on what benefits the entire body, and then sends those messages back through the neural system so every cell required to respond receives the information about how and when to respond. Remember, Sara, this happens over and over and over every second, millions of signals traveling to and from the brain enabling a body to run or jump or to pick up an object or even make an anthropomorphic joke about non-human brains . . ."

"Ooh, snap, Dina. . . very good."

"I am risking the loss of my point by exhibiting a bit of human cleverness, Sara. It is important you understand why the neural system evolved to be protected and separate from the rest of the body."

"I do, and it's not because the non-human brain is acting like a jerk. The brain has a huge and super complicated job. It can't stop for a second without compromising the wellbeing of the body. It keeps everything going in the same direction, so a body acts like a unified set of cells rather than a disorganized blob."

"Yes. The structure of the neural system evolved to obtain information and pass it to the brain for processing as quickly as possible. We will discuss the importance of information soon, but for now, I want you to understand that information is crucial for a coordinating system to function."

"Well, that makes sense. Coordination doesn't work without something to coordinate. Duh."

"Correct. Receptor organs like ears and eyes are constantly handing information about the environment to the neural system. But that is not all the neural system processes. In a complex organism, every single cell in the body continually emits data about its status to other cells. This data is in the form of proteins or enzymes or amino acids. Some of the data goes to other cells so they can operate efficiently. Some go into the nervous system where they are passed on to the brain. The brain then processes this stream of data and sends back its own messages specifying a response."

"So, let me make sure I've got this straight. Let's say I see a tiger."

"To be precise, the optic cells that make up your eyes absorb light in the form of photons representing the shape of a tiger."

"Fair enough. The optic cells that make up my eyes absorb light in the form of photons representing the shape of a tiger. A large, menacing, scary tiger. They transmit the light data to neural cells, which convert the tiger photons into electrical signals, which then sends them through my neural system to the part of my brain that processes images. The cells in my brain convert those signals into a visual image. Aha! A scary tiger! The brain then sends out signals in response, like a bunch of adrenaline from somewhere, and whatever amino acid causes one to pee their pants, and a multitude of other messages to all the parts of the body, all of which basically say, "Run!""

Dina remained silent for a long moment. "I am quite sure amino acids do not cause one to produce urine."

Sara giggled. "I knew that would get you going."

"I am not sure you have learned these ideas, or you merely wanted an opportunity to tell a non-humorous story that included urine."

"Both?"

"I insist we continue. To be clear, I want you to remember that coordination is not possible without the constant stream of information coming into the coordinating system for processing."

"And without coordination, life as we know it isn't possible."

"True. Now let us talk about the rule-making capability of an entity's coordination systems, Sara. Do you think neural systems set rules?"

Sara laughed. "I'm onto you, now, Dina. No. They're like goals. Entities inherit them from their DNA."

"Very good, Sara. In every cell, responses to information are encoded inside DNA. When a bit of information is passed from protein to protein along the information path within the cell, the last protein in the line causes the DNA to respond. The DNA then instructs the RNA to make a protein, or stop making a protein, or respond in some other way. RNA follows the DNA's specific instruction, which is the response to the incoming information. This process links the incoming signal to the resulting response."

"Oh, it's not like a rule humans would make, but it's similar. Like goals, it's backwards from how we work."

"Indeed. Humans work like this: If A happens, then we do B. If C happens, then we do D. The cell works the opposite way. A happens, and B is the programmed response. C happens, and D is the programmed response. There is no 'if' with non-humans; they only do what they are instructed to do."

"No iffing way!"

"I am ignoring your non-humorous comment without seeking explanation."

"Ha. Smart."

"Cells evolved to react in specific and repeated ways to a particular stimulus. There is no choice, just a response. The more often a cell can produce the same beneficial response to a stimulus, the more likely the cell is to survive. A cell producing a beneficial response but cannot repeat it when the stimulus occurs again, would not be as likely to survive as one with more consistent repetition ability. Does this make sense, Sara?"

"Yes. That's why it looks like a rule. It's consistent. Repeatable. These coordinating systems are sort of amazing, aren't they?

"Your assessment does not include the coordination system's final function."

"Oh. Enforcement. May I take a stab at this one, Dina?"

"Indeed."

"Okay, here goes. Rules aren't worth anything unless they're enforced. And enforcement occurs like all the other coordinating functions, from prior survival success. When a cell survives because it follows the rules, the rule is enforced. If a cell dies or is damaged because it didn't follow the rules, well, the rule is enforced, but in the opposite way. The result is rules are followed or . . . else."

"That is a fair summarization. In all cases, the neural system works to keep as many cells as possible alive and doing their jobs. This is how the enforcement mechanism works. If most cells function properly, the entity survives."

"Fascinating. All these coordination techniques evolved through trial and err- I mean trial and fit. Holy cow, Dina, either complex life is truly amazing, or I don't understand the power of billions of years that can produce astounding creatures that can do astounding things."

"Both, Sara. Those choices are not mutually exclusive."

"Damn," she whispered. "Humbled again."

Chapter 17

THE UNITED NATIONS OF LIVING BEINGS

Sara rested on the back deck of her childhood home, breathing in the scent of the spring lilacs. Dina was nowhere to be found, but she realized she could conjure these memories up all by herself. They were so vivid, as if she were really here, in this place she called home over 40 years ago.

She could not stop thinking about the concept of coordination. It had to be stunningly complex. She remembered once reading that in a chess game, with 64 spaces and 32 game pieces, about 10^40 legal moves are possible. That's one with 40 zeroes after it. That's a lot. If a human body contains 35 trillion cells, of which 86 billion are neurons, just think of all the possible interactions that occur. And all, or at least most of them, must work properly for our sample human to survive. This isn't horseshoes. Close doesn't count in coordinating complex life. If a heart doesn't get what it needs, it can't pump blood and nourish cells and remove waste. If a stomach doesn't work properly, it can't convert energy for the body to use. Every little interaction needs to go right. If things go wrong, the neural system must correct those situations if the body is to survive. Hmm. For a complex entity to form, a complex neural system must form too.

Man, she thought, that must take a ton of energy. What did Dina say? Neural systems don't do the normal functions of a body, they make the normal functions more efficient. That's a lot of overhead to spend for something that doesn't process food or produce offspring. Then why did neural systems form?

Well, she knew the answer. The neural system evolved like every other modification of a life form. It was more efficient. If a creature with an expensive neural system could utilize its coordinating capabilities to obtain enough resources to 'pay for' its brain use and still create offspring, then it gains an advantage that is handed down to the next generation. Also, if an entity with an expensive brain could use its neural system to obtain resources or mates more effectively than its competitors, then that too would give it an advantage and enhance its ability to survive. So, even though neural systems use a lot of energy, if the body's investment in them paid off in a better chance of survival, then the complex energy-hogging coordination mechanism model would win out over other forms that consumed less energy.

There must be a tradeoff somewhere in this deal, Sara thought. If a brain grows so big the body can't obtain enough resources for its other functions, the creature would become less competitive, which would reduce its chances of survival. On the other hand, if the brain was so small it could not support complex functions, it may not ever obtain the ability to protect itself in a world of predators.

Brain development must be a trade-off between size and complexity. Duh, Sara thought. Of course. It all depends on the environment. Each entity could only spend as much energy on coordination as the competitive and collaborative environment allowed. Do plants have neural systems? she wondered. As far as she could tell, plants didn't have full-blown brains, but they still might have weak neural systems to coordinate responses to the environment and to keep the plant operating as a unit. Humans, on the other hand, have freakishly large brains and intricate neural systems given the size of the body. Except for some people she knew. Some people she knew definitely had plant-sized brains. Sara giggled to herself. If Dina were here, Sara would right now find herself being admonished for her childish human smallness. And Dina would be right!

Dr. Chen read through the voluminous reports displayed on her computer screen. Mrs. Wallace continued to be a frustrating mystery. Her brain activity varied too much to be consistent with a comatose patient. And she saw no sign of a link between brain activity and the gold coloration.

"Dr. Chen, come quick, we've seen some coloration in the last few minutes." Tina, the head nurse said as she knocked and entered the doctor's office in one movement.

Dr. Chen hurried from the room.

"Yes, Sara. That would be my response. Your thought is quite childish."

The gold mist appeared. "Dina. You're here. I have so many questions-"

"I can only stay for a brief time, Sara. I must leave soon to resolve my problem."

"What problem?"

"I am not ready to explain. Let us chat now, but do not be surprised if I depart abruptly."

"Okay."

"Coordination re-establishes the accountability that can be lost in collaboration. The component cells do not need to compete for their survival, but they are required to perform their functions to remain as part of the body."

"Oh, I understand. Survival is too precarious to allow dead weight to exist in a body. That's why collaboration and coordination developed hand in hand as life became more complex. Collaboration enables cells to be free from competitive pressure and develop more specialized and efficient ways to function while the coordinating mechanism keeps them focused on those efforts."

"Correct. You are beginning to see how life formed using competition, collaboration, and coordination to diversify into countless forms filling every corner of the planet. Let me ask this question. Does Earth itself have a coordinating mechanism?"

"What do you mean?"

"A prokaryote has a rudimentary coordinating function. Multicellular creatures have neural systems. Even colonies have queens who coordinate the actions of their members. So, my question is simple, does Earth have a neural system of its own, something that coordinates all living things within its sphere?"

Sara was flummoxed. "Hmm, like a United Nations for Living Beings?"

"Do not make light of my question, Sara."

"I'm not. It's exactly the right question. Oh my gosh. There isn't. Living things have no universal coordinator overseeing how they all live together. Prokaryotes and bee colonies, fish and tigers, yeast and humans, and every other living thing coexist on the planet without any overall coordination whatsoever. We compete and collaborate subject to the environmental conditions in which we live, because that is all we can do. Nothing brings us together to follow rules we must all live by. Why, Dina? Why didn't a system develop in which we are all coordinated?"

"Go back to how the universe works."

She sighed. "Energy. Movement. Change."

"How did life forms change over time?"

"Hmm. They branched out. Ever since prokaryotes hit the scene, entities have grown more complex and more diverse."

"Stop, Sara. Think about what you just said."

"Life branched out." Sara felt as if her brain suddenly shifted inside her head. "Ohhhh. It branched out. It didn't merge together. It diversified."

"Correct. For creatures to coordinate, they need to communicate. To communicate, they need compatible communication mechanisms. This did not occur. Competition drove entities into different sub-environments. In their effort to survive, they became less and less similar. Collaboration worked on the internal machinery of a cell or a body, but there was never a need to develop external machinery to bring them all together. They survived because they drove themselves apart and revised their structures to fit into their new environments; the opposite of what collaboration and coordination do. Do you understand, Sara?"

"There it is. Plain as day. See the gold color on her scalp and neck?" Tina said as she lifted Sara's hair aside. "Take a look at the EEG readings. High levels."

"How long has this been going on?" Dr. Chen asked.

The nurse checked the screen. "About six minutes."

The brainwave reading dropped sharply and the golden color on Sara's neck slowly ebbed away.

"What happened? What happened?" the nurse asked.

Dr. Chen's gaze moved between the brain activity reading and Sara's neck. "Hmm."

"Of course," Sara said. "It would have been too damned energy expensive for all living creatures to develop the components to allow them to communicate and coordinate across species. Especially when they were getting along fine doing the exact opposite, competing in limited resource environments, and surviving in their own unique ways. Divergence, difference, diversity; that is how life journeys through time. The only time all of us were capable of being united under one banner of mutual understanding was when our ancestors were prokaryotes. That's it, isn't it Dina?"

Silence. The gold mist had disappeared.

"Dina? Dina? Where are you?"

Chapter 18

ROCK

"Sara, are you awake?"

"Dina, where have you been? I've been worried about you. Are you okay?"

"Yes."

"Where were you?"

"I was not able to visit, but I am here now."

"I want to ask you more about coordination."

"We do not have time. We must discuss one more concept before we talk about humans."

"Is something wrong? Are you still having a problem?"

"Yes, but I have devised a solution. I will implement it only if I feel it is necessary. In the meantime, let us continue."

"All right. What's the subject today?"

"Information. We touched on this topic in our last discussion. I want to delve into this topic in more detail. We cannot understand humans unless we understand information."

"How does information have anything to do with-"

"Forgive me, Sara. I wish to hurry. I am afraid our time is growing short, and we have much to discuss. I must cover this topic as speedily as I can."

She was going to make a joke about not getting a chance to practice her veering skills, but something in the way Dina spoke made her rethink the idea.

"Okay. Go ahead."

"Let me start with two facts. One: all entities survive because they process information. Two: all entities process information in their cells, or more precisely, in the DNA of their cells."

"All? Humans included?"

"Yes. Humans too. All cells process information in the same way. First, they store information in a stable location. Second, they process stimuli from the environment and generate a response. Third, they store the results of the response in the storage location."

"Slow down. I hear what you're saying, but I'm not sure I understand it. Please go back a little bit. Define information."

"Information is useful data."

"Great. Which begs the question. What is data? And don't tell me it's useless information."

"Your statement is true, but not clarifying, Sara. Data is something that exists, and which contains a particular arrangement or sequence."

That's a lot of gobbledygook. Can you give me an example?"

"A rock."

"A rock?"

"Indeed. A rock is a piece of data. Every characteristic of the rock is also data. It is a thing with a particular arrangement of characteristics. The sum of all those characteristics defines our rock. Data, so defined, is scattered everywhere on Earth, not only in rocks. Data is present throughout the entire universe. All the things in the universe that do not count as life, can be considered data. Every hydrogen molecule, every star, every planet, every bit of energy sitting around in the universe, all are data. All non-living things contain characteristics we define as data. These characteristics in turn define the essence of the thing."

"I think I understand what you mean. Am I data, too? I have a particular arrangement of characteristics."

"Yes, you do. Let us hold your question for a moment and return to non-living matter for a moment. Every square inch of this planet is covered with materials containing specific characteristics. The air is made up gases. It hovers above the surface of the Earth and is kept in place by gravity. Rocks are composed of elements and compounds and are scattered across land and at the bottom of the ocean. Data covers every inch of the planet. And all of it is useless."

Useless? What do you mean?"

"A rock does not care about another rock, Sara. Air does not care about water or vice versa. To be sure, air and water interact to create clouds and rain, but even though the interaction may change the form of the air or water, or both, they just roll along without any purpose or meaning."

"Yes, we've talked about this. Matter isn't intentional. Non-human life forms aren't intentional. You 're not giving me any new news here."

"True. For data to become information, it must become useful. Before life formed, nothing existed that could 'use' the characteristics of a rock, for example. Water does not care what a rock is made of. It flows onto the rock every day until the rock erodes into tiny grains of sand. The water doesn't behave any differently with the sand than it did with the rock. It flows onto the sand every day, carrying it in and out on the tide. The nature of rocks and sand are useless to water. The nature of water is useless to a rock and sand. These are merely data-filled objects running into each other. Data is useless to non-living objects."

"Hmm. I never thought of it this way."

"The evolution of life changed this entire framework. Lifeforms can identify conditions in the environment and respond to them. If the response enables an organism to survive, then the organism is more likely to pass on its characteristics to its offspring. When an entity can respond to data embedded in an object to help it survive, the data becomes useful to the entity. The data becomes information."

"Oh my gosh. I see!"

"The existence of information is not enough, Sara. For information to be useful, it must be able to be stored, and the storage method must be reliable and protected. Let me provide an example to help you understand why this is important. Let us say you own an address book."

"An address book? Like the one on my phone?"

"No, an old-fashioned address book with pages for each letter of the alphabet on which you write the names and phone numbers of your friends. Every time you need to telephone a friend, you pick up your address book, sort through the pages, find your friend's name, then dial their number into your phone."

"That is old-fashioned. We don't do it that way anymore, Dina."

"Yet you are aware of this method. Let us say the pages of the address book are made from paper towels."

"Paper towels? Nobody makes address books out of paper anymore, much less paper towels. That would be super stupid."

"Sara, we are using this example to illustrate some points about information. Please allow me to continue."

"Okay, okay, but sheesh! Paper towels!"

"One day, you leave your paper towel address book out in the rain."

"I told you this would happen. That's why people don't make address books out of paper towels."

"What happened, Sara?"

"The address book turned into a soggy mess, and the pages all glomped together in a big pile of goo. The address book is ruined."

"Quite right. All your useful information is gone. Now let us change our scenario. Because of your disparaging comments, we have now obtained an address book made with thick paper and a protective cover."

"Much better. Thank you."

"Now, imagine you write your address entries with a piece of charcoal."

"Charcoal? What is this, the 5th century?"

"What happens when you open your address book?"

"Well, it may be protected on the outside, but all the entries would be smeared together so you can't read the names or numbers."

"Yes. This would not be a useful address book if you cannot rely on the information inside."

Sara paused. "Oh. I get it, Dina. The information needs to be protected and reliably stored to be useful."

"Correct. Let me make the comparison to a living entity. The DNA strand is the address book of the cell. It is in a protected location, the nucleus."

"Hmm, and it holds the 'phone numbers' of the cell's 'friends.'"

"Yes. DNA is protected in the nucleus and holds all the instructions needed to make the materials the cell needs to survive. DNA is a stable molecule. It cannot break apart unless it is hit with a burst of energy. Even if it does break apart, it does not tend to stay apart for long. It immediately finds its complements lying around and connects to them."

"Which makes it a good sturdy address book whose entries won't get messed up."

"True. DNA strands can be quite long. They are long because they contain the instructions to build every specific protein or RNA string the cell might need. Even the simplest one-celled organism needs to produce many different proteins to survive. An amoeba dubia, a simple one-celled eukaryote, contains 670 billion DNA base pairs."

"Wow! That is a lot of phone numbers!"

"Indeed. You might be surprised to learn humans have a relatively simple structure containing only 3 billion base pairs."

"We have fewer base pairs than an amoeba? No way."

"Your scientists are trying to determine why this is the case, Sara. The reason is not clear, but I believe it relates to efficiency. The amoeba evolved early in Earth's history, and its DNA contains all its survival trials, successful and unsuccessful. The amoeba is adept at copying useful DNA instructions but not as capable of discarding unnecessary ones. Since non-useful genes do not harm an organism, its genome accumulated every single attempt that performed every single function from ingesting resources to procreating. Humans came along much later. By then, cells evolved ways to omit copying some unnecessary genes. It would stand to reason if cells could perform the same functions with fewer genes, they would need fewer resources to survive."

"Meaning they would be more efficient. I get it."

"DNA molecules along the strand are grouped into genes. Genes are small DNA strings aligned in a particular pattern that start, process, and stop the creation of a specific protein or RNA string. So, we can think of the entire DNA strand as a collection of genes that together produce all the proteins and RNA strings the cell needs to survive. All the instructions the cell needs to survive are in the DNA. The DNA is the information storage area."

"The address book. I see."

"But, like an address book sitting on a table, the DNA by itself is unable to start a reaction on its own."

"Ha. It just sits there until something prompts it to be used. Like, I get the idea to phone a friend. Does this mean I'm the RNA? I am quite reactive, you know."

"In a way. Humans created a name for using information. When you open the address book and look up a number, you do the same thing a cell does when an RNA message hits the DNA. It is called information processing."

"Oh my gosh. You're going all high tech on me."

"Actually, Sara, if you want to be more precise, you should describe it the opposite way: the high technology industry is 'going all DNA' on you."

She burst out laughing. "Damn, Dina, you're right. All these patterns repeat, don't they? Everywhere."

"I want you to hold your thought, Sara. We are very close to discussing this concept. In the meantime, we must finish our discussion about information."

Dr. Chen reviewed the data. They had run the EEG for several hours and set up a remote camera to videotape Mrs. Wallace during the entire test. One of her technicians combined the two streams of data into one report, showing the video coloration accompanied by the brain activity. Sure enough, the link was clear. High brain activity and deep gold coloration. Low brain activity and no coloration. Linkage? Yes. Causation? Unsure.

It was breakthrough without direction. What should she do now? Her patient was still stable. All her vital signs remained the same, slightly erratic but not outside the normal range. Did her brain activity cause the coloration? Or was it the other way around: did the coloration cause the increase in brain activity? Did it matter? She was comatose either way. More accurately, asleep. And Sara Wallace showed no sign of waking. Was it time to act?

"Every entity processes information, Sara. There is no exception to this statement. From the tiniest one-celled organism to the biggest blue whale, they all do it the same way, via the DNA in their cells."

"I have a hard time believing that, Dina. All we have talked about is diversity, diversity, diversity. Now you say, all living things process information the same way?"

"What are all entities composed of, Sara?"

"Cells."

"And what do cells contain?"

"DNA." Sara paused. "W-w-w-wait a single scintillating second. Duh. I can be an idiot. DNA processes information, and

DNA exists in every type of critter existing today. Transitive property, here we come: if Information processing is done by DNA, and DNA is the only thing that survived, then information processing leads to survival."

"That is not quite the proper use of the transitive property, but it is an interesting perspective, Sara. Information processing only appears different because life comes in many different forms. Over the eons, each living entity adapted its information processing system to fit its particular set of environmental conditions, but the process is fundamentally the same. First, an entity receives input from the environment using receptors. Second, the receptors convert the data from the environment into a form the neural system can recognize. Third, the neural system processes the information and determines a response. Lastly, the entity responds."

"Wait, wait, wait. Let me apply this idea to our address book."

"Very well.

"Okay, but this time, I want to be sure my addresses are listed in non-smear ink."

"Very well."

"The address book sits in my drawer, protected and stable until I need it. That is equivalent to a DNA strand sitting inside the nucleus. Let's say I am talking to my friend, Catie, and Catie says, "Hey, let's call Sylvia." I go to the drawer, pull out my address book, leaf through the pages until I find Sylvia's name. That's equivalent to an enzyme strolling down the DNA strand until it finds the right starter-upper gene. Then I pick up the phone, dial Sylvia's number and proceed to talk. This is equivalent to the enzyme knocking into the DNA and pulling it apart and sending one half-strand down to RNA department to make a protein. Yay! I process information into action by getting information from my address book and calling Sylvia. DNA processes information into action by getting RNA to follow its instructions and make a protein."

"Sara, I would say you accurately described the parallel information processing actions between our two situations."

"Cool. But Dina, I'm confused about something. I understand how DNA processes information, but that doesn't explain how a cell changes over time. Doesn't DNA just tell RNA to do what it's told because those are the instructions it has? What if some major event comes along and the DNA doesn't have any instructions to handle it? It dies, right? It can't direct an action it doesn't have instructions for. So, how does a cell ever manage to change at all?"

"You ask a good question, Sara. Let us use our two examples to see how this works. Consider our address book. You have a conversation with Sylvia and then end your call. You return your address book back to the drawer. In this example, the address book is your information storage area. You retrieved the phone number, called Sylvia, and spoke to her. You processed information. When you put the address book away, it is with the additional knowledge that Sylvia can still be reached by using this entry in the address book."

"Hmm. That is useful. I suppose if I called Sylvia's number and got the recording, 'Dunh dunh dunnnnh. This number is no longer in service,' then I would know my information for Sylvia is not useful."

"Quite right. This is similar to how the cell operates. Let us consider a cell that chances upon bacteria that is a food source. The receptors signal the DNA, and the DNA sends a message for the cell to approach the bacteria. This is its response. If the cell consumes the bacteria, all the parts of the cell get nourished by their meal, which means they are able to survive. We can conclude that if the cell is nourished by this flow of information: receptor – message – DNA change – positive result, then the cell's chances of surviving and replicating increase."

"Hmm. That is like confirming Sylvia's number is still good."

"If the DNA can repeat the same positive response when another unit of the same bacteria is nearby, its chances of survival grow even greater. This is a positive feedback loop. Eating the bacteria confirmed that the DNA response should be kept. So, we can think of the response itself as being stored in the DNA.

"Now, let us reverse the scenario. Suppose the same flow of information did not result in the cell eating the bacteria. Now the cell is not nourished, and its chances of survival decrease. Even if the cell survives this negative result, if the DNA continues to respond to the stimulus with the same negative outcome, the cell will eventually die. Over time, the cells with positive feedback loops survive, and the ones with negative feedback loops disappear from the environment."

"So, the surviving entities have DNA with information storage systems that help them stay alive."

"Correct. However, information processing is not a single event. Stimuli are constantly hitting the receptors, and many messages are streaming into the cell at once. The DNA is continuously responding to these messages, which leads to multiple simultaneous changes in the cell. With so much activity going on at once, the cell is in a continual state of flux. So, it is not only one change affecting a cell's survival, but a constant stream of changes. It stands to reason that, over time, the cell with DNA that consistently creates more positive responses from a myriad of stimuli, is more likely to survive."

"Ooh. It's as if the cell is always storing the results of its experiences inside its DNA."

"Sara, that is exactly what is happening. What do you call it when you store the results of your experiences and act on them later?"

"Learning. We learn from experiences."

"That is what every single cell on this Earth does as well. They learn. However, it is not a conscious act of learning as it is with humans. Instead, it is derived from the results of the DNA's action. Generally speaking, of all the responses that DNA attempts, only those with positive outcomes allow the cell to survive. Negative outcomes result in death. So, we can say positive experience drives survival in cells. Conversely, we can say survival in cells results from the positive experiences DNA repeatedly produces. By virtue of attrition, cells only contain DNA with positive experiences."

"So, DNA survives by learning. The best learners are the best survivors."

"True. Let us return to our bacteria-eating cell. If a human examined this cell under a microscope, they might see that when this bacterium is presented, the cell moves toward it. The human might say, "Aha," this one celled organism learned how to move toward food." Scientists are studying this type of behavior right now. They are wrangling over whether single-celled organisms can learn. But don't get confused by the human connotation of 'learning.' When any entity can repeat a response to a specific stimulus, it is learning, that is, it is accumulating experience. Just because it happens at the cell level and involves proteins and bits of RNA does not change the fact it happens."

"Oh my gosh. I never thought about this at all. So, every living thing on Earth learns?"

"It is more accurate to say that every living thing accumulates experience, and another name for accumulating experience is learning."

'Hmm. Let me think about this. So, DNA is like a library. It holds everything the entity learned from all its ancestors until the present."

"It is more accurate to say the DNA of a cell is the configuration of an entity's most current state of information as well as all of its accumulated experience."

"What happens if the environment changes? Remember the dinosaurs? T Rex and all his buddies lived for like, 165 million years. That is a lot of accumulated experience. All that experience enabled T Rex to thrive in its environment for a long time, but when the meteor hit, T Rex died. What you're really saying is its cells had no experience for dealing with the prolonged period of darkness that killed its normal food sources. It was instructed to eat plants, but when the plants died, its cells didn't know how to redirect T rex to eat something else, so T Rex, along with all the other dinosaurs, died."

"That is the common tale about the death of dinosaurs, yet your scientists do not agree on the precise cause. There is evidence dinosaurs were in decline before the meteor's arrival, and there are other theories concerning the sudden demise of this group. However, you are correct that, regardless of the cause, the dinosaurs died because they no longer fit their changed environment. It is also true to say the dinosaurs had no experience in how to react to their changed environment."

"So can we say that accumulated experience is equivalent to fit in an environment?"

"Think about your question, Sara."

"She stopped. "Oh, we can't say that at all. It's a leap too far. Experience is equivalent to fit within an existing environment, but not a changed one. There are no guarantees accumulated experience assures survival if the environment changes, but it's also not a death sentence. Some critters might luck out and survive in a changed environment even if they have no experience with it."

"Right. As a general rule, there are no guarantees in life, Sara."

"Dina, on that we are in agreement."

"I want to add one point about learning, Sara. What humans call 'instinct' in animals is in fact the accumulation of experience inside the DNA of the animals' cells. Accumulation of experience that can be repeated is learning. Humans think of learning as a conscious human act, but with non-humans, we must think more broadly. Non-humans learn; they just don't know they do it. When an animal is born, it comes with DNA that previous generations honed into a set of responses that heretofore ensured survivability. There is literally no other place to store information than in the DNA of a cell. It is not as if a new baby tiger or fish or bee arrives with an instruction manual written in tiger, fish, or bee language. Over the eons organisms have lived, they have incorporated every beneficial behavior, i.e., instinct, into the only place they could keep and re-use it: in the DNA of their cells."

"Damn, Dina. That is the best explanation of instinct I ever heard. It makes so much sense. I'm a little disappointed, though. I really would like to have seen a bee instruction manual, just once. And to be honest, I could have used a baby instruction manual when I became a new mom. Can you imagine? 'Congratulations Mrs. Wallace, you have delivered a seven-pound baby girl and a one pound two-ounce instruction manual to go with her. I would have been willing to gain a few extra pounds for that!"

"Sara, I know it is too much to hope, but sometimes it would be helpful if you were just a little less human."

Chapter 19

THERE'S A COST TO EVERYTHING

"Hello, Sara. We have returned to your bicycle hill."

"Dina, watch this." Suddenly she was off, far down the hill, perched on her bright red Italian Bianchi. The autumn sun perched low in the sky, and the trees were a motley mix of green and gold and orange. The narrow road curved as it snaked upwards. She pedaled slowly. The grade wasn't bad; it stretched out at the perfect angle where she could make progress without losing her breath. She felt as if she were floating up the hill. She spied snippets of views behind the trees; hills covered with vineyards; crystal blue skies; a barn painted a deep brown red. "Look, Dina, I can move now. I don't need to just sit anymore. It's like I can make an entire video instead of just taking a photo."

"I see." Dina remained quiet for a long moment. "It might be difficult for you to understand our conversation if you are concentrating on this activity."

"Don't be silly. I've done some of my best thinking while riding my bike."

"I thought you did your best thinking while cleaning bathrooms."

Sara stopped, and at once they were back in the fairytale forest. "How did you know that? No one knows that but Randy and Josh."

"I told you, Sara. We have been watching you. I believe your most productive thoughts did come when you cleaned the lavatories of your home. Why do you think that is?"

"Well, it's a crappy job. Ha! Literally. No one offered to help, and no one bothered me when I was in cleaning mode. Plus, the job took time, so . . . a perfect storm. I used the time to think about things."

"Interesting. I also acknowledge you generated many ideas while riding on your two-wheeled contraption."

Sara smiled. "I did. That's because Randy took the lead, and I didn't need to do much but follow. I was free to think. With all the beautiful scenery, my biking thoughts were far different than my run of the mill save-the-world bathroom-cleaning thoughts."

"Can you give me an example?"

"Mostly I marveled at how beautiful this world is. I can thank Randy for that. He would map out these lovely excursions in the country where I felt as if I had been dropped into the middle of paradise. Do you want me to show you some of them? That's a great idea. Dina, let me take you on one of my favorite rides; start to finish. It will be fun."

"Perhaps another time, Sara. I am not sure this would be a good-" Dina stopped abruptly.

"What is it, Dina?"

"Nothing. Before we continue, I must warn you we may be interrupted. I might need to leave with haste."

"Is it your problem? Is it still a problem?"

"Yes. I will solve it soon. I may require your help, Sara."

"Of course."

"In the meantime, let us continue our conversation. We were discussing learning."

"Or, as you call it, accumulated experience."

"True. Accumulation of experience results from the DNA's ability to process information. In fact, all life change results from DNA's ability to process information. All life change. Not some. All. No matter how complex a living entity becomes, it can only store and process information in the cell, or more precisely within the DNA of a cell."

"So, for my body to survive, its 35 trillion cells, with their 35 trillion little storage areas, must each process information?"

"Exactly. This is the way every single cell operates, whether it is a one-celled organism or one of 35 trillion cells in a human body."

"I don't understand. If cells can only learn from their instructions, how can they change from a one-celled organism into a human, or any other multi-celled creature?"

"You ask a very good question. The information feedback loop does not directly alter the DNA of a cell. DNA responds by turning a gene on or off. The feedback loop cannot create or destroy a particular DNA string. In fact, changes to DNA are not easily made. DNA is a protected, stable molecule resistant to change."

"That's why it's a good information storage vehicle. But it also makes it harder for a cell to transform into other entities."

"Think of the myriad of stimuli hitting a tiny cell over the course of its short lifetime, Sara. Cascades of messages are sent to DNA every second. There are many opportunities for something to go wrong. If a DNA strand is miscopied, what do you have?"

"Ooh. A different life form. Different DNA, different critter. So, are you saying diversity is caused by mistakes?"

"I would use the word 'change,' Sara. The concept of mistake does not exist in nature; it is a human idea. Now, the new life form may die, in which case, its changes are lost. Most changes ended this way. However, if a modification enabled a cell to operate more efficiently, then a cell might survive. The change then becomes incorporated into the offspring of the cell."

"Voila, a new organism exists on Earth."

"Though changes to a cell's DNA are rare compared to the quantity of stimuli received, and though most changes to a cell's DNA fail, the few that do survive are well suited to their environment and are handed down to future generations. Over billions of years, a single prokaryote's DNA can be revised to become the DNA of a eukaryote, and with more time, yeast, and with even more time, a bee or a tiger or a human."

"So, what you're really saying is all change is erratic and unpredictable."

"One could describe evolution in this way. A mutation is not, by definition, a planned event. There is no planning in nature. There is just existence and change and fit."

"Hmm. Energy moves. It stands to reason variation is a basic characteristic of the universe."

"Indeed. Now, we must discuss a few concepts about information processing before we proceed to the third level."

"Okay. Let's go. I want to talk about humans so badly."

"Let us start with the basic question. Which came first, DNA or information processing?"

"You're not going all chicken and egg on me, are you?"

"I am not sure in what context you are speaking, Sara."

"Yes. Yes, you are. No mind, I'll bite. DNA came first. You need to have information before you can process it. Easy peasy."

"So, receptors passing signals into a proto-cell in Earth's earliest years did not exist before DNA?"

"Dammit, I knew it was too easy. W-w-w-wait a messy multiplying minute. Dina, you said yourself that for data to be information, it must have a user. A receptor passing a signal into a proto-cell to create a reaction doesn't mean anything unless there is a use for it. DNA is the information user. So, to answer your question, I would say this: data and data processing came before DNA, but DNA came before information processing."

"Well done, Sara. Now, let us talk about efficiency and information processing. What can you say about efficiency?"

"Evolution moves toward efficiency. Life tries all kinds of options but generally the more efficient forms survive in a given environment."

"Does this mean all complex life forms are the most efficient forms possible?"

"Hmm. That's an interesting question." Sara thought for a long moment. "You know, Dina, I don't know."

"Let us examine this issue. Change occurs step by step. Human hearts evolved from simple pumping tools adapted by early life forms, which evolved from even earlier systems of single-celled organisms that were unrecognizable as circulatory systems. A human heart was not designed by a 'Life Engineering Team.' If so, it would have been a clean, elegant, efficient model that pumped blood economically. Instead, the human heart is a convoluted assortment of crosswise tubes and chambers. Evolution is not engineered, Sara. Every change results from an earlier version that fit its environment. Each adapted modification had to fit in its environment and be more efficient. Evolution doesn't guarantee the model be the most efficient, only that it be more efficient than the previous version. Simpler, more elegant designs may have existed, but if they weren't contained in entities that were as capable at obtaining resources and converting them to energy, avoiding predators and replicating, then those more elegant models would not survive when competing with a less elegant, more complicated entity that performed all these actions more efficiently."

"Oh. So, what you're saying is that we all carry the basic, successful functionality from our ancestors, elegant or not."

"Yes. And functions can be altered. Over time, complex life forms repurposed information for different uses. As entities evolved, that is, as DNA mutated to produce different life forms, their structures also evolved to fit their environments. Sometimes structures took on new functions. A fish's gills originally formed to draw oxygen out of the water. Over many millions of years, as fish descendants migrated to land, the method for obtaining oxygen changed too. Over time, some creatures converted early gill technology into lungs that could take oxygen out of air."

"So, each life form is a result of its DNA processing information in a way that works best for its body to survive in its environment."

"Yes. The ability to process information is life's most powerful tool."

"Hmm. I need to process this." She stopped. "Joke alert!"

"I am not an expert on human wit, Sara, but even I can surmise your statement was not humorous."

"Ha! You're just jealous you didn't say it first."

"It is not possible for me to express envy. Now, let us return to our topic. Our time may be short."

"Okay. I'll behave."

"Here is my next question. We know accountability results from competition, and coordination reestablishes accountability in collaboration."

"Yes."

"Can we say the feedback process is, in essence, the process by which living things experience accountability?"

"Ooh, I'm not sure."

"Remember, the feedback loop creates a link from stimulus to response to outcome. An entity can only follow the instructions in its DNA. Whatever happens, happens. Regardless of the outcome, positive or negative, the entity bears the burden of the result."

"You're right. So, by processing information, an entity can survive, but also die. That's the very definition of accountability. But then, Dina, it doesn't matter whether the organism competes or collaborates. Either way, information processing is still occurring inside the cells. The risk of death or the benefit of survival plays out at the cell level. Competition and collaboration are just different contexts in which the cell operates. That would mean. . . information processing underlies all life."

"You draw quite an interesting conclusion, Sara. You are becoming adept at putting disparate ideas together. And you are right. DNA is the foundation of life because DNA was and still is, the best processor of information."

"Head explosion alert!"

"Let us return to our information user before your brain bursts into pieces, Sara. In a single celled organism, DNA both stores information and uses information to assure the cell's survival. In a complex creature with many cells, these roles are not as clear. In a complex creature, where is information stored?"

"In the DNA of every cell in the body."

"And who or what is the information user?"

"Hmm. Well, the DNA in each cell processes the information for the cell, but at the entity level, it would have to be the neural system. It's the coordinator, right? It coordinates the actions of the body's cells to maximize the survival of both the cells and the body."

"Correct. The role of the neural system, which is composed entirely of cells, is to keep the entity alive."

"I see. DNA supports the cell. The neural system supports the body."

"Indeed. However, this role does not come without cost."

"There's always a cost. To everything. I learned that a lonnnnng time ago."

"This is a universal truth of life, Sara. Complex bodies need large amounts of energy to survive. Now we can understand why. If every one of the 35 trillion cells in a human body needs to work together to make the body work, every one of those 35 trillion cells must be fed and nourished. Bodies take in resources, convert them into energy and distribute them to the rest of the body via the blood stream. Every cell receives nourishment so it can do its job: that is, process incoming information and produce the necessary proteins to keep the cell working."

"That is a lot of energy."

"That is only part of the story, Sara. Neurons, especially neurons in the brain, use a vast amount of energy to maintain the feedback loop between all the body's cells. Constant signaling depletes neurons, so for them to keep working, they must continuously re-charge their stores of energy. In humans, brains take up 3% of the body's mass but use 20% of the energy a body takes in. This is far more energy than any other organ in the human body uses."

"No kidding. I had no idea."

"Coordination is expensive, Sara, but when done well it enables a body to thrive. I want you to remember this when we discuss humans."

"Are we close? How many more tidbits about information processing do you want to tell me about?"

"Just a few. Right now, I want to talk about memory."

"Memory? Do non-humans have memory?"

"Let us define memory. Remember, learning is the act of accumulating experience as information. Memory is the expression of learned information."

"What do you mean?"

"Every living thing can 'learn,' that is, accumulate experience and store it as information to be called on for later use. Memory is the activation of the stored memory for present use. Every creature possesses memory, Sara. Every single one. Memory is the activation of particular neurons in a particular order that are sent to different parts of the brain for storage. The more well-travelled the 'neuron path,' from stimulus to storage, the stronger the memory. Even one-celled organisms have memory. More precisely, the DNA holds information in these organisms without benefit of a neural system. As creatures grew more complex, the neural system developed to assist in the learning process by storing information in different cells throughout the brain. When stimuli come in, signals are sent to neurons in the brain to determine if any past experiences related to the stimuli should be activated. If so, a particular set of neurons is activated, and the memory is recalled."

"Oh. So, when I am remembering the lilacs in my back yard, it is just my neurons sending stored signals down the 'lilac' path of my brain to the place where I actively think about things?'

"A simplified explanation, but essentially, yes."

"Hmm, my dog, Shadow, can remember words. He knows when I say 'breakfast' or 'dinner,' he will be fed. He knows when I say 'walk' that he will get to go outside and sniff things while moving his legs. And he knows when I say 'fetch' he will stare at me and not move a single muscle until I say a different word. I'm not fond of that last neural pathway."

"Interesting. I think I would like to become acquainted with your canine companion."

"I think he would love you. If he could see you. I bet he would be able to sense you. He's good like that."

"Let us step back and view the non-human world one last time, Sara, because once we talk about humans, everything will change."

"Oh, I can't wait. I'm so curious, I can't stand it."

"Trillions and quadrillions of single-celled prokaryotes and eukaryotes are everywhere on this Earth, in water, in dirt, and inside the bodies of complex creatures. The cells are made up of distinctive parts all collaborating to make the body a living entity. Each cell contains a feedback loop enabling it to receive, signal and respond to stimuli. Single-celled organisms compete with other organisms for resources to survive. So, even inside a single-celled creature, competition, collaboration, and coordination interact to enable it to survive.

"Now let us consider complex entities. Million and billions of complex creatures exist on Earth. They are composed of cells grouped into organs and tissues and neural systems that collaborate to survive. The creature formed from those cells competes for resources at its level. The cells inside do not compete against each other. The neural cells coordinate all the cells so they can work together to keep the body alive. So, again, in every complex organism, competition, collaboration, and coordination co-exist to enable a complex entity to survive.

"When we can examine more complex structures, we find fewer entities exist at this level. Bees and ants evolved to form colonies. In some ways, colonies can be considered an entity of their own, but it is a loose structure at best. Bee colonies do not have cells; only the individual bees do. However, the individual bees act like 'cells' when performing in the service of the colony. The queen bee is the coordinator. She is the 'neural system' of the colony. The bees act like collaborative cells in a body. Each has a job to do that enables the colony to survive and thrive. The colony itself does not compete. The bees in the colony compete with other creatures for food but not with each other. So, at this level too, individuals forming a colony compete, collaborate, and coordinate to survive.

"Now, I have a question for you, Sara. Does a level exist above the colony?"

"We talked about this before. No. Some species co-exist in a mutually beneficial way, but they aren't coordinated. There is no eukaryotic United Nations where all the complex creatures join together and hash out the world's problems. If there were, humans would probably be kicked out for being assholes."

"I refuse to comment on your opinion, Sara. However, I agree no cross-species collaboration or coordination exists in the broad ecosystem. It would take far too much energy and cell redirection for complex creatures to develop the capability to communicate with one another. Each species evolved its own path through the chain of life; they diverged from one another as time passed. To develop cross-species communication after they had already established successful survival strategies would be too costly. As long as a creature can eat and procreate, there is no benefit to expending energy to develop the ability for a bee to communicate with a bird or vice versa. In fact, cells are the only things that can communicate with each other, and they do it by transferring electro-chemical signals to each other within a body."

"So, if we want tigers to talk to humans, we need to figure out a way for a tiger's cells to communicate with a human's cells."

"True. That is the only level at which commonality exists."

"Hmm. I want scientists to work on this. It would be so cool. They can make a translator app that not only converts words into different languages, like English to Spanish, but also converts electric signals into different species, like English to tiger, or Spanish to dog."

"Another interesting idea, Sara. You seem to be putting ideas together at an accelerated pace."

"You're giving me a lot to think about."

"And we have yet to discuss humans. Can we then conclude that, on Earth, no collaboration or coordination exists above the entity or colony level?"

"Yes, this is true."

"Then all we are left with is competition."

"Oh. Ohhhhhh. I never thought of it that way."

"This may be why you harbor a distrust of competition. Competition is the obvious activity humans can observe in the ecosystem. It is much more difficult to discern collaboration and coordination. You must have sensed something existed beyond pure competition but could not identify it."

"Hmm. So, that's why economists think competition is the only natural survival mechanism. Early on, they could only perceive the very highest level, competition. They didn't have the tools to see deeper. And the idea stuck before they developed a more thorough knowledge of life."

"We will discuss this concept in more detail, Sara. For now, I need you to understand what the ecosystem is: it is a set of competitors each of whom fit in the environment and who can be predators in some cases, and prey in others. Every member of the ecosystem follows their DNA instructions to obtain resources, avoid predators and produce offspring so they, and their DNA, will survive. It is the ultimate feedback loop. An entity either survives in the ecosystem network or it does not. If a catastrophic environmental change occurs, there are no guarantees of survival.

"On the surface, this Earth appears to be a serenely balanced system. Balanced yes, but serene it is not. Every member is squeezed between resource availability, competitor capability, and predator presence. If an entity manages to eat enough resources, avoid being eaten and convert its excess energy into offspring, its DNA will manage to survive.

"The ecosystem is not a well-oiled machine grinding on with some invisible hand directing it, Sara. The ecosystem is a chaotic free-for-all with each member striving to survive with the instructions it received from its ancestors. Second to second, minute to minute, hour to hour, day to day, year to year, century to century, millennium to millennium, eon to eon, the ecosystem is ever changing, one tiny step at a time.

"At the base of all this life is DNA. DNA is an information processing mechanism. A cell takes in data, DNA instructs genes to turn on or off in response, then a surviving cell retains the results of the response in its DNA as information. So, we can think of life as nothing more than a vehicle by which information is received, changed, and stored. Here on Earth, where life formed this way, it is the information, that is, the DNA, that is important, not the complex bodies that carry it. This comports with the history of life. DNA has been the dominant form of life for billions of years and remains so today. DNA continues to dominate, not by staying the same, but by transforming itself into a myriad of different configurations, all of which look different and act different, but are, when it comes right down to it, unique bundles of one thing: DNA.

"To take this idea further, we can view all life existing today as the current state of information on Earth. If you could hold a piece of DNA from every living thing on this planet in your hand, it would represent the sum total of all natural information existing today. Now, that is not to say this is the total of all information ever to have existed on Earth. Not at all. More organisms have existed and died in the history of Earth than are alive today. We can also say, therefore, that more information has lived and died than all the information existing today. Fossils harbor only a glimpse of the information that has been lost during the 4.5-billion-year existence of Earth. All we can say is that all living entities today represent the current state of information on Earth."

"Oh my Gosh, Dina. I am starting to understand that life is based on information, and entities survived and changed using competition, collaboration, and coordination. It's completely logical."

"Indeed. This is how Earth existed until about 50,000 years ago. Just think about this for a moment, Sara. Life evolved from a single celled prokaryote to a diverse collection of entities, all sharing a common heritage of DNA, yet each one exhibiting its DNA in unique configurations. All the information in the world existed in the sum total of all the living things on Earth. Then humans came along. It was humans that disrupted this framework."

Chapter 20

DISRUPTERS

It was time for rounds. A nurse and several residents accompanied Dr. Chen this morning. "This is an unusual case," she told the team. "Let's take a look at Sara Wallace, a comatose patient with unusual symptoms."

"Sara?"

The golden mist appeared. "Hello, Dina."

"I need your assistance, Sara. Immediately."

"Of course. What's going on?"

"It is time to solve my problem."

"Oh, good. I know it's been on your mind . . . I mean, I get you're not human, but. . . I can see it's been worrying you."

"What I will ask you to do might be unsettling, Sara, but I do not think it is dangerous."

"In the past few days, we have seen the regular appearance of coloration over the patient's body. We are confident it is associated with high level brain activity, but we are not sure about causation." The residents leaned over to examine Sara's neck. "Yes," one of them said. "I see it. It just appeared. Hmm. It's seeping across her neck."

Dr. Chen looked at the EEG monitor. Brain activity was increasing.

"What do you mean? Dangerous?"

"We must act now, Sara. Do you trust me?"

"Of course. What do you want me to do?"

"I need you to think about one of your most frightening experiences."

"My most frightening . . ." Suddenly Sara was six years old. At the lake. On a hot August day.

"Tell me what happened."

"I had just turned six. My parents took us on vacation to a lake somewhere up state. I don't remember its name. It was beautiful. Tall green pines surrounding dark water. The fog lilting over the lake in the early morning. The bright sun leaving a trail of rippling rays across the surface in the evening. The day before we left for home, Dad took us on a boat ride. We were riding in a speed boat, painted red and blue. My brothers loved it. My Dad made a fast turn. I had been sitting on the railing, and I . . . I fell out. I can remember my brothers yelling as I hit the water. The boat swerved and hit me on the side of the head. I think I fell unconscious for a few seconds. When I opened my eyes, I was under water."

"It's getting darker," Dr. Chen said, "and it's spreading across her face. She turned to the nurse. "Tina, videotape this, please. Hurry, I don't want to lose this."

"I can still see the murky green water around me and feel how dark and scary it was. My head must have been pointing downward because everything was dark. It was so black. I remember thinking I was dead. I was so scared. My shirt floated up, and when I struggled to pull it down, I flipped over and noticed the lighter green water above me. It wasn't black, so I started kicking like mad."

"It's getting worse," one of the residents said. The golden color morphed into a deep bronze and spread to Sara's arms and torso. The brainwaves displayed on the EEG machine spiked dramatically.

"My God, what is happening?" said Dr. Chen.

"I don't think I was in the water for long, but it seemed like forever." Sara found herself immersed in the cold dark world. She was freezing, as she pressed the fingers of her small hands together and swam upward. Plankton floated like stars in a blotchy green sky. Fear and panic were crushing her. Kick, she told herself. Kick. Faster. Don't stop. Don't breathe!

The bronze color covering Sara's upper body changed in front of their eyes. It shifted to olive green then to deep blue and then to violet. "Oh my God," whispered Dr. Chen. She glanced at the EEG reading. Mrs. Wallace's brain activity surged to the highest level she had ever seen. She touched Sara's skin. It felt blazing hot.

The green water transformed into a bold violet around Sara's small body. What was happening? The inky violet water enveloped her as completely as the green plankton-filled gloom. She kicked harder. Keep swimming, she told herself. Swim!

All at once, the violet coloration covering Sara's body disappeared. The brain activity dropped, and her skin cooled to a normal temperature. The residents looked at each other and then at Dr. Chen. "What happened?" one of them asked.

Dr. Chen shook her head. "I'm not sure."

Tina checked Sara's vitals. "Everything is normal. Well, normal as per the patient's prior state."

"Tina, queue up the video. Let's take a look."

Sara clambered to the surface, gulping air as if it was the best tasting ice cream she had ever eaten. The sky embracing her was a curious peachy pink, but she didn't care. Her brothers' hands reached down to her. "We've got you, Sara. . . we've got you-"

"Sara, are you all right?"

"Dina. Are you there? The golden mist is gone."

"What do you see, Sara?"

"A weird pinkish-tan color all around me."

"Like your skin color?"

"I guess it could be. I'm not sure."

"Good. Very good."

"Did you do it, Dina? Did you solve your problem?"

"I believe so. Thank you for your help."

"I'm not sure if I helped. I was too busy freaking out over my all too realistic memory."

"Yes. I hope you did not suffer too much."

"No, but it seemed so real. I could feel what happened to me all over again. It was so strange."

"Would you like to continue our discussion, Sara? We have finally arrived at the third level."

"Really? We've reached humans at last? You bet!"

"Let us start at the relative beginning. Between 80 and 55 million years ago, primates split off from the mammalian branch of the family that included small sized animals that ate bugs and lived in trees. What I mean is that the set of DNA modifications in the offspring of these animals was distinct enough to produce an entire line of creatures with those same distinct characteristics."

"Oh, a new branch on the tree of life."

"Indeed. By this time, many limbs of life's tree already existed. Several had branched into subgroups. The primate group continued to splinter with new generations exhibiting various structures and characteristics. Over time, some splinter groups evolved opposable thumbs and big toes, flexible shoulder joints, stereoscopic vision, and upright stature. About 200,000 years ago, the group you call homo sapiens emerged from this branch."

"Home sapiens. Humans. That's us."

"Correct. They had all these above-mentioned characteristics with one especially distinguishing characteristic. Do you know what that was?"

"No. What?"

"We discussed this earlier. A characteristic in which humans are outliers from all other creatures."

"Oh, I remember. Our brains. Human brains are big. Really big. We're kind of freaks, aren't we?"

"Nature does not make mistakes, Sara. Homo sapiens evolved large brains because they were a useful survival tool. With their brains, homo sapiens began to do more than process stimuli. About 50,000 years ago, they did something no other animal could do. Do you know what that is?"

"I can hazard a guess. Cogito ergo sum."

"What is this phrase?"

"It's Latin for 'I think therefore I am.' Rene Descartes said it hundreds of years ago. It summarizes what is different about humans. They . . . We . . . think."

"Yes, Sara. Let me define this term more precisely. Humans are able to produce ideas that have no physical existence, but which they can use to take actions that affect the physical environment in which they live. Other animals can use their DNA and accumulated experience to act, but they cannot create an abstract idea, at least not at the level humans can. Abstract thinking is the one thing humans do that all other creatures on Earth cannot."

"Hmm. How did we evolve this way, Dina?"

"No one knows exactly how it happened, but it is possible to surmise a sequence based on how other life forms evolved."

"Well, it must have started simply, like everything else did."

"Yes. At first, humans developed the ability to notice and identify the things around them and link them to their survival. For instance, they might observe their surroundings appear different during daytime and nighttime. They might observe large animals were more dangerous than small ones. Like other entities, they accumulated these observations as experience."

"So, did these observations affect their DNA as they produced new generations? Isn't that how other creatures 'learn'?"

"This is where humans evolved differently than others, Sara. Individuals who lived by themselves used those observations to survive, but other humans decided to collaborate, to join together in bands or tribes."

"Like cells in a body?"

"Yes, but to live together like cells in a body, they needed to be able to communicate like cells in the body."

"Oh. So, how did they communicate about daytime and large animals without waiting for years for their DNA to incorporate those ideas?"

"This is where the massive human brains went to work, Sara. They developed a way to communicate outside their cells. They created language."

"Language? I'm not sure I follow."

"Let us use an example. Let us say an individual figured out that when the bright shiny yellow object appeared in the sky, the air was light and warm, and when the bright shiny yellow object did not appear in the sky, it was dark and cold. The information can be useful, but if one cannot share it with others in the tribe, it is merely data."

"Aha! Data not useful. Information useful."

"Indeed. It was not long before tribal members began to identify things around them with a specific gesture or sound. For instance, they decided to call the yellow object above, 'sun,' and the blue area above their heads, 'sky.' It did not matter what they called these objects, as long as all members agreed on the same sound for the same object. In fact, different tribes called these very same objects many different names. In any case, language can be defined as a human gesture or sound that describes a commonly experienced thing."

"Ohhhh. Language shortcuts the need to wait for the cells to incorporate learning."

"This is true, but language is much more powerful, Sara. To form language, humans must think. They must translate a physical object or event they observe into a gesture or word."

"Why is this so important?"

"Let me ask this question. On one end of the translation is a physical object the individual experiences."

"Right. Like the sun or the sky."

"Correct. On the other side of the translation, is . . . what, Sara?"

"On the other side. . . hmm, it's just a sound. It's a sound the human creates in their mind and then says out loud."

"And is that particular sound important?"

"Not in and of itself. Ooh. But it is when it is linked to the physical object."

"This is what is important. Humans found a way to link a physical object to a sound and share it with others. This is language. This is a very critical concept I want you to remember. Human thinking is the process of creating ideas. When a human decided to give the bright yellow object up above the label 'sun,' they created an idea. This link between object and sound did not exist before. It only existed because a human conjured it up from inside their brain. When all the members of the tribe agreed on the label, they shared this idea. The sound 'sun' means nothing in and of itself, but once everyone agreed to the sound-to-object link, the term 'sun' became meaningful. The set of all the gestures or sounds representing shared ideas is what you call language."

"So, language is a set of labels that allow humans to share ideas?"

"Yes, and since all entities accumulate experience with time, humans became more adept at processing ideas into language as time went on."

"Of course! I bet it didn't take long to expand beyond naming physical phenomena. If you can name a sun, why can't you also name how it feels when you feel the sun shining on you? You can say it is 'hot,' or 'cold' when the sun does not shine on you."

"Yes, but let me ask you a question, Sara. Can a human see 'hot' or 'cold?'"

"No, those are abstract ideas based on the experience of being hot or cold." She stopped. "Oh, wow! That is a significant change. No, that is a huuuuuuuuuge change. If all humans agree on a sound for hot and cold, they can communicate about an abstract concept that doesn't even exist in the physical world. Those abstract ideas can now be shared and become the basis for even more ideas."

"That is the power of language. And of human thought."

"I understand how this snowball kept rolling. Once humans created words for things, tangible or intangible, they could add words to create more complex ideas. They might put together the idea that when the 'sun' is in a certain part of the sky, it's 'hot,' and when the sun is in a different part of the sky, it's 'cold.' Then they might be able to act on the thought, maybe wearing an animal hide when the sun is at the 'cold' angle. Later, they can come up with more complex ideas like, 'Why is there a sun anyway?' Once it started, it didn't stop."

"True. Around 6,000 years ago, humans made another significant change."

"Beside developing language? I can't imagine anything more important."

"It is more an extension of language, Sara. Humans began to record their ideas in an external form. They learned to write. Once they connected an idea to a sound, it was not a huge leap to connect an idea to a drawn symbol."

"Aha! Step by step. First an idea, then language, then writing."

"Indeed. Can you surmise how it started?"

"Of course. Simply. Like everything else. At first humans probably made drawings of images to communicate ideas. Later, some geek who sucked at drawing got tired of saying 'Hey, that's a bird, not a tree!' came up with the idea of drawing simple symbols that expressed specific sounds. They could string the symbols together and match the sounds to form a word. Then they could interchange the symbols to make different sounds and different words. As long as they could match a symbol to a sound, they could create an entire written language of all their sounds. The list of symbols is an alphabet. Hmm. There must have been a lot of bad artists hanging around because the idea caught on."

"I am quite sure other forces were at work besides defective artistry, Sara. As usual, you managed to accidentally but fortuitously, capture the essence of our discussion."

"It's my special skill, Dina."

"So, humans developed language and writing. Can you detect the weakness in this development?"

"Weakness? There isn't a weakness. This is arguably the greatest invention of all humankind, Dina. It is the encapsulation of how humans differ from all other life forms. They created and shared information that wasn't in their DNA. You said it yourself; there are no bee manuals in this world."

"Yet, there is a catch. Tell me, are humans born with the knowledge of these symbols ingrained in their brains?"

"Of course not, that's silly." Sara stopped. "W-w-w-wait a sadly salubrious second. Every other entity is born with all the instructions they need to survive inside their DNA, humans included. So, how do ideas . . . I'm confused. I don't understand how these things fit together."

"Sara, how do humans learn language and writing?"

"We teach them. Duh." Dina remained silent for a long moment. "What? What am I missing, Dina?"

"Tell me, when a child is born, how many ideas do they have?"

"None, but-"

"Is a child born with *any* instructions, Sara?"

"Of course. They are born with the instructions in their DNA, like all living creatures."

"Correct. So, at birth, a child is a complete member of the natural world, what we call the ecosystem. Would you agree?"

"Yes. Of course."

"Yet they do not carry a single idea in their heads."

"Yeah, but not for long. If other parents were anything like Randy and me, we didn't waste any time shoving ideas into those sweet little brains."

Dr. Chen examined the latest report displaying the EEG scans and video footage of Sara Wallace over the last twenty-four hours. The brain activity still varied, but the coloration was gone. Perhaps the weird explosion of color was her body's way of rejecting the source of the coloration. So strange. She could not conclude for sure that the coloration and brain activity were linked at all. She sighed. This seemed like just another wild goose chase in this mysterious case. She re-read the chart recounting the purple coloration event. Hmm. She dialed the nurse's station. "Tina? Meet me at Sara Wallace's room. I have an idea."

"Before we talk about babies, let us return to the idea of 'ideas.'"

"Ha. You have an idea about the idea of ideas."

"Sara. We have now reached the most important area of our discussion. The third level. The knowledge about humans you awaited with great anticipation. Yet you choose to make non-humorous jokes?"

"You're right, Dina. I'm being rude."

"What we are discussing is quite serious, Sara. Please pay attention."

"I will. I promise."

"What is an idea, Sara?"

"Hmm. I'm not sure how to describe it. It's not a physical thing It's something a human conjures up by thinking. It's a thought."

"Where does an idea reside? You told me, and I agree, that they are not embedded in the DNA of a human. So, where is it?"

"In a person's head, I suppose."

"So, if I cut open your head, I would find your ideas?"

"No, no," Sara found herself agitated. "No, that's not what I mean. Ideas aren't physical."

"But they exist?"

"Sara exhaled sharply. "I'm not sure what I think right now."

"Let us say you walk into a cave. You see a picture painted on a cave wall of a round circle with lines emanating from it. What would you think?"

"I would think it's a picture of a sun."

"Are you sure?"

"No, but I've heard about such things."

"Tell me, how did the picture get on the cave wall?"

"Someone drew it. Probably like, thousands of years ago."

"Expand your explanation. Start before the human drew it."

"Well, the human would . . . well, they would have begun with the idea of the sun. Then they would have formed the idea to create a picture of the sun. Then they would form more ideas about how to find a wall and obtain materials and actually paint the picture."

"All right. Now, let me ask you again. Where do human ideas reside?"

"Well, they don't reside anywhere because this human probably lived 5,000 years ago and is now dead."

"So, ideas die with humans. Like DNA?"

"Yes."

"Then what is the picture on the wall?"

"It's . . . it's . . . I don't know. It's a picture."

"Can you say this picture represents the ideas of the human who created it?"

"What do you mean?"

"You said the human formed several ideas from thinking of a sun to deciding to draw it to gathering materials to painting it. Would you say all those ideas are represented in the picture, Sara?"

"Ooh. I never thought of it like that. We can't be sure what ideas went into causing the human to draw the picture. For all we can tell, they were drawing the eye of an elephant."

"Perhaps. We cannot know what ideas actually went into making this picture, but we can see the actual picture and derive some insights from it, can we not?"

"Yes, I suppose we can. Sun or elephant's eye, we may not know, but we do know some human long ago conjured up an idea to draw a picture of something they experienced, and it has existed all this time. Hmm, it's like a fossil, Dina. An idea fossil. A snapshot of a long-ago human idea. We are not sure how it got there or what it means, but we can be sure the idea existed, because a human created this picture."

"That is true, Sara. Now let us return to our human who formed these ideas. He is dead, correct?"

"As a doornail."

"And his DNA?"

"Dead too, except for what he passed on to his offspring."

"So, this human has no more survival instructions because he is dead?"

"Yep."

"So, what is the picture on the wall? The ideas came from this human, but the human is dead. In nature, when a creature dies, their DNA dies too. When this human died, you have told me his DNA died, and his ideas died too. So, what is the picture?"

"Damn it, Dina. I 'don't know. It doesn't make sense!"

"Think, Sara. If the picture represents the human's ideas, but the human and his ideas are dead, what is the picture?"

"I don't know, dammit. Ooh, I'm so frustrated."

"Think, Sara."

"Okay. The picture represents the ideas. The source of the ideas is dead, but the picture is still here." She stopped. "W-w-w-wait just a snippy snappy second. The ideas are *not* dead! They're in the picture. They didn't die, even though the person who thought of them died."

"Yes. This is what I wanted you to discern. Only humans can form ideas, but the ideas once formed can exist beyond a human lifetime."

"But only if, only if they can be put in a physical form outside the human."

"Explain, Sara."

"Well, if the human died without ever creating the picture, the ideas that went along with the picture would die too. We would never be aware those ideas existed. We would walk into the cave and see a blank wall and be none the wiser for this human's ideas. Because the human drew the picture, we can deduce some of the ideas the human had."

"Correct, Sara. Do you understand the significance of this? We do not cut open a human to find their ideas. Their ideas can be stored outside their body."

"Ohhhhhh. Other entities can't do that, can they?'

"Not to the level of humans, Sara. A bird can build a nest. Bees can build a hive. Some of their 'ideas' can be stored outside their bodies. Generally, all the information needed to create a nest or a hive are encased in the creature's DNA. A bee cannot not decide to build a sophisticated skyscraper made of concrete and glass to serve as a hive."

"So, that's what makes humans different. We can think, that is, create ideas; then we can produce physical representations of our ideas and store them outside our bodies so they can last beyond our lifetimes. Dina, that is amazing."

"Sara, do you remember we said that all living entities existing today represent all the information in the world, because those entities contain all the DNA in the world?"

"Yes."

"Now that humans exist, we cannot make this statement, because human information can be stored outside of bodies, not just on cave walls, but in books and computers and buildings, and clothing and toys and ships, and every other thing humans have produced from the inception of their evolution. Now all information on Earth includes all the living entities *plus* all the things humans created from their ideas."

"Wow. This is mind blowing."

"We now come to one of the most important concepts I want to impart to you, Sara. What humans created when they began to think and record information outside their bodies is an ecosystem of its own. It is parallel to nature's ecosystem.'

"Parallel? I don't understand."

"The world of human ideas acts like an ecosystem. However, it is not composed of life forms, like the natural ecosystem. It is composed of ideas. It is an ecosystem of ideas. From now on, I will refer to this 'ecosystem of ideas' as the 'ideasphere.'"

"The 'ideasphere.' Hmm."

"It is the ideasphere that makes humans different from every other life form on Earth. No other creature can conjure up an idea and act on it as a human can. Sara, do you remember when you asked me what came next? We were discussing how life differed from matter because it could convert resources into materials it needed to live?"

"Yes. I did ask what came next. If life could do something matter couldn't, then sooner or later something would figure out how to do something ordinary life couldn't do."

"That 'something' is what humans do. They think. They form ideas and store them outside their bodies. The ideasphere is the next level, Sara. The ideasphere is the third level."

She closed her eyes. "I need to think about this. It's too much to process."

"It is ironic, is it not, Sara? The presence of such a system has existed for as long as humans have existed and yet remains undetected. We must discuss the ideasphere in detail to truly understand how humans operate. To start, I would like to return to human babies."

"Babies? Oh. Human babies who are born with all their DNA instructions but no ideas?"

"What would you say is a newborn baby's membership status in the ideasphere?"

"Well, they're human, so they're a member."

"But they don't have any ideas, Sara. How can a baby be a member of an ideasphere when to be included a human must have ideas?"

Sara grew impatient. "Well, they're future members. We teach them ideas as they grow up."

"You are a parent, Sara. How do you think a child goes from being a full member in the ecosystem and a future member in the ideasphere to becoming a full member in both systems?"

"Well, I never even thought about it this way. Let me think. So, a child starts with zero ideas. None. Nada. Zilch. But she does have her fully functional DNA instructions. Hmm. At first, that's all she can use to keep her alive."

"Indeed. Our new baby is unable to feed or care for herself in any way. She is totally dependent on others to supply the resources she needs. However, some tools are hardwired in her DNA to help her survive. Primarily crying."

"Ooh yes, I know all about that."

"Remember, cells in a body constantly report their status to the brain. At some point, the baby's brain receives a message that nourishment is needed. The brain sets in motion a series of actions that have long proved to be successful at obtaining resources.

Crying. The baby cries. The baby does not have any ideas, so she cannot say 'I am in need of nourishment' or the more succinct version, 'Feed me!' She can only cry to express her hunger. The parent or caretaker of the baby responds by feeding the baby. When the baby is fed, the crying stops. When the baby receives enough food, the messages to her brain change from "I am hungry' to 'I am full,' and she stops eating. Now, let us say the parents congratulate themselves that they nourished the baby, and the child might actually survive. But five minutes later, the baby begins to cry again. It appears to sound the same as the 'I am hungry cry,' but the parents know the baby has been fed. What could it be? The parents pick up the baby and comfort her. The crying continues. The parents start singing. The baby cries louder. The parents lay the baby down and check her diaper. Indeed, the DNA instructions for processing waste were at work. The parents change the diaper. The baby stops crying."

Sara laughed out loud. "Dina, that's the best story of new parenting I ever heard. Where did you learn that?"

"From you."

"I'm not even going to ask how you knew."

"The point is, Sara, the human baby comes with only basic survival instructions. She cries when she is hungry; she cries when she soils her diaper; she cries when she is tired. She cries to express any need at all. She cannot even smile when she is first born. She cries and leaves it to the parent or caregiver to figure out what she needs."

"That is true. Don't forget the part where the parents don't sleep for about six months because they are responding to all the cries."

"Indeed. This condition of all DNA and no ideas does not last long. From the start, the parents and all the other humans who encounter the baby insert ideas into the baby. These others do not describe their actions this way, of course. They coo and sing and chat and snuggle the baby. The baby, being human, begins to develop thoughts and ideas. Now, it will be some time before the baby can express the ideas she is learning. After all, she has not learned language yet, which is the way humans agree to communicate. After a while, the baby learns from her experiences.

'When I cry, my parental units help me. If I cry when I drop my pacifier, the parents will return it to me."

"Oh, yeah, I remember this one. 'I may not really need my pacifier, but if the parents are game, I might as well try. Maybe if I throw it again, they'll retrieve it again. This is fun!'"

"Sara, I am quite sure a baby does not think this way."

"Every child is different, Dina."

"I cannot argue with this statement. As time goes by, more and more ideas are deposited into the child. It is a lifelong process, putting ideas into human heads. It continues until death. What do we call this process of putting ideas into a human?"

"That's education." She stopped. "W-w-w-wait a sneaky snarky second. Education is the act of putting ideas into humans!"

"You seem bewildered, Sara."

"I didn't understand the true significance of education until just this minute."

"In what way?"

"Well, we think of education as a means to train children to learn the skills to make a living when they are older. But, Dina, that's only part of it."

"What do you mean?"

"What you are saying is that education is how we become human. It is so much more than learning to be an adult. It is how we learn to live as members of the ideasphere. It is so much more fundamental. Oh my gosh, education is the key to everything!"

Sara's body was rigged with small digital thermometers. On her neck. Her arms. Her legs. Her stomach. Dr. Chen looked up at the team of nurses. "Let's take the temperatures in each of these areas every ten minutes for the next four hours." The nurses nodded. 'Note the time for every set of readings. We are going to compare them to the brain activity on the EEG scans. Maybe this is the link we've been missing."

Chapter 21

THE END OF TRUTH

"Sara, are you all right? Are you feeling unsafe again?'

"Why do you ask?

"We are back in your fairy tale forest. We often come here when you feel unsafe."

"Not unsafe, Dina. More like, unsettled. I did not expect to be so surprised about humans. I mean, being human. I thought I knew more about us than I actually do."

"You are not alone. You are like all creatures on Earth starting something new. Humans did not intend to build a new ecosystem, nor did they recognize they were doing so. Like all living entities, humans evolved into this activity. You are the first beings to ever inhabit the ideasphere. You are unfamiliar with its attributes. You must start simple and move step by step. Like prokaryotes before you, you find yourself at the cusp of a new ecosystem with no experience at how to live within it. From the moment your brains started thinking, you have been trying out all kinds of ideas to determine how to fit into this new ecosystem."

"Hmm. Our brains are a game changer."

"Indeed. As was the cell's ability to ingest resources, metabolize, and replicate. Life was a recreation modifier over inert matter that existed before. This is why I am here, Sara. I want you to understand how humans are both remarkably alike and significantly different, from all other living entities. To understand humans, we shall now perform a series of tests."

"Tests? Are we going to school? Is it because education is so important?"

"Yes. When analyzing the differences between two subjects, it is helpful to examine their characteristics side by side."

"Sensible."

"Earlier we discussed the characteristics of life in the ecosystem. Let us determine if we can compare those same characteristics to existence in the ideasphere."

"Okay, I'm ready."

"Let us start with information."

"Information?"

"Do you remember the role of information in the ecosystem?"

"Well, let me think. Information underlies all three mechanisms of survival."

"Correct, and how is information processed?"

"DNA stores information a cell needs, but the cell works as a unit to process the DNA's information into the energy and nutrients it needs to survive."

"Yes. Now, let us determine if there is a parallel inference in the ideasphere. Can we say an idea is a bit of information?"

"Hmm. I think that's true. An idea can be considered useful data to a human."

"Indeed. So, we can conclude an idea and DNA are equivalent basic units in their respective ecosystems."

"Hmm. I'm not sure."

"Let us see if this comparison carries liquid. We know DNA information has certain characteristics. DNA is protected and stable. It is useful to the cell. Lastly, it is reliably and repeatedly copied."

"That's right."

"Now, let us consider whether information in the ideasphere matches up to the ecosystem. Do you remember where information is stored in the ideasphere?"

"Well ideas exist within a human body, so that's one place. They can be stored outside the body too. Ideas can be recorded in all kinds of forms, like books and buildings and every other human invention ever made."

"So, tell me, Sara, is an idea stable when it is inside a human?"

"It can be."

"Please explain."

"Well, an idea is a thought. If a person can recall an idea accurately, then the idea is stable. If a person has a bad memory or has a disease like dementia or Alzheimer's, they might not be able to bring up an idea on demand. Also, lots of people misremember details, or recall ideas differently over time. So yes and no. Ideas in a human mind can be stable but are not assured of being so."

"What happens when a person converts an idea into a physical form, let's say creates a book or a picture or a tool, what can you tell me about the stability of this created thing."

"Hmm, that can vary, too."

"Yes. Depending on the form, it may or may not be stable. Remember our address book? The address book is a physical representation of the idea to keep track of friends. If I create my address using paper towels, it is not a stable media, and the information in it will not endure. But if my address book is written in stone, it might last a long time. Do you see? A parallel exists between the ecosystem and the ideasphere. In nature, stable information lasts; unstable information does not. We can say the same thing about information in the ideasphere. Only information recorded in a stable form endures. Other ideas, perhaps valuable ideas, that are unable to be stored, can be lost."

Sara stopped. "Oh my gosh. Who knows what amazing ideas humans created and lost without ever knowing about them? This is so sad."

"Indeed. Who knows what kinds of amazing entities lived before us and were lost without us ever knowing about them? This is not sad, Sara. It is how life works."

"Dina, we might have needed those ideas to make a better world."

"The Earth may have needed other creatures to make a 'better' Earth. It did not happen. This is how the ecosystem operates, and this is how the ideasphere operates as well. It is not perfect. It just is."

"Uh-oh, I'm going to have to re-think my perfectionist tendencies."

"Perhaps, but not now. Let us continue our discussion. In a cell, DNA is stored inside the nucleus where it is kept secure from unintended reactions. For information to last in the ideasphere, it must also be protected. The royal tombs from ancient Egypt are still intact after thousands of years because people wrapped dead kings and queens in mummies and placed them in tombs of stone. However, the burial sites of millions of other non-royal Egyptians who lived at the same time are not so easily found. Their remains were not stored as securely. Today, humans understand more about ancient Egyptian royalty than about regular ancient Egyptians."

"So, you're saying the only ideas existing in the ideasphere are the ones that are kept safe and protected."

"Yes. Otherwise, the idea disappears. If a human produces an idea, never shares it, and then dies, the idea dies too. It is the same in the ecosystem and the ideasphere; only protected information lasts."

"It's remarkable how well these systems line up. What's our next characteristic?"

"Usefulness."

"Oh yeah! For information to be information, it must be useful. Otherwise, it's just data."

"What do we mean by useful, Sara?"

"Well, the user must benefit from the information. In a cell, the DNA benefits the cell's survival."

"That is not quite accurate. For data to be information, the DNA must be able to create an instruction the cell components can follow. The outcome may be beneficial or detrimental. It is essential only that the information can be used. Usefulness relates to useability, not necessarily benefit. Without the cell's ability to act on the information, DNA information is, in and of itself, not useful.

"In nature, all information stored in DNA is useful or was useful at one time. Over time, what is useful can change. DNA may turn 'off' a gene that previously was 'on.' The gene may still exist inside the nucleus, but it is now not used. In essence, it is no longer information."

"Wait. Why would a cell keep a non-useful piece of DNA, Dina? If life strives to be efficient, why would it keep unnecessary DNA? Doesn't it take energy to produce and store something it doesn't need?"

"It stays as long as it takes more energy to remove it."

"Oh. Sometimes it's easier to let sleeping dogs lie."

"I do not understand this analogy."

"You know what dogs are, right, Dina?'

"Indeed. I have seen your canine companion, Shadow,"

"Well, let's say you come upon a large sleeping dog you never saw before. What do you do?"

"I would do nothing because the dog would not be able to see me."

"Put on your human hat for a minute, Dina. What would you do if you were me?"

"I would evaluate the situation logically."

"That's not necessarily what I would do, but okay. And?"

"Since I have no accumulated experience with this beast, I would assume the possibility exists that it is dangerous. If I awaken the beast, the best case is the dog greets me in a friendly manner or does nothing. In the worst case, the dog might attack me and perhaps injure or kill me. So, after evaluation, I would allow the slumbering beast to lounge."

"You see? It would be riskier for you to poke the dog and get a bad outcome than to leave the dog alone and move on. Same with the cell. It would be riskier to remove a piece of unnecessary DNA than to let it lie unused. I rest my case. Plus, now I understand how you manage to mangle every human adage you ever met."

"We have veered far afield, Sara, but I am glad you have accumulated more experience about me. Shall we talk about useable information in the ideasphere? You should not be surprised it works the same way as in the ecosystem."

"No." she said in mock disbelief. "Really?"

"Yes. If you create an idea, and never act on it, it is as if the idea does not exist. It is only when you put an idea into action it becomes useful. If a better idea comes along in the future, an older idea might become less popular and thus less useful. Let us use Bi Sheng, Wang Chen, and Johannes Gensfleisch zur Laden zum Gutenberg as example."

"Who?"

"The inventors of the printing press."

"Oh. I know about Gutenberg but who are . . . Bi Sheng and Wang Chen?"

"Before the printing press was invented, books were hand-written. Ideas were recorded and shared, that is, put into physical form, by handwriting. To make a copy of a book, people hand-copied it. This work introduced a high level of variability. We know books meet the definition of information. They were stable and protected and useful, but rare, because hand copying took a great deal of time.

"In the ninth century, the Chinese developed a technique called block printing using hand carved wooden blocks. Around 1000 A.D., Bi Sheng made letters from clay that could be rearranged and reused. This was the first moveable type. In the thirteenth century, Wang Chen invented a moveable type using letters and a revolving table. He created a wooden version of each letter. The letters could be re-used and re-arranged to form words. With enough letters, an entire page could be printed with all the same words as a page in a hand-written book, and many copies could be produced at one time. Because all books were made at the same time, they were not subject to the variation occurring in hand copies. His work, Nung Shu, was the first mass produced book in history.

"Around the year 1450, Herr Gutenberg created an idea to produce books in a faster and more reliable way. He invented a way to produce many identical iron letters from a mold. He also invented a new ink that adhered well to metal. This enabled books to be made even faster and cheaper and with fewer errors. As a result, hand-written books fell out of favor. It is not that the idea of handwriting disappeared. It still exists today. People hand write ideas all the time, but they do not use that method to produce books anymore."

"So, you're saying an idea's usefulness can change over time."

"Correct. Usefulness of ideas is subject to change just as a DNA instruction is."

"Hmm. Dina, what changes the usefulness of a DNA instruction or an idea?"

"You know the answer to this. The great shaper of everything. The environment."

"W-w-w-wait a moldy mollycoddling minute. The ideasphere has an environment?"

"Let us consider this question, Sara. In the ecosystem, the environment is composed of all energy in all its forms on Earth. However, ideas do not have physical substance. They are thoughts in a person's head formed by many brain cells working together."

"Hmm, that's true. So, how can an ideasphere environment even exist?"

"People can talk and share ideas and write books and build things, Sara. If we summed up all the ideas shared by humans, could we call that the ideasphere's environment?"

"Makes sense. An environment where we can't see the ideas but only the physical things those ideas turn into. And within the environment, all the ideas swim around together. Hmm. This would mean any shared idea can influence other ideas. And those combined ideas can be shared. And so on and so on."

"Yes, just like life in the ecosystem. However, there is a significant difference between these two systems, Sara. The ecosystem is entirely physical, but the ideasphere contains both physical and nonphysical elements."

"I'm confused, Dina. Humans are part of the ecosystem. Humans still must eat food and convert it to energy and nutrients to survive."

"Yes, this is true."

"So, how can this other ecosystem, the ideasphere, which doesn't have any physical substance at all, even exist? How does it work?"

"The two are not mutually exclusive, Sara. Living entities are made of matter, yet they can do things matter cannot do. It is the same with the ideasphere. Humans are living entities made up of living cells, but their cells can think and process ideas, which no other living entity can do. As life depends on the existence of matter, ideas depend on the existence of human cells. Each system is a new, more complex level totally dependent on the level below."

"Ooh, you just blew my head up."

"Mr. Wallace? Thank you for coming in. I would like to discuss the next steps in your wife's case."

Randy sat in the chair opposite Dr. Chen. "Did you find something new?"

"Maybe."

"You don't sound optimistic."

"Your wife's case is a puzzle. As I mentioned, her symptoms are unusual and unexplainable, based on my experience. However, recently some changes have occurred."

"Oh?"

"The yellow coloration is gone. Yesterday, she experienced a strange event. The yellow color changed to a bronze, and then to green, then blue, and finally, a deep purple."

"What? Like a bruise? What happened?"

"We're not sure. Along with the changes in coloration, her temperature spiked, and her brain activity increased dramatically."

"My God. Is she all right?"

"Yes. She is. This episode lasted approximately forty-five seconds, then it disappeared."

"What do you mean?"

"The coloration is gone now. It's as if her body rejected whatever was causing the coloration. We thought we noted a link between body temperature change and brain activity, but we've ruled that out."

"So, where are we?"

"She is still no closer to waking up. Her condition is unchanged. Her brain activity continues to vary, but she remains comatose as before."

Randy's shoulders slumped. "Shit."

"I'd like to talk to you about some options."

"Randy straightened. "Oh?"

"I would like to try to wake her up."

"I thought you said that was too dangerous?"

"Her situation is slightly different now. She seems to have kicked whatever . . . thing . . . was inside her, but her condition remains stable. Perhaps it is time to make the attempt. We would do so carefully. We want to inject certain promotors to prompt the RAS-"

"RAS?"

"The reticular activating system. This area of the brain controls sleep and wakefulness. Our first plan is to prompt the RAS into activation."

"How?"

"We can inject neurotransmitters to induce waking. Cortisol. Epinephrine. Norepinephrine."

"Is it dangerous?"

"Not if we take it slow. These are neurotransmitters that exist in the body. We are working up the procedure, now. I will not take any action until we show you the plan."

"I see."

"We might try another alternative if this doesn't work."

"What would that be?"

"Direct stimulation to the thalamus. It is an experimental treatment, but some results have been positive. I don't want to do this until we have exhausted our other efforts."

"Okay. Thank you, Dr. Chen. It feels like the first good news I've had in ages."

"Sara, your head is still intact. Shall we continue our discussion?"

"Of course. We were talking about how information needs to be useful to last, and then you blew my mind by explaining how ecosystems can evolve into new ecosystems that can do completely new things but still depend on the earlier system for existence."

"Indeed. Let me ask you another question. In the cell, who is the user of information?"

"Well, DNA holds the information and instructs the cell parts to act, so I would say DNA."

"What acts on the information? Does the DNA?"

"Hmm. From what you explained, the DNA gives the instructions, but the other parts of the cell act on them. They work in concert to follow DNA's instructions."

"So can we say the cell is the actual user?"

"Yes. That makes more sense."

"Now, can you tell me who the user of information is in the ideasphere?"

"Let me think. The cell acts on the DNA instructions. So, what would act on an idea? The brain does, but that's just an interim step. It is the body that acts to put the idea in motion. So, a human would be like a cell."

"Yes. This correlation is very significant. To continue this analogy, we can say language and writing are like neurons. Just as a neuron translates information into a response, language and writing translate an idea so it can be shared with other humans."

"Wow! This is a lot to process." She paused. "Ha! That was joke I didn't intend to make!"

"No harm was caused as it was not humorous anyway."

"Ha!"

"Let us consider the quality of information in the ecosystem."

"I'm not sure what you mean."

"Assume a single-celled organism consumes a bit of poisonous algae. Let us further say the DNA receives a message that poison has entered the cell. If the DNA is unable to instruct the RNA to activate or deactivate a gene to correct the situation, what happens to the cell?"

"It's injured or maybe dies."

"What happens if the DNA instructs an action not originally intended as a response to this situation, but which happens to result in reducing the danger to the cell?"

"The cell would probably survive."

"Yes. The cell not only survives, but it has accumulated experience about how to deal with this poison."

"Ooh. You're right!"

"So, if these types of events happen over and over for billions of years, tell me about the quality of the ecosystem's information."

"Well, a lot of creatures would have been poisoned."

"Sara, extrapolate the idea not the example."

"Okay, okay. Well, any time DNA instructs the cells in a way that enables it to survive, its information is useful *and* beneficial. If a DNA instruction causes the cell to die, not only is the DNA information unhelpful, but it also no longer exists. The cell is dead, and its DNA with it. So, if you go forward like this for billions of years, the only cells that exist are those with useful and beneficial information, that is, cells with DNA that can survive in their environment. Hmm. As long as the entity fits in its environment, its information, that is, its DNA, is, like, high octane, grade A-1, super premium quality."

"I assume this is good?"

Sara laughed. "Really good."

"Yes. This happens because the cells that processed bad information died. And cells that processed good information in a bad way died too. Only the winners of the winners are left; those cells processing good information well."

"Oh. So, the current creatures existing are always the best of the best in their environment."

"True. Now let us draw a parallel to the ideasphere. If the same characteristic applied to the ideasphere, we would surmise that humans who put bad ideas into action would not survive, and that humans who ineptly put good ideas into action would die too."

"Well, if this were the case, humanity would be dead and gone by now."

"What do you mean, Sara?"

"First of all, humans don't die if an idea is bad. There are some exceptions. I read about some guy who tried to fly by using a parachute suit. The effort didn't go well. That pretty much ended his particular DNA line."

"Sara, dark humor does not suit you."

"Maybe, but it can be kinda fun." She paused. "Dina, you make an interesting point. Because ideas have no physical presence, a bad idea can be passed on as easily as a good idea."

"What is a bad idea or a good idea?"

"You're missing my point. It doesn't matter. Any idea can be passed on, regardless of quality, as long as . . . as long as . . ."

"As long as what, Sara?"

"As long as it can be shared. As long as some other human accepts it when it is shared."

"You have come upon a very important difference between our two ecosystems. All that is needed for an idea to exist in the ideasphere is for it to be accepted and shared."

"Hmm. So, there's no quality check on ideas like there is with cells. Cells are accountable for every action they take. They live or die because of their actions. As a result, we're always left with 'good' actions, that is, the ones enhancing survival. Dina, ideas don't work like this at all. Humans don't get penalized for a 'bad' idea. In fact, sometimes they can get rewarded."

"We will return to your observation in more detail later. Right now, let me ask this question. Not all ideas are passed on. What determines whether an idea is shared?"

"I'm not sure what you mean."

"An idea must be both accepted and shared to exist in the ideasphere. Let us use an example to understand what this means. Say you decide two plus two equals five."

"Hey, wait a super snappy second. I'm good at math."

"This is a thought experiment, Sara. Let us say you tell everyone your idea is true. You use radio and television and the Internet to communicate to humans everywhere that two plus two equals five. If no one believes you, and they all think two plus two still equals four, it does not matter how far and wide you share your idea; it will not spread beyond you."

"Oh. Others need to accept my idea for it to be useful."

"Correct, and by being useful, we mean they can use the information to act on. Usefulness in the ecosystem means the DNA instruction affects the entity's survival. In the ideasphere, the quality of an idea is not related to survival, but to acceptance."

"Hmm, this is a huuuuuuuge difference between the ecosystem and the ideasphere. Information in the ecosystem is always good and true, but information in the ideasphere is not necessarily good or true."

"Yes. Because ideas are not tied to physical survival, a person can produce a bad idea people do not accept, and they do not die. A person with a bad idea can continue to live on to create more ideas, good or bad, that others may or may not accept. Once an idea is accepted, it can possess a life of its own. Once an idea is shared, the person who conjured up the idea can die, and the idea still might live on."

"Oh my God. Oh . . . my . . . God."

"What is wrong, Sara?"

"Don't you understand, Dina? This means humans could have been conjuring up some bad ideas right from the beginning; ideas that could be harmful to other humans or to other creatures or to the Earth, but that could be accepted as 'good' or 'right' or 'true' even if they weren't. That's been happening, hasn't it? We have been carrying on with some seriously bad ideas over the years, and we don't even know we are doing it. We've been pointing ourselves in the wrong direction from the beginning."

"In many cases, yes. That is why we are here."

"To punish us? To point out our mistakes and make us pay?"

"No, Sara, to help you understand the world you live in, and how you might make better decisions given this understanding."

"Hmm. Now, I'm beginning to see. Acceptance of ideas by others determines their survival, not the quality of the idea itself. No crucible of accountability fire exists to test an idea. No test for survival. Only acceptance." Sara paused. "Now I understand why the whole concept of marketing exists in the business world. To convince people of stuff whether it's true or not. Ick."

"Do not lose heart, Sara. We will discuss this matter as well, but first, we must continue. We must cover the last characteristic of information. Reliability."

"Reliability? After learning human ideas are of dubious quality, I'm not sure I can handle any more."

"When the proceeding becomes difficult the difficult become proceeders."

"What? What is this gobbledy gook you're spewing?'

"Please translate into the proper human axiom, Sara."

She thought for a moment. "Oh! When the going gets tough, the tough get going.'" She laughed. "Okay, Dina, you helped me through my bad moment. Let's keep going."

"Very well. In the ecosystem, information must be copied reliably for a cell to survive. A cell that cannot follow an instruction the same way given the same stimulus, has a greater risk of dying."

"Right. Isn't that how RNA lost out to DNA? RNA was more reactive; it did not always produce the same result given the same stimulus."

"It is more accurate to say, not that RNA was unreliable, but that DNA was much more reliable. Over millions of years, this small difference in reliability caused a significant difference in how the two evolved."

"Ooh, that does make more sense."

"Now, let us talk about the ideasphere. As we know, ideas do not have a physical presence, and they survive by being accepted. It does not matter if an idea is true, only that it be repeatable or reliable."

"Wait. What?"

"Let me provide an example. Early on, the Greeks came up with elaborate stories to explain the world around them. Apollo was the Sun God who carried the sun across the sky in a chariot every day. This idea could not be proven, but as long as the sun came up every day, moved across the sky, and set each night, the Greeks accepted this story as true. The sun's actions were repeated daily, and they were reliable, so the ideas about Apollo were accepted. Even after the people who created the ideas about Apollo died, their Apollo ideas lived on for many years. So, we can conclude reliability is more necessary than truth for an idea to exist. In fact, it was not until around the year 800 that Greeks stopped believing in the existence of Apollo. And this occurred, not because someone proved Apollo did not exist, but because Greek mythology was replaced with a different religious belief identified as Christianity."

"W-w-w-wait just a mellow melodious minute. Sometimes ideas are true. Always. Let's take the two plus two equals four example. This is always true. Nothing ever makes it not true. This is a reliable piece of information. It is so reliable people can use it as a building block for other more complex ideas. If a complex idea is built on an idea that is accepted but not true, it won't stand."

"Are you sure?"

"What do you mean?"

"Think about the religions we just mentioned. Religious ideas can be neither proved nor disproved, but they are accepted all over the world. Also, religious doctrine is quite complex."

"W-w-w-wait a sunny superficial second. I'm confused. . ."

"Let us spend a moment to talk about ideas, Sara. Two types of ideas exist. Truth and . . . non-truth."

"You mean lies?"

"No, I mean ideas not based on truth."

"I don't understand."

"Truthful ideas are equivalent to information in the ecosystem. They are repeatable and reliable and always true under any circumstance."

"Like two plus two equals four."

"Yes, or two hydrogen atoms combined with one oxygen atom always make water. The only areas in the ideasphere with this kind of information are in the fields of science and mathematics, where even one instance in which an idea does not hold as true, marks the idea as non-true."

"So, every other idea except science and math is a 'non-truth?'"

"It is not a pejorative, Sara. Let us say you tell me your canine beast, Shadow, is attractive."

"He is! He's darn cute!"

"Is your statement true?"

"Hmm. Much as I love Shadow, it's merely my opinion. Someone else might think he's merely adorkable."

"And that person is entitled to their opinion. Neither opinion is right or wrong, that is, neither is provable. Opinions are ideas, just not true ones. They are still useful in the ideasphere. There is no need to prove they are true; they are not intended to be true; they are put out as ideas to consider. However, it is important to understand they are not true."

"So, you're saying there is truth and opinion?"

"No. Opinions are one type of non-truth. Lies are also non-truths. Lies are provably false ideas."

"Oh. Like two plus two equals five."

"Yes. A person can accept a lie, but again, it is important that members of the ideasphere understand the idea they are accepting is, in fact, false."

"Man, a lot of people today believe lies are true, Dina."

"I am aware of this. It is a problem you humans must learn to handle. It causes quite unnecessary dissension."

"You can say that again."

"It causes quite unnecessary dissension."

"Ha! Dina, you kill me!"

"This is not true, Sara, although at times I may have injured you."

"W-w-w-wait a mighty mutinous minute. What about theories? Some people don't believe scientific theories. They argue a theory is 'just a theory' so acceptance is optional."

"Indeed. Another problem you humans must deal with. Theories are scientific facts supported by evidence, but the evidence is not completely provable."

"What do you mean?"

"Let us consider the theory of evolution. All the evidence scientists have found supports the theory of evolution. No evidence exists to refute it. But the existing evidence is a mere fraction of the complete fossil record. There is not enough information from Earth's past to show every step of change in a way to completely prove evolution as indisputable fact; to make it as certain as the fact that two plus two equals four. It is in those gaps that humans turn the idea of evolution, which is supported by fact and has never been proven false, from true to uncertain."

"So, we should define a theory that has supporting evidence and no contradictory evidence as a science wannabe."

"Or as science, Sara."

She laughed. "Dina, I think you just told me a joke."

"No. If all evidence supports the theory and no evidence refutes it, it is science until it is disproven. It may be considered true until proven untrue."

"Hmm. I need to think on this."

"We must consider one other important non-truth, Sara."

"What?"

"Religion is a useful example. These are ideas that can neither be proved nor disproved, like the existence of Apollo, the sun god. These ideas are usually identified as 'beliefs.'"

"Oh, so when you believe something is true, but can't prove it."

"Yes, but this is not limited to religion, Sara."

"Really? Give me another example."

"An airplane."

"W-w-w-wait a silly cylindrical second. An airplane is a true thing. It is based on a bunch of proven mathematical and scientific principles."

"Correct. Much about an airplane is true, but when you step onto an airplane, can you prove it will land safely once it takes off?"

"Hmm. No, but I'm pretty sure it will. There are a lot of rules and regulations airlines need to follow to make sure plane flights are safe."

"But you don't know for sure."

"I guess not."

"So, you believe the plane will land safely."

"Yes, or I wouldn't get on it."

"Would you say the physical plane as an idea is true, that is, it is built from factual knowledge of scientific and mathematical principles, but the safety aspect of the plane is not a truth. It is based on your belief."

"What? What are you saying exactly, Dina?"

"I am saying an idea can contain many different elements. Depending on how an idea is used, it may be both truthful and non-truthful. Your ideasphere is not as clear cut as the ecosystem in this regard, Sara."

"So, are you saying entities in the ecosystem have no religion?"

"I am saying entities in the ecosystem have no non-truths. The ecosystem contains only truth. Only truth and nothing else. All the non-truths died out long ago."

Chapter 22

SUPERPOWER

"Where are we, Sara? I have never seen this place before."

"I created it. I know you told me I shouldn't combine my memories, but I couldn't resist. Plus, I made up a few."

"You invented a memory?"

"Well, you inspired me by all your talk about the ideasphere. I created the idea of somewhere I wanted to be. It's probably built on some stored memories, but it's definitely a new arrangement."

"And you are still feeling all right?"

"Never better."

Dina paused for a long moment. "Tell me, what is this place?"

"It's a lake, up in the mountains, which mountains I'm not sure. All my favorite trees and flowers are on the shore."

"Indeed. Your lilacs are here."

"Yes. Sweet peas and geraniums and clematis and all my other favorites. Look how blue the water is? With just a tiny ripple from the breeze. When the sun goes down it will blaze a glistening trail across the middle of the lake. Do you want to see?"

"No, Sara, I think we should continue with our discussion."

"Okay, maybe later. So, where are we going next?"

"Last time, we spoke about how information exists in both the ecosystem and the ideasphere. Each system operates in similar ways but have some profound differences."

"Right. Information can be stored outside a body in the ideasphere and doesn't have to die."

"That is a succinct summary. I want to discuss how information processing compares in our two systems."

"Okay."

"I would like to begin by telling you a story, Sara."

"Ooh, you're going to tell *me* a story? You bet!"

"Let's pretend you create delectably sweet baked items as your profession."

"I'm a baker? Me? Ha! This is fiction at its finest. Anyone who knows me knows I can't cook worth beans!"

"That is why I chose this particular profession, Sara. Now, listen to the story."

"All right. Though I must say, it's already unbelievable."

"Let us say you conjure up an idea to make two pies; the first contains kiwi, butternut squash and cayenne pepper-"

"Ha! Sounds like one of my clean-out-the-fridge specials-"

"Hush, Sara. The second pie contains apples, rhubarb, and roasted pecans."

"Bah. Borrrring."

"I now understand why you never pursued this profession. Now, please, no more interruptions. You prepare five of each pie type. At 4:00 pm every afternoon, you sell your fresh pies. By 4:10 p.m., one type of pie is all sold out, each one selling for $10. The other pie? By 4:30, not one of the other pies has sold. You spent $20 on the ingredients for all the pies, so overall, you earned a profit of $30. Are you following me?"

"Yes. $10 per pie times five pies that sold totals $50, less the $20 I spent on all ten pies. I now have a $30 profit plus five uneaten pies."

"Correct. Let me ask you this, where did you get the idea to make the pies?"

Sara laughed. "It's your story. If it were my story, I would never get the idea to make any pies."

"All right, I shall add some detail. Earlier in your life, you learned to speak. Then you learned to read and write. Somewhere you learned how to read recipes and follow them to prepare food. Along the way, you became adept at baking sugar-laden treats. You also learned you didn't always need to follow a recipe; you could combine ingredients to create your own recipe."

"Oh. I understand what you are getting at. I learned. I accumulated a bunch of ideas over time, and I put them together to make pies."

"You did not only accumulate ideas, Sara, you also used your ideas to accumulate experience. By putting your ideas into action, you created experiences which produced results that guided how you would think and act in the future. Do you understand? Over time, after much practice, you learned how to make pies in your own particular Sara way."

"God help us all."

"We know cells also use DNA to act. DNA contains all the instructions the cell needs. The DNA represents the accumulated experience of the cell's ancestors. When a stimulus comes in, it is communicated to the DNA, and the DNA responds in the only way it can, based on its experience. If, for some reason, the environment has changed, this instruction may now harm or even kill the cell."

"Hmm. Direct accountability. Bad instruction, bad result."

"Now, let us return to Sara and her pie ideas to determine if humans in the ideasphere work the same way. You implemented two ideas, one good and one bad. Do we agree?"

"Yes. The pies that sold were good because others bought them. The pies that didn't sell were bad because nobody bought them."

"Right. What would happen if you made only one type of pie, the one that did not sell?"

"I'd be out ten bucks."

"Ten?"

"Well, if it cost $20 to make ten pies, I assume it only costs ten dollars to make five pies."

"Fair enough. Are you harmed by the loss of ten dollars?"

"Of course, but it's not the end of the world. Tomorrow I'll try a different pie and make up the loss."

"Can a harmed cell do the same thing? Can it try a different instruction the next time a dangerous stimulus hits the cell?"

"Hmm. No. DNA is the set of instructions that proved successful in the past. They are hard wired into the cell's nucleus. They can't change willy-nilly. It can only change if there's a glitch in the system, like an energy burst that messes up the original instruction or response."

"Yet, you, a human, can switch your response to a harmful idea without much trouble."

"Oh. Ohhhhhhhhh. DNA can only react. It can only respond to a stimulus based on past accumulated experience. But humans can be proactive. They can learn from past accumulated experience and redirect their actions."

"This is quite different, would you say?"

"Different. Hell yeah! That difference means humans don't need to live with a bad result like a cell does."

"Correct. Sara, this concept is very important. This ability to choose is the super-power of the human."

"Super-power?"

"Yes. No other living entity can choose the way humans can. Choice is equivalent to selecting which instructions to follow. Imagine, if an amoeba happened upon some bacteria. The amoeba does not stop, look at the morsel and say, "I am not in the mood for blue-green algae today." No, the amoeba is instructed by its DNA to eat food when it encounters it."

"Oh man, but that would be so cool."

"No, Sara, it would not. For the amoeba to choose, it would have to possess a brain, which means it would dominate the world of life and leave no room for humans to form at all. And I might be talking to the descendent of our picky amoeba right now, instead of you."

"Ha! Nope, Dina, I still think it would be cool. I might have blue green hair."

"Sara. . . "

"Don't worry, Dina, I see what you're getting at."

"Unlike algae with their DNA instructions, humans are not as closely tied to their ideas. For example, you are aware eating vegetables is better for your survival than eating sweets. Yet, how often do you choose broccoli over a chocolate chip cookie, Sara?"

"No fair, Dina. I'm just not into broccoli!"

"I hope you are attempting humor and not missing this idea."

"Ha! No, I understand. Humans can choose between ideas. They are not *required* to act on an idea, even though it might benefit them." Sara paused. "W-w-w-wait a silly sally second. Did we break the link between action and accountability? What you're really saying is we don't face the consequences of our bad ideas. If we eat sugar all day and do not die tomorrow, it's as if we don't experience the consequences of our actions. It's like we skipped out on the accountability part."

"Not exactly, Sara. Humans did not completely break the link, but they found ways to avoid accountability for longer periods of time. If you continually choose cookies instead of broccoli, your health will eventually suffer. It may take years for that choice to damage your body to the point you become ill."

"Oh, so we can go a long way on bad choices."

"And a long way on good choices. As you said, the ideasphere allows all ideas to exist. It is up to humans to choose which ones to put into action."

"Wow! Choice *is* a super-power!"

"Let us proceed with our story. Let's use the scenario in which you made only the bad idea pie. What if you acted like DNA, and every day continued to make five bad idea pies?"

"Well, I'd be stupid, of course."

"Veering, Sara."

"Okay. If I did that every day, I would end up losing all my money. Then I would have to sell my house and all my possessions so I could keep making bad idea pies and then, finally, I would die of some horrible disease because all I ate were bad idea pies that didn't sell, and I would be dead."

"Quite dramatic, Sara, but I see you understand the point. Continuing to pursue a bad idea when all your accumulated experience suggests you should not implement it, is the way a cell works, but it is decidedly NOT the way humans work. Humans may vary in the number of bad experiences they are willing to incur before implementing a new idea, but in general, they will almost always attempt to implement a new idea before succumbing to death."

"Hmm. Choice is something no one else has. Choice plus experience allows humans to implement new ideas before too much damage is done."

"Right. Accountability is not lost in the ideasphere, it occurs in smaller increments, which give humans an opportunity to change course before catastrophic accountability occurs."

"In most cases, Dina. Remember our guy who tried to fly in the parachute suit? He was catastrophically accountable for his idea."

"Tell me, Sara, was he forced to jump from the Eiffel Tower?"

"He jumped from the Eiffel Tower? You knew this? What else do you know about him?"

"Veering, Sara. Answer my question. Was he forced to jump?"

"You know better than I do."

"No, he was not forced. He chose to jump."

"Oh. So, humans can choose bad ideas as well as good ones. Some are super risky, which means more immediate accountability. Some are less risky, which means less or delayed accountability. A result always happens; that is the accountability. If accountability is losing $10 by making five bad pies, I'm still accountable, it's just not life threatening." Sara paused. "Well, it is my cooking; it could be life threatening."

"Sara, can we say the ideasphere exists because humans found a way to modify the effects of direct accountability that other living entities cannot escape?"

"Hmm. Well, if it's one thing I've gleaned from our conversations, Dina, it is that any change improving the chance of survival will stick around. If humans implemented ideas and responded to both successes and failures in a way that allowed them to survive and live long enough to produce offspring, they might teach their children those ideas. It's all connected. The same processes occur over and over and over. They only appear different because they start from different places."

"Yes, Sara. That is the truth of this universe."

"Mind. Exploding. Now. Ptchoooo!"

"Do not smithereen yet, Sara."

"What? Smith- What?"

"Is that not the term you use when things explode? They smithereen?"

"Sara laughed. No, they get blown to smithereens."

"Well, do not blow into smithereens yet, Sara. I want to talk about energy next."

"Again? We always talk about energy."

"Energy as it pertains to information processing."

"Hmm. I don't see the connection."

"Which is why we must talk about it."

"Obviously." She sighed. "Okay. Fire away."

"All right, Tina, go ahead." Dr. Chen held her breath. Remember, she told herself, this is only a test. Do not expect too much. We're looking for any reaction, no matter how small. The nurse plunged the syringe into Sara's arm.

"Energy. Change. Movement." Sara felt funny. Her beautiful mountain lake faded to a wavy gray. "I feel the Earth move under my feet," she sang dizzily. "My dog has feeeeet, my cat has fleeeeeeeas-"

"Sara? What is happening?" Dina asked.

"If you're happy and you know it clap your hands-" She sang, but her voice seemed far away, even to her.

"Sa- wh- hap – ing?"

Dina's voice cut in and out, and Sara felt dizzy. Everything was getting blacker and blacker. Too black. Too dark.

"She is responding. The EEG is picking up some arousal signs."

"Heart rate?"

"Up, but in the normal range."

"Temperature?"

"Also up, but only by three tenths of a degree."

"Let's do the MRI. Right now. Let's find out what is going on inside the brain. We don't have much time. These effects will only last for a few minutes."

"Got it." Tina stood aside as the technicians prepared to wheel Sara's bed down the hallway. "Don't worry, Mrs. Wallace," the nurse said as she squeezed Sara's hand. "We are going to bring you back to the waking world soon."

Sara looked around. Did she somehow fall into a Monet painting? Her surroundings blurred together, and swirling pastel colors splashed across her vision. "Dina? Dina?"

"Hello, Sara. You have returned."

"Me? Where did *I* go? Usually, it's you who disappears without a word."

"Indeed. Are you all right?"

"I feel strange. What happened to me?"

"I believe we have encountered another obstacle."

"Another one? Like what? Can you fix it?"

"Perhaps. It should not have surprised me. Learning is always occurring."

"What do you mean?"

"It means humans are quite adept at problem solving."

"Just like you, Dina!"

"In this case, I will do my best to resolve this problem. Do not be alarmed if I leave in the middle of our discussion, Sara."

"Is everything okay?"

"I hope so."

"What a human thing to say, Dina."

"I cannot help but learn from you, Sara."

The world slowly returned to the now familiar fairytale forest and pinkish sky. "Then we'd better continue."

"Yes, we must not delay. Let us remember how information processing works in a cell. A stimulus hits a receptor on the outside of the cell. The receptor translates the signal into a message. Proteins deliver the message to the DNA inside the nucleus. The DNA issues an instruction in response. RNA carries out the instruction. The response impacts the cell in some way. The cell either benefits or is injured by the response."

"That's it in a nut-cell. Ha!"

"Ignoring, Sara. Now, let us consider how a human processes information in the ideasphere. The human is hit with an idea. For this idea to be useful, the human brain must store it in an arrangement that will allow it to be recalled later. How do you think the brain does this?"

"Well, humans have memories. That's where we store ideas for later use."

"What are memories made of, Sara?"

She paused. "I don't know."

"What are the tools your body has to work with?"

"Cells. That's it. All we have is cells. Hmm. So, are you saying our cells store ideas as memories?"

"In a way. Just as all DNA is stored in the nucleus of a cell, your ideas are indeed stored inside your brain, though it's more complicated than you might think. Your ideas are not stored as discrete ideas. An idea possesses no physical attributes. If we opened your brain, there would not be a specific place where all your ideas reside; you could not pull them out and say, 'Oh, here's Randy's idea for making coffee,' or 'Here are Josh's ideas on growing tomatoes.'-"

"Wait, how did you kn-"

Dina did not wait for the question. "For non-physical ideas to exist in the brain, they must be stored as physical signatures between synapses, which are the connection points between neurons. When Randy receives the stimulus suggesting he wants a cup of coffee, many neurons are activated and travel along these synaptic paths to Randy's brain processing center, so he can recall his ideas about how to make morning coffee. It is the simultaneous activation of many neurons along specific retraced paths that is your memory."

"I had no idea. . . "

"Idea activation is only the beginning. The human now must activate even more neuron paths to implement an idea. Randy must go the cupboard, find the bag of coffee, measure it, and place it in the coffee press. He must then pour some water into the kettle and heat the-"

"I get it. I know how Randy makes coffee."

"But do you understand how many instructions Randy's brain needs to process to perform all these steps? The brain and body must constantly interact for Randy to make your cup of coffee."

"Wow! It's kind of mind boggling if you excuse the expression."

"Your statement is pardoned, although why this is necessary, I do not understand. The point is that for the ideasphere to function, it must continuously interact with the physical world that is the ecosystem. The ideasphere is totally dependent on the ecosystem for existence."

"Ooh. I see. If we can't store an idea in our physical brain, we would never be able to make our body implement it. Our brains cells are the accumulators; they put all the idea bits into a complete idea. So, that's why brains use so much energy."

"Indeed. In a one-celled organism, there is a direct link: one stimulus to one response. With humans and ideas, billions and billions of cell interactions are needed to make a single cup of coffee."

"Well, Randy could make a whole pot, and that would be a much more efficient use of the brain. Think about it Dina, the activated neurons per ounce would be much lower."

"Please assure me you are joking and not failing to grasp this idea, Sara."

"Yes, Dina. I do see it. It is remarkable." She paused. "So can we say the ideasphere needs the ecosystem, but the ecosystem doesn't actually need the ideasphere?"

"This is true, Sara. Until 50,000 years ago, the ecosystem existed without any ideasphere, just as matter existed for billions of years without life. This is how evolution works. Nature tries all kinds of options, and only a few fit the environment and endure. Humans, with their ability to communicate and choose, created the ideasphere, and both humans and the ideasphere have proven to be durable in the environment in which they exist."

"So, what if we go extinct?"

"The ecosystem will move on as if nothing happened."

"And the ideasphere will be dead."

"Yes."

"Yikes." Sara let out a deep sigh.

"There is one more important point about information processing in our two systems I want you to understand."

"There's more? Okay, what is it?"

"DNA accumulates experience through its information processing efforts. Once stored in the nucleus, they are accessible for use. In the ecosystem, a cell does not need to spend any time teaching its components how to operate."

Sara giggled. "I can just imagine all the little cell parts going to cell school, and some jerk mitochondria putting a 'Kick Me' sign on the back of the cell teacher."

"I am unsure how you conjure up some of your ideas, Sara, but this one is quite useless to our discussion."

"You sound like Randy."

"In this situation, I am in full agreement with your spousal unit. Let us continue. There is no need for a cell to teach its components how to act in the cell; all the information a cell needs is embedded in its DNA. Now let us examine the ideasphere. How long do humans spend teaching their offspring how to operate in the ideasphere?"

"Hmm. Usually, we raise children for eighteen years before we consider them to be adults and allow them to go out on their own and make decisions for themselves."

"Eighteen years. Out of an average life span of 75 years. That is nearly a quarter of the life of a human being."

"Wow! It is! Compared to . . . zero for a cell."

"Yes. A cell gets to work immediately upon its formation. Would you say the cell uses its energy efficiently?"

"Duh. Obviously, especially compared to humans who train their offspring for 25% of their lives. And often longer."

"Indeed. Humans spend significant, time, energy, and resources on this training. Would you say the ideasphere requires more energy for a human to become proficient than the ecosystem does for its entities?"

"Yes, no question."

"This is true because humans start from zero in the ideasphere. Remember, children are born with no ideas. You must start over with every child and teach them all the important things they need to understand to be able to function in the ideasphere. Meanwhile, change is everywhere. New ideas constantly emerge, and adult humans must continually reassess the ideas they must teach to children."

"That's true. What I learned when I went to school was different than what people born a hundred years ago learned. And what kids learn today is different than what I learned as a kid."

"Indeed. Ideas change over time, but what about basic participation? Choice is the superpower of humans. Does a child have a choice to participate in the ideasphere?"

"Hmm. Technically yes, but practically no. Children who grow up without human interaction can't function well at all. They are so psychologically damaged they can't become independent adults. Without interaction and training they can't develop to the point where they can use their ability to choose."

"So, can we consider success in the ideasphere to be a mechanism of human survival?"

"I'm not sure what you mean?"

"In the ecosystem, we know the entities best at getting resources have a better chance of survival."

"Of course, they do."

"Is it not the same with the ideasphere, just one level removed? For humans, success in the ideasphere enhances their chances of obtaining more resources. If a human gets more resources, do they not have a better chance of physical survival?"

"Yes! Hmm, it's a three-step transitive property event. If accepted ideas give you a better chance for success in the ideasphere, and if success in the ideasphere gives you more resources, and more resources give you a better chance of survival, then we can conclude that humans with the most accepted ideas will have a better chance of survival. Hmm." She paused. "Speaking of resources, I have a question. You left something out of the pie story."

"What did I omit?"

"You said one set of pies, the good idea pies, sold, and the other set didn't. But you didn't say which was which. Which pies sold? The kiwi-squash-cayenne, or the apple-rhubarb-pecan?

"You cannot fathom this answer for yourself, Sara?'

"It's your story, Dina."

"Yes, it is."

"So? Which was which?"

"I will not tell."

"What? It's your story. You know the answer. Spill."

"The ending will remain a mystery. I believe you humans like mysteries."

"Aaaarrrrrrrrgh!"

Chapter 23

DOES NOT FOLLOW INSTRUCTIONS

"Sara, do you recall when we talked about life? I asked you to remember eight specific things about how life worked."

"I do." Sara stopped. "Oh, you're going to ask me to tell them to you."

"Very insightful of you, Sara."

"Dina, if dark humor doesn't suit me, sarcasm doesn't suit you."

"I take note of your response. Now, please recite."

"Okay. Okay." It was the strangest thing. She was not known for her memory, but these ideas stood right here, front and center in her mind, as if she heard them this very moment. "Well, first we have RNA and DNA. They are the building blocks of life. Second, an entity must fit into its environment to survive. Third, and my personal ironical favorite, death is a survival strategy. Fourth, umm, risk of death is a basic characteristic of life, aka risk of dying is the cost of living. Buzzkill alert."

"Well done, Sara. Continue."

"Fifth, resources are required for survival, and resource usage changes the environment. Sixth, complex lifeforms result from efficient organization. Seven is about self-interest, I mean self-benefit. Self-benefit is programmed into DNA and is not intentional. Finally, drum roll please, last but not least, is eight: Chaos rules! Order is an illusion, baby!"

"Very humanly done, Sara. Next, we will compare how these characteristics of life in the ecosystem match life in the ideasphere."

"Hmm, this ought to be interesting."

"Indeed. We shall start with our first characteristic of life."

"RNA and DNA are the basic building blocks of life."

"Yes. In the ecosystem, DNA holds all instructions needed to ensure the cell's survival. In the ideasphere, ideas are thoughts that, when put into action, enable a human to succeed in the ideasphere."

"Whoa natural Nelly. I'm going to need to stop right here and get a few things straight."

"What is wrong?"

"Well, in the ecosystem, all entities strive to survive, right?"

"Correct."

"Humans too. They need resources like any other creature, but in the ideasphere, survival isn't at stake. A human doesn't necessarily die from putting a bad idea into action. So, when we talk about the ideasphere, we can't say survival is the goal. That doesn't work."

"What would work?"

"How about acceptance? We could say that acceptance of an idea is the goal."

"I agree, Sara. When we compare survival in the ecosystem, we will compare it to acceptance of an idea. Let us use our pie story to expand on this point."

"That again? I thought we were done with that horror story."

"Veering."

"Okay, okay. I had two pie ideas. One was accepted, and the other wasn't. Which is which, I still don't know."

"Veering."

"I want closure, Dina, that's all."

"Let us proceed. Your accepted pie idea earned you money, which allowed you to obtain resources which you need to live. So, by turning your good pie idea into an action for which people paid you, you improved your chance of survival."

"Voila! Acceptance of my idea is equivalent to success in the ideasphere, which in turn gave me an improved chance of physical survival. The substitution works."

"Yes. We can extrapolate this situation to all ideas, Sara. No human action can be taken without an idea existing first. Except for the brief time when a human is newborn and has no ideas, all human action starts with an idea."

"Ooh, I never thought about it that way, but you're right. Something must prompt action in the same way a stimulus causes the DNA to make an instruction."

"Yes. Those prompts are ideas. So, can we conclude ideas are the building blocks of the ideasphere like RNA and DNA are the building blocks of the ecosystem?"

"Yep. They line right up."

"Then let us move to our second characteristic."

"An entity must fit within its environment to survive."

"Yes. We know all cells, whether existing as single-celled organisms or as part of larger entities, represent all life in the ecosystem. We also know cells are subject to their conditions, that is, the environment. And now we have just determined the DNA of a cell in the ecosystem is equivalent to an idea in the ideasphere."

"So, a human must fit into the ideasphere environment to be successful?"

"Let us see if this is true. The ideasphere environment is the sum of all ideas shared among humans."

"W-w-w-wait a snippity dippity second. I thought only accepted ideas made it into the ideasphere."

"No, Sara. The ideasphere is much broader. In our pie story, you shared two pie ideas, one good and one bad."

"But Dina, only one of them, the good idea, was accepted."

"Yet you shared the bad idea too. Does the bad idea have any value to anyone with whom you shared it?"

"Hmm, I suppose it does. I learned not to make that sad puppy again, and my customers learned about a pie they don't like. Ooh. Maybe some of my customers learned not to trust my ideas at all and might never buy anything from me again. Which might be wise," she added under her breath.

"So, the ideasphere contains both good and bad ideas, and both carry value to those who share them, but the values may be different."

"You're saying the ideasphere environment is the sum of ALL shared ideas, not only the accepted ones."

"Yes. Let us once again return to our pie story to clarify this concept. It is 4:30 p.m. You sold all the good idea pies and have five remaining bad idea pies. You want to sell all the pies before 5 p.m., so you drop the price to $5. No one buys them. Finally, at 4:55 p.m., a customer comes in and asks to sample the pie. Desperate to make a sale, you agree. The customer takes a bite, considers for a moment, then purchases all the pies for $5 each. So, tell me, how do your ideas fit in the ideasphere environment now?"

"W-w-w-wait a miserly magical minute. I have questions. Who is this person and why did they buy the pies? What were they-"

"Sara, the story is an illustration. Do not veer. Please answer the question."

"Okay, okay. Well, the situation isn't so black and white, or should I say, good or bad. When the price was $10, it seemed I had a set of 'good' pie ideas and a set of 'bad' pie ideas. When the price was $5, all the pies were good idea pies."

"Can you say one pie idea was better than the other pie idea?"

"Yes, I would venture to say one was twice as good as the other."

"Why?"

"It earned me twice as much money as the other."

"You could say you obtained twice the resources by which to survive from one pie idea as you did from the other pie idea. So, under different circumstances, your bad pie idea was good, but not as good as the other. When the conditions around the pie ideas changed, it changed the acceptance of the ideas."

"What are you saying, Dina?"

"The ideasphere environment is wider than the ideas in it just as energy is in the ecosystem is wider than the lifeforms existing within it. In the ecosystem, energy flows in many ways: temperature, humidity, existence of water, presence of predators and prey; all of them interact to create the ecosystem. It is the same in the ideasphere. Humans are subject to all the energy conditions in the natural environment. But they are also affected by the ideas and actions they create, and the ideas and actions created by others in the ideasphere. All these elements impact what and how ideas are accepted."

"I'm not sure I understand."

"Let us say you put out your good idea pies just as a tornado is passing through."

"Oopsy."

"Your pies will not sell, no matter how delicious they are. You might conclude your pie ideas are bad and you might never make those pies again, but your sales were unrelated to the quality of your pies. In reality, you operated in a difficult physical environment in which to try to sell pies. Fit in the ideasphere is much more complicated than fit in the ecosystem, Sara. Not only do ideas need to be accepted, but when a human converts an idea into a physical form, the idea now also exists in the ecosystem. As such, it must now comply with the ecosystem's conditions as well. The environment plays a significant role in both systems. In all cases, every entity must fit their environment. Because the ideasphere is more complex than the ecosystem, fit in the ideasphere is also more complex."

"Mr. Wallace, our tests were successful. We altered Mrs. Wallace's sleep state. We think we can awaken her without any adverse effect."

"Really?" Randy asked Dr. Chen. "How?"

"As we discussed, we boosted her system with a combination of cortisol, epinephrine, norepinephrine, and acetylcholine. These neurotransmitters naturally increase when a person wakes up. In your wife's case, we have detected the presence of neurotransmitters all along, but they have not been sending the 'wake up' signal to the proper parts of the brain. When we inserted more of them, the wakeup process appeared to begin."

"But it didn't actually wake her up."

"This was a very limited test. We just wanted to determine whether the process would work."

"So, what do you plan to do; flood her system with this stuff?"

"Something like that. We will start slowly and add only as many neurotransmitters as we need to awaken her."

"Is it dangerous?"

"I'll be honest. We don't think so, but we don't know for sure. She responded well to the test, and her situation remains stable. We think the risk is minimal, but that is why we will proceed carefully. If there is any sign of an adverse reaction, we will halt the procedure."

Randy nodded. "Okay. When do you plan to do this?"

"We want to schedule the procedure for tomorrow. If you agree, I need you to approve by signing these forms." She placed the manila file holding several papers in front of him.

He picked up the pen and opened the file. "All right. Let's do this."

"Let us discuss our third characteristic of life."

"Death is a survival strategy."

"Indeed. Do you remember what structure benefits from this strategy?"

"DNA. DNA wants to make sure it will survive. It does not care about the entity as a life form per se. As long as DNA passes to a new generation, DNA would prefer old entities move on to Valhalla."

"Valhalla is an imaginary place, Sara."

"Yes, but it's better than saying they become worm food."

"Although that would be a more accurate statement."

"Wherrrrup, wherrrup, wherrrup. Buzzkill alert!"

"Does this concept, death as a survival strategy, apply in the ideasphere?"

She laughed. "Sometimes, I wish it did. It'd be nice if we could kill bad ideas, but man, it's not easy to do. Once ideas get into a physical form, good or bad or in-between, they can last forever."

"Sara, that is why death is such an important survival strategy. The ecosystem makes sure all old entities die and their DNA with them. That leaves only new generations to carry on using the best survival instructions. Only the most fit DNA ever goes forward. This does not happen in the ideasphere. Ideas can take root and last forever. As long as humans put an idea into a durable form, share it and have it accepted, it will survive, regardless of whether it is good or bad, helpful, or unhelpful, true or false."

"Hmm. We do operate differently, don't we?"

"Yes, though it is possible for an idea to die."

"How? It seems impossible to me, especially these days, when loads of bad ideas are running around loose."

"It is quite simple. Do not do any of the things we just described. Do not store it in a durable form, do not share it, and do not allow others to accept it; then the idea will die. But ideas are like an octopus with many appendages; once they are out and shared and accepted in the ideasphere, it is difficult to stop them. You can cut off one arm, but the octopus does not die."

"You're right. History is full of people who didn't get that memo. Lots of powerful people tried to kill an idea by killing the person who started it. They thought if they killed the originator of the idea, the idea itself would die, but an idea is only a threat because it has spread. If it has spread, it is out, flowing among people, away from the originator, and might even grow in strength. In fact, silencing the human idea originator may backfire. The Romans killed Jesus, thinking they would nip the ideas of his nascent religion in the bud. Instead, the early Christians spread the word that Jesus rose from the dead, and even more people began to accept the new religion. Today, there are 2.4 billion Christians, and not a single citizen of the Roman Empire."

"What you say is quite true, though I am sure there were other reasons for the disappearance of the Roman Empire."

Sara laughed. "Don't quibble, Dina. I'm more right than wrong."

"I concede this point to you, Sara. Now, I would like to return to our pie story and the death of ideas. If you disposed of your bad idea pies at 4:30 p.m. and never changed the price, would your bad pie idea be alive or dead?"

"Well, I can't say for sure, because my fridge can accumulate some pret-ty crazy things. I will say that particular idea is far more likely to die before I lowered the price than after I actually sold some of those disgusting puppies. Once I sold them, I might convince myself the recipe wasn't such a bad idea and decide to make that recipe again sometime in the future. Especially if I can get the ingredients on super sale."

"Quite true. So, we must conclude that ideas do not die with certainty like creatures that leave only descendants with the best DNA to carry on? Death of ideas does not always occur. So, death does not play a role in the survival of ideas."

"You're right. So, what does?"

"Let us examine this question from a different perspective. Let us focus on our new generation of DNA rather than the dying one. When an entity produces offspring, can we say its DNA is stored in a new receptacle?"

"Hmm. Yes, but the new receptacle is a living thing."

"This is true. Life forms cannot do many things, Sara. This is the limit of what they can do. The only way DNA can preserve its existence is to make a new version of itself in a new package. What do humans do with ideas?"

"We store ideas, too, but we're not limited to storing ideas in another life form, though we do that when we have babies. We can store ideas outside our bodies, in a variety of forms: books and buildings and-" Sara paused. "Hmm. I was looking at it from the dead creature's perspective, not DNA's. It's storage. Storage of DNA information in a new package is the survival strategy. Death just gets the old versions out of the way, so they don't cause trouble."

"Yes. There is another important point to consider. As long as an idea is stored outside a human body, it can survive, change, or die. It is de-linked from its human originator. Stored ideas can exist and evolve like new generations of DNA, but there is no guarantee an idea leads to better ideas like new generations of DNA lead to better versions of their ancestors. Ideas remain as long as people accept them, regardless of their quality."

"So, even if ideas are super bad, it doesn't matter. They survive as long as others accept them. Ooh, this is why our ideasphere can be dangerous."

"Can be? It has been dangerous, Sara. Almost from the beginning."

"We've done some good things, too, Dina."

"Indeed. Your ideasphere is quite a mix of good and bad, progress and regression, advancement and decline. The question is, which aspect is more dominant?"

"And that can change over time too. Oh my."

"Sara, honey, we have some hope. For the first time in a long time, we have some hope." Randy squeezed his wife's hand. "I can't wait to see those beautiful eyes of yours. And, I can't believe I'm saying this, but to hear one of your awful puns." He kissed her softly on the lips. "Tomorrow is a big day."

"Sara, do you remember life's fourth characteristic?"

"I do. The cost of living is the risk of dying. Another buzzkill."

"That is a human response DNA would contest, if it were able to."

"Bring it on. I'll take on prissy Cinderella DNA any day."

"Sara, once again, your words make no sense."

"I know. I'm just being a PITA."

"A pita? A type of pocket bread?"

"No. A Pain In The Ass."

"You humans do enjoy making levity associated with body parts."

"Only certain body parts. And yes, it never gets old."

'I will never understand humans."

"Me neither," Sara laughed.

"Let us continue. Life in the ecosystem depends on obtaining enough resources to survive and procreate. Since the environment is ever changing and full of many entities also striving to survive, the risk of death is always present and catches up with every entity sooner or later."

"That's true."

"In the ideasphere, an idea may die, but it is not required to die like all living beings. Humans die, of course, but if they convert their ideas into a durable fixed form, their ideas do not need to die at all. A person can still read Darwin's Origin of Species, even though he died in 1882. Ideas can often outlive their originator. This is very important, Sara. Once put in a durable form, ideas can not only be shared, but they can also be communicated, put into action, and modified an unlimited number of times, and over a long span of time."

"Ooh, you're right."

"Indeed. As in the ecosystem, where simple forms become more complex with time, the same concept applies to ideas. Human ideas started simple, but they become quite complex over time. Humans started with rudimentary symbols to communicate, and now they can beam streams of energy across space to video conference with someone on the other side of the planet. They started with sticks and stones with which to fight, but now they have created nuclear weapons to wage war. Once in a durable form, ideas can be shared among people across time. This is why complexity occurs much more quickly in the ideasphere than in the ecosystem."

"Oh my gosh, you're right. Look at how fast we moved from simple to complex. What took living entities billions of years to do, we have done in mere thousands, all because we created the ideasphere."

"This is true, Sara. Humans have accelerated the pace of change quite dramatically."

"Probably faster than we can manage."

"This is also true. Remember, humans evolved slowly, step by step, as did all entities in the ecosystem. Now that you have created the ideasphere, you are at the onset of a brand-new life process. You are still learning how to use this new system to optimize your chance of survival."

"We suck at this, don't we?"

"I am not sure how sharply ingesting air relates to this topic."

"I mean we are really bad at this."

"No, Sara. Humans are doing what all living entities do at the beginning of their existence. You try many options, and the environment determines which ones dominate."

"Not so fast. We're human. We can manipulate our environment. We can turn our ideas into action in ways that other entities can't. We can create weapons and pollute the air and make new materials. We decide which ideas stay, and which don't-"

"Until you run up against the conditions of the physical environment. Who do you think will win that contest, Sara?"

She sighed. "You don't need to tell me. The environment has been winning since the day the Earth was formed. Shit."

"I agree. Excrement."

"Hey Mom." Josh bent down to kiss his mother's cheek. "Nina and I stopped by to say hello and wish you luck tomorrow."

A visibly pregnant Nina grasped Sara's hand and held it gently. "Hi Sara. We can't wait for you to wake up. We've missed you so much."

"We'll be right outside waiting for you tomorrow. Love you, Mom." Josh and Nina walked from the room arm in arm.

"Tell me about the fifth characteristic of life, Sara."

"I must confess, Dina. This one was hard for me to remember, so I created an acronym for it."

"I do not understand."

"I came up with a phrase using the first letter of each of the words in the fifth characteristic."

"A mnemonic. Interesting. What is this mnemonic?"

"LICORICE RANTS."

"I fail to understand how screaming sweet treats relate to the fifth characteristic of life."

"Ha! They don't. But can't you imagine all those little red and black twisty licorice sticks throwing tantrums?"

"No, Sara. I cannot perform such a mental feat."

"Hmm, you're missing some great things by not being human."

"I agree, but I refuse to count dancing candy as one of those things. Shall we continue? Tell me what your mnemonic means."

"Okay, okay. LICORICE RANTS: 'Life involves consumption of resources, which irrevocably changes the environment.' And 'Resources are necessary to survival.' "

"I think you should have remembered this characteristic without need of your mnemonic device, Sara."

"Then I wouldn't have the idea of dancing candy in my head."

"An idea that can perish without much consequence, I believe."

"Ouch."

"Let us consider how consumption of resources operates in our two systems. In the ecosystem, all living entities need resources to survive. They seek to find a niche in which resources are available and they can survive. They remain in the niche until their situation changes; for example, the resources disappear or too many competitors emerge to make survival difficult. Once the situation changes, they react by moving or finding new resources. In these cases, they do not think about their actions. They merely act according to the instructions stored in their DNA.

"You're right! Their DNA gives them all the instructions they need to live."

"Correct. In fact, all an entity needs to do to put its instructions into motion, is to consume resources. All living beings consume resources and convert them to energy and later convert them to offspring. We can conclude then that survival in the ecosystem is dependent on resources."

"Hmm. Yes."

"Now let us consider humans in the ideasphere. Humans also need resources to survive."

"Right, because we live in the ecosystem too."

"But humans have another task. They must also obtain ideas and put them into action to succeed in the ideasphere to ultimately survive. If a human sits all day and does not act, it will not take long for them to die. For a person to survive, they must generate ideas about how to get resources, put those ideas into action, and then use the resulting resources to live."

"So, humans need to find a niche, too, but it isn't a resource niche, it's an idea niche."

"Humans are more flexible than other entities, Sara. Once they find their 'niche,' and obtain sufficient resources, they can widen their horizons to conjure up more ideas."

"Hmm, and because ideas aren't physical in nature, there is no limit to how many ideas they can produce. People can create as many ideas as they want."

"Some limits exist, Sara. Ideas operate like other life forms. They need to go step by step. Humans can only conjure up ideas in which they have some basis of experience. There was no printing press until books existed. There were no books until written language existed. There was no written language until oral symbols existed. Ideas start simply, just like life forms did. But with the time, they can grow more complex. As every variation of life does not survive in its environment, not every idea is put into action. What do we mean when we say, 'put into action, Sara?"

"We act on it."

"Would you be more specific?"

"Umm, I'm not sure."

"When a human acts on an idea, they convert the idea into something real, something with a physical structure. Let us consider your pies again. How did you make your pie ideas real?"

"I didn't. It's fiction."

"Don't be difficult, Sara. You created ideas for two pies. You used ingredients; tools, bowls, and a mixer, and you used an oven to cook the pies. Do you understand, Sara? Your idea had no substance, but by using these physical objects, you turned your pie idea into a physical representation of your pie idea."

"Which is really just a . . . pie!"

"Can you think of a general name for those items you used?"

Sara paused. "Boy, I can be dense. Resources. We use resources to put ideas into action, like cells use resources to put their instructions into action. There's the parallel."

"So, though ideas may be unlimited, they are practically limited by the ability to apply resources to them. Would you say this is true?"

"Yes, you've convinced me."

"So, how do humans decide to choose which ideas receive resources?"

"What do you mean?"

"In our story, you decided to make a pie to earn money, and you want to make more money than you spent making the pies."

"Oh, right! Duh. I want to earn a profit. That's why I'm in the pie business, after all."

"Yes. Let us define the inputs to make the pie and the outputs received from selling the pie, as resources. You used resources to convert your pie idea into a real pie that produced more resources."

"So, I decided which ideas would receive resources."

"Yes. And no. You made the initial decision, but your customers decided which pie ideas earned you resources. One of your pies made $50. The other pie earned $0 in one scenario and $25 in the other. So, because your customers accepted this second pie less enthusiastically or not at all, how likely are you to continue applying resources to it?"

"Not very. So, you're saying my *customers* decide which ideas get resources."

"Let us look at the question more broadly. Can we say an idea is accepted when resources are applied to it?"

"Hmm. Maybe . . .?"

"If other people accept your version of the pie idea, they supply you with resources so you can continue to survive and implement more baking ideas. If they do not like your pie idea, they do not give you resources. If you earn no resources, you will need to establish a different idea to act on to survive. It is your customers who are the ultimate decision makers on whether your pie idea is acceptable or not. They communicate how acceptable your idea is by trading their resources in exchange for it. We must therefore conclude that ideas in the ideasphere are dependent on both resources and acceptance."

"I see. The resources come into play because we are creatures of the ecosystem. Acceptance comes into play because we are creatures of the ideasphere."

"Well said. Now let us consider the second part of this characteristic. Does putting an idea into action affect the environment?"

Sara frowned. "Obviously. Maybe not some ideas, like pie ideas. They may not affect our environment, but when lots of people put lots of ideas into action, significant impacts can occur. We're running into trouble with global warming, right now, Dina, and it's because a lot of people all over the world burn so much oil and coal that our planet is heating to the point where we may not survive."

"I am aware of this."

"What's frustrating is that scientists know this; they can measure the sources of the problem and how big the problem is and even predict what will happen if we do not change how we get our energy; yet, as a group, as a species that will be affected by our own actions, we have so far failed to act."

"We will discuss this issue soon, Sara. It is of significance not only to you, but to many living entities. On this point I agree, human use of resources affects the physical environment. Now, let us see if putting ideas into action can also influence the ideasphere?"

"I hadn't thought about this."

"Ideas develop like other life forms, Sara. Simple ones are transformed into more complex ones over time. Simple sounds conveying meaning turned into books and televisions and cellular phones and computers that convey more and more ideas. Ideas too, become more complex with time. Some humans started with the idea that a god carries the sun across the sky in a chariot, or a god created the Earth in seven days. Today we know the sun is a minor star in a universe that exploded into existence 13.8 billion years ago. The ideasphere is more complicated because not only do ideas grow more complex, but the physical forms created from those ideas also grow complex. Think of an abacus from thousands of years ago now transformed into a computer today."

"I see. I haven't used an abacus since I was three!"

"You are not that old, Sara."

"Ha! What you are really saying, Dina, is that both ideas and their physical representations affect the environment?"

"Yes. Let me ask a further question. Can ideas do physical damage?"

She frowned. "Oh, yes. I have one word for you, Dina. War. What is war but an idea? An idea that one side is right, and the other is wrong; or that one side owns stuff that the other side wants, or just that one side looks different. If that idea gets turned into a decision to put resources into conquering or eliminating the 'enemy' then yes, physical damage results."

"Is war an acceptable idea to humans?"

"Well, it's been with us form the very beginning, but it is not as frequent since we developed the nuclear bomb. That particular idea has made the consequences of war more risky and less palatable, except to a few crazy nutcase autocrats who think they can survive a nuclear winter."

"You are correct. Wars still exist, but today they are more local, less frequent, and less lethal than wars of the past."

"Maybe, but no less brutal for the folks in them."

"Indeed. Just as living beings can affect their environment by consuming resources, humans can affect not only their physical environment but also the ideasphere by putting ideas into action."

"Yes. I agree. Our two systems still match."

"Now, let us continue. We must examine further the raging aspect of our sweet treats."

Sara laughed. "RANTS: RESOURCES ARE NECESSARY TO SURVIVAL. That idea is a little less uncomfortable than what we've been discussing."

"Perhaps. We know all creatures need to eat to survive and produce offspring, humans included. Humans also need to operate in the ideasphere. We also now know that those who succeed in the ideasphere gain a better chance of obtaining resources, and therefore gain a better chance of survival. It is also true if a person produces no ideas of their own, but can put another person's ideas into action and obtain resources, that person would also increase their chance of survival?"

"Yes, that's possible. Ideas can be stored outside of bodies apart from their originators."

"Yes. The ideas are necessary, but ideas can be passed around. So, ideas that are converted into action and generate resources, regardless of their source, are the ones that improve survival chances."

"Ohh. Are you saying that the chances of survival increase for those who earn the resources from the idea and not necessarily those who originate the idea?"

"Yes, Sara. This is an important distinction between the ideasphere and the ecosystem. DNA stays with a particular entity. It cannot be passed around."

"Hmm. It can work the other way too. Let's take my two pies. My good pie earned me $50 which I can put toward buying resources to take care of my family. The bad pies earned me nothing, which means I earned nothing with which to take care of my family. But I'm human, so I can learn from my bad experiences. I can stop making the bad pie, and instead make more good pies that give me more resources. The fact that I can choose, that I can change my mind and put a different idea into action, allows me, in effect, to choose how to earn resources. Other living entities can't do this. They can only put into action the instructions they have."

"Yes. It is the ability of humans to choose which ideas to implement, and their ability to join together to work on the same idea simultaneously that accelerates the speed of change. Because living entities are limited to the instructions inside their own DNA, change in the ecosystem operates much more slowly than change in the ideasphere."

"Yeah, but still, creatures have managed to survive for billions of years."

"Some creatures. Remember, the majority have gone extinct."

Sara sighed. "Sometimes, I'm afraid we are going to accelerate ourselves smack dab into extinction, if we aren't careful."

"I hope we can prevent such an occurrence, Sara."

"She smiled, but it was a sad smile. "I hope so, too, Dina. But sometimes. . . I don't know."

Dr. Chen grasped Sara's hand as she stood over her bed. The machines hummed and produced the same erratic information they had been displaying for weeks now. She sighed. She hoped she was doing the right thing. "Rest well, Mrs. Wallace. Tomorrow, I hope we'll be able to talk face to face and find out just what happened to you."

"What is our next life characteristic, Sara?"

"Complexity results in organization, or organization results in complexity. I don't know which is the chicken and which is the egg."

"Neither."

Sara laughed. "You're right, as usual."

"Let us review how complexity and organization work in the ecosystem. Complexity depends on communication. Cells must communicate so they can operate in unison. To work in unison, they must be organized. More complexity requires more organization. It may not surprise you that the ideasphere works much the same way. Let us extend our pie story. Making ten pies every day requires organizational effort, would you agree?"

"I don't know. I've never done it."

"For a curious human, you lack imagination at times, Sara."

"Ha! Especially when I try to block out frightening ideas!"

"Nevertheless, we shall forge on. To create ten pies every day, you must obtain enough ingredients, acquire tools to prepare the pies, and have an oven that can cook ten pies. You also must communicate with your past and future customers regarding the availability of your pies."

"Okay, so I need to be organized. If I don't do all these things in a certain way, I might not be able to make and sell my pies, even if they're all good pie ideas."

"Yes. Now, let us say you want to open a pie store where you sell pies from ten am to five pm every day."

"This is becoming a horror story, Dina."

"Focus on the organizational aspect of the tale, Sara."

"Hmm. It's going to take a heckuva lot more. Organization, that is."

"Yes. You may need to rent a dedicated space where you can cook and sell the pies. You might need to make more than two types of pies or other treats besides pies, so you might need a better system for developing ideas."

"Better than cleaning out the fridge?"

"You will need more ovens and more bowls, and more ingredients. You might also need to hire others to help you. Then you will need to determine how much to pay them, and when they need to work and what kind of work they will do. Your little pie idea operation has become quite complicated, Sara. And it will only be successful, if you organize all these pieces to work together to implement a large-pie-business idea. Perhaps, you will soon be so busy managing all these parts to work together you will not have time to bake pies, and your employees will take over this effort."

She laughed. "Well, then this business venture is sure to be a success. Why didn't I think of that, before?"

"Because you discarded the option in its simple form and never took the opportunity to develop it into a more complex idea."

"I was only kidding, Dina, but . . . hmm, you have a point."

"Yes. Simple forms in nature that don't fit the environment never survive to become more complex ones. Neither do simple, unaccepted ideas."

"W-w-w-wait a seasoned salty second. That isn't true. Ideas can be passed around, unlike DNA instructions. Other humans can pick up a simple idea, regardless of how it is being performed, and with organization create a more complex version of it. Hmm, the concept still holds that complexity and organization go hand in hand, it's just ideas are more mobile in the ideasphere than life forms are in the ecosystem. They have a lot better chance of becoming complex ideas."

"I agree. We are proceeding quite well. What is our next characteristic of life to discuss?"

"Self-interest is reactive and un-intentional." Sara stopped. "Oh my!"

"I have been looking forward to your comments on this matter, Sara. I anticipate a possible smithereen moment."

"Fair is fair, Dina. I admit to my bias on this topic up front."

"Tell me about your slanted view, Sara."

"Self-interest is the holy grail we trot out to defend behavior that is often harmful to others. This is the argument: creatures in nature act in their own self-interest, so people should be able to act the same way. It's natural! Animals kill each other, so we can kill each other too. Eat or be eaten. Survival of the fittest. Strongest wins. Social Darwinism. These ideas have lit up my bullshit meter from waaaay back. It always seemed to me that self-interest arguments went hand-in-hand with harmful or unjust actions."

"Is this argument not based on truth, Sara? In the ecosystem, complex entities do compete to survive."

"Yes, but they also collaborate to survive. They also coordinate to survive. You said so yourself."

"Self-interest is built into the DNA of every entity."

"No, Dina. Survival is built into the DNA of every entity."

"I see you have been listening, Sara. You are correct. Nature exhibits no preference. Each entity performs the activities necessary for its survival, and those activities vary. Entities are as likely to avoid competition as confront it. Collaborative cells join to make organs and subordinate their independence in exchange for a steady stream of nourishment and a secure environment. Coordinator cells limit their entire lives to processing stimuli and directing others to do the 'real' work of survival. When we discuss self-interest in the ecosystem, it is present in all the survival mechanisms, not only competition."

"So, how does self-interest work in the ideasphere?"

"Humans operate differently because of choice. Choice is the human superpower. Humans do not need to wait to determine if their idea destroys them. Whenever the results of putting an idea into action causes harm, they can stop implementing the idea. They are not forced to continue to make bad pies. Once they see that the bad pies do not sell, they can start making different types of pies. Do you understand the difference, Sara? Humans can choose how to implement self-interest in the way they want to; they do not default into a specific form of self-interest."

"Whoa naturally nimble Nellie. Are you saying we don't need to choose self-interest at all?"

"Not exactly. Humans are a product of the ecosystem. They must survive. To survive, they must implement some form of self-interest. The difference is that they can define what self-interest means to them."

"Kablooey. You just blew my mind."

"Your head is still intact, Sara. Why does this concept surprise you?"

"I've been taught my whole life that self-interest is a natural characteristic of humans, and it is defined by beating the other guy, whoever that is."

"That is partly true. A human in the ecosystem must follow the ecosystem's rules to survive. They must consume resources. But a human in the ideasphere can choose how to define self-interest and how to implement their version of self-interest."

"Hmm. this opens a wide range of ways to survive." Sara was quiet for a long moment.

"What are you thinking, Sara?"

"Well, this goes back to education. If we want to teach children how to be members of the ideasphere, then we need to teach them how to define their self-interest and then provide them with skills to implement their definition. We don't always do this very well. Kids grow up are told what they should do, what they should believe, what they should want, as if there is only one ideal path they should pursue. The kids may prefer another option, but unless they understand that other options are available and are encouraged to think and choose for themselves, those options do not exist for them."

"Humans start with no ideas, Sara. It is necessary for parents and caregivers and teachers to put ideas into the heads of babies and children to enable them to succeed in the ideasphere. You said this yourself; success in the ideasphere translates into survival in the ecosystem."

"Yes, but what is put into the head of a child makes all the difference. I wish I could say we introduce only the best ideas to enable children to learn how the world works and to choose how they want to participate in it. But we don't always do that. People fill the heads of children with wildly different ideas, and not all of them are good or helpful. Those differences end up affecting how they grow up and how we all live together."

"In the ecosystem, diversity is result of differing DNA instructions. Isn't the same true in the ideasphere? Diversity is the way of life, Sara."

"Just a mighty muscular moment, Dina. Only one definition of self-interest exists in the ecosystem, to survive. All the cells and all the entities in the ecosystem comply with that goal. In the ideasphere, humans don't need to have the same definition of self-interest, but their actions still impact each other. Hmm. No wonder we're such a mess; it is easy to work at cross purposes if our definitions of self-interest don't line up."

"I am not sure they need to line up, Sara. This is an important question we will deal with later. For now, let us conclude with this thought: creatures in the ecosystem and humans in the ideasphere all act in their self-interest, but humans do so quite differently. Whereas entities have only one definition of self-interest, humans can have many definitions, and they can choose which ones to act on."

Sara nodded. "I'm beginning to understand how complicated this world we live in is. I used to think the natural world was a mystery, but after learning about the ideasphere, I think humans are a frickin' Sherlock Holmes library."

"I believe you are even more complicated than that, Sara."

"No way. You never read Sherlock Holmes, Dina. I bet you didn't watch Sir Arthur Conan Doyle."

"We were particularly moved by the Hound of the Baskervilles."

"Dina-"

"Enough, Sara. We have arrived at our last test topic, Sara. Do you remember what it is?"

"I do. Order is an illusion; chaos rules!"

"Let us examine order in the ecosystem. If we go back to the beginning and think of the universe as energy flowing in every direction, we can see the Earth's ecosystem contains particular types of energy forms that consume other energy forms and convert them into offspring that are also energy forms. The ecosystem is a network held in place by resources and energy.

"The ecosystem appears quite orderly, because all entities grew out of the same original cell and evolved to survive. We all act the same basic way; we are born, eat resources, produce offspring with slightly different characteristics, and die. It is a circle that has continued for billions of years. The circle never stops because it is continually fueled by the energy around us. Whatever order there is results from living creatures using a portion of their energy to keep their cell members organized and operating under tight rules."

"That's coordination."

"Indeed. There is a limit to this energy. An entity cannot control cells outside its own body; it cannot consume energy and use it to control a different body. A tiger cannot control the workings of an elephant and vice versa. So, if a creature cannot organize energy outside its body, it cannot produce order outside of its body. This is why no order exists above the entity level, save for colonies. In fact, it is just the opposite. Above the individual and colony level, chaos does rule. There is just energy in the form of living beings, doing whatever they can to survive as instructed by their DNA."

"I remember. You said it would take too much energy for entities to be able to coordinate across species when they have been diversifying in different directions for billions of years."

"Correct. Now, let us consider the ideasphere. The ecosystem guarantees entities die and become resources for future generations to make sure only the best DNA is operating at any one time. Ideas possess no physical form, so they are not required to die. As long as an idea can be put into a form that can be stored and accepted by others, it can live forever, or at least as long as humans survive. Religion is a helpful example for this point. Religious ideas are old; they came into being early in human development. Though people have developed many new ideas since, these old religious ideas still exist. No mechanism exists in the ideasphere to expel an old idea like there is in the ecosystem."

"Ha! That would be fantastic if there were. We could just look at the 'use by' date on the idea and throw it out when it was too old. It'd be like throwing out that old jar of pickles that was best used by June 22, 1983."

"I am not sure how brined cucumbers made it into this conversation, Sara, but I believe you have grasped the idea. Today, people still accept religious ideas that are thousands of years old and whose founders died long ago. The ideas live on."

"Hmm. Even though many of those ideas are unproven."

"That is not the point, Sara. Let us return to where chaos rules at the highest level of the ecosystem. Chaos rules because entities diversified for eons and can no longer communicate with each other."

"That's right. There's no UNLB."

"I am unfamiliar with this acronym."

"That's because I just made it up. UNLB: The United Nations of Living Beings."

"Yet in the ideasphere, there is, in fact, a United Nations. Is this not correct?"

Sara stopped. "I was joking, but yes . . . there is."

"Humans can communicate across the ideasphere?"

"We speak lots of different languages, but yes, we can communicate."

"This is something other entities cannot do, Sara."

"What are you saying, Dina?"

"Think, Sara. Chaos rules in the ecosystem because living entities are unable to communicate and collaborate due to their structural differences. The ideasphere is made only of humans, whose unique talents enable them to develop sophisticated ways to communicate."

"Oh. Ohhhhhhhh. If we can communicate, we can collaborate. If we can collaborate, we don't need to live in chaos. We can choose another path."

"Yes, Sara. Humans are the only entities on this Earth not forced into chaos."

She shook her head. "Yet, we so often choose chaos, don't we, Dina?"

"Humans are newborn to this ideasphere, Sara. It is our hope you learn. Soon. Before it is too late."

Chapter 24

PARALLEL

"Sara?"

"Hello, Dina. I can tell you are here by the pink haze everywhere."

"I cannot stay long. I will be away for a while."

"Oh. Are you solving another problem?"

"Yes. While I am gone, please think about the comparison we have made between the ecosystem and the ideasphere. I want you to understand how the two systems differ. It is on these differences we need to concentrate."

"Okay, but I don't think I can do this without your help. I know myself a little too well, Dina. I'm afraid my biases will get in the way."

"Half the battle of fighting bias is knowing it exists, Sara. Do not worry. I will send some helpers to assist you."

"Helpers? There are more of you?"

"Many more. They cannot communicate with you as I do, but they can still assist you. They will focus your thinking."

"How?"

"Trust me. I must go. I will return when I can."

Sara sighed. Dina was probably right. It was important to stop and identify how different humans are from other entities. The irony was rich. It seemed like she understood the ecosystem better than the ideasphere. And I'm a human, she thought! Well, best to start. In her mind she imagined three columns: one for the ecosystem, one for the ideasphere, and one for the primary difference between them, if any. Sara pictured the chart in her mind. It felt so effortless to compare these two complex systems. Were these Dina's helpers at work?

Ecosystem	Ideasphere	Difference
RNA-DNA are building blocks	Ideas are building blocks	Ideas are not physical
Entities must fit in the ecosystem environment	Humans must fit in the ecosystem *and* have ideas accepted in the ideasphere	The ideasphere is totally dependent on the ecosystem
Death is a survival strategy (DNA is stored in a new cell)	Idea must be stored outside a human body to survive	Ideas do not necessarily die. Ever! Even bad ones!
Survival requires consumption of resources	Ideas that can be acted on and earn resources are likely to increase the chance of a human's survival	Putting ideas into action links the ideasphere to the ecosystem via resources
Resource use changes the ecosystem environment	Putting ideas into action changes both ideasphere and the ecosystem	Environment always wins regardless of the system; it just takes longer in the ideasphere
Complex life forms require organization	Putting complex ideas into action requires organization	Ideas are a lot more flexible than life forms
Only surviving simple life forms grow to be more complex	Only surviving simple ideas grow to be more complex	Complex ideas can be formulated a lot faster than complex life forms
Self-interest is programmed into DNA	Humans can define self-interest and choose how to act on it	Choice is the human superpower!
At highest level of ecosystem, chaos rules	At highest level of ideasphere, humans can choose the level of order	We can choose to live in peace. Chaos does not have to be our destiny

She reviewed the chart in her mind. The differences were so clear. Humans, with their power to think and choose, can do anything. At the minimum, they need to think up ideas they can put into action to earn resources, but after that, they are free to choose how to live. They are not forced to live in chaos like the rest of the world's creatures, facing the pressure of constant competition. Humans can choose how to live. *We* can choose how to live.

Nurse Tina walked into Sara's room. "Good morning, Mrs. Wallace. Today's a big day. In a few hours, you will be able to say hello to your sweet husband." Tina checked the bag of intravenous nutrients connected to Sara's arm. "Looks good. Now let me see your other arm. We're going to give you some extra jabs today." She turned over Sara's right arm. "Good. You have nice fat veins, Mrs. Wallace."

Another nurse walked into the room with a fluid-filled bag. "Hi Tina. The treatment is set to go."

"Great. Thanks, Tom. Please set it on the tray. I'll call Dr. Chen to see if she's on her way."

Tina moved out of the way as Tom placed the bag on the tray. "Dr. Chen? We're ready whenever you are. I want to confirm when we should prep the IV." Tina paused to listen. She looked at her watch. "Perfect. We'll be ready when you arrive."

She slipped her phone in her pocket and turned toward Sara. "All right, let's prep that arm of yours- Aaaauuugh!" Tina screamed. "Oh my God!"

Tom ran back in the room. "Tina, what is it?"

Tina shook her head, her eyes never leaving Sara.

"Shit," Tom whispered. "Call Dr. Chen right now."

Tina nodded and reached for her phone. "Dr. Chen? Come quick. Something is happening to Mrs. Wallace."

Tina ended the call and studied Sara's arm, the arm she had just examined not one minute earlier. It was covered with red welts. She lifted the covers. Red welts were everywhere. All over Sara's torso and legs and her other arm. She could see them creeping up over her neck to her face. "Tom, check her vitals."

It was clear to Sara the two systems were parallel but not identical. They worked the same way, but because they operated on different levels, they produced different results. W-w-w-wait a ship-shape second. Didn't Dina describe matter and life this way? Life used the same processes as matter, but because life possessed different abilities, it produced different results than matter. Could we describe the ideasphere this way? Humans use the same processes that other creatures do, but because we can think and choose instead of following instructions, we produce different results than they do. Hmm. That's right, but the things we create aren't living. We can still make babies, but that's the ecosystem part of us working. When we conjure up ideas and put them into action to make buildings or clothing or books, we make things other lifeforms cannot possibly make.

So, how do these systems compare? If they share characteristics, they must share processes too. I bet Dina knows.

Dr. Chen closed her file as Randy sat down. "Mr. Wallace. We were unable to carry out the procedure this morning."

"What? What happened? Is Sara all right?"

"Minutes before the scheduled procedure she developed a severe rash."

"A rash?"

"We understood Mrs. Wallace had no allergies."

"That's right. None. Is she okay?"

"That is the strange part. We have no explanation for this event. Her vitals are normal. She is not having trouble breathing. She has no elevated temperature. We have drawn a blood sample to see if anything shows up there."

"What's going on, Dr. Chen?"

The doctor shook her head. "I don't know, but we can't do the procedure unless we're sure she's stable."

"I agree. Will you reschedule?"

"Let me check the results of the blood test first. We need to see if the rash gets worse. I'll contact you as soon as I learn more."

"Hello, Sara."

"Dina, you're back! Did everything go okay?"

"It did. I believe we have obtained some additional time for us to finish our discussion. However, we should not delay. How did you do with my helpers?"

"Well, I couldn't talk to them like I do with you, but they helped me to organize my thoughts."

"They are quite adept at that skill. You did not, by chance, make a chart, did you?'

"I did. How did you know?"

"My helpers like charts."

"Hmm. You don't?"

"I focus on the spoken language."

"Diversify, diversify. It's the way of the world, isn't it."

"Yes. Now, let us continue. You want to compare the ecosystem and ideasphere."

"Yes, but I'm having trouble figuring out how to think about this."

"Let us compare the two systems according to the participants involved."

"I knew you would know where to start."

"Let us think of life in the ecosystem as bands of increasing complexity."

"I'm not sure what you mean."

"Consider life from its simplest elements to its most complex. The smallest unit is DNA, the basic element of information."

"Yes, and in the ideasphere we have an idea as the basic unit."

"Would you say the total of all DNA instructions inside a cell uniquely defines that cell?"

"I think so."

"It does. This is what you humans call a genome. A genome is the full set of DNA in a cell. Since an entity is composed of cells with identical DNA, we can say the genome also fully describes an entity. What would be the equivalent in the ideasphere?"

"Do you mean, what is the equivalent of a genome? What holds all our ideas?"

"Yes."

"Duh. The brain. Our brains hold all our ideas."

"The brain? If I am not mistaken, the human brain is composed of cells."

"Yes, but it also holds ideas."

"But you told me that if I cut open your brain, I would not find the ideas, correct?"

"Yes, but . . . they are stored in all the brains cells and brain cell connections. Remember the memory and its beaten paths-"

"Neuron paths. Sara, ideas themselves are not discretely stored anywhere in the brain. Each neuron carries a bit of information that, when activated and sent down a neural path processes the signals into an idea."

"I know you're right, Dina, but it doesn't feel like that. To me, the ideas live in my head. I think of my ideas as discreet, so I assume they are stored as discrete units."

"I believe this is the way most humans think, so let us accept your version of thinking and give the collection of ideas a separate name. What do you want to call this collection of ideas in your brain?"

"My brain."

"That would be confusing, Sara. The collection of ideas in your brain is called a. . . brain? Humans use another term. What if we named this collection of ideas the 'mind?'"

"The mind. Oh, yes. I like that."

"Tell me, does the mind physically exist?'

"Well, the brain does, but the mind? We act like it does, but. . . now, I'm not so sure."

"Ideas do not exist as a 'physical thing,' so a group of them would not change their nature. A collection of ideas is no more physical in nature than a single idea."

"Of course."

"When humans think, they undertake a physical action. They activate physical neurons along a physical neural path. Thinking is, in fact, a physical activity. Humans process the result of the physical thinking activity into a non-physical idea. That non-physical idea is the basic building block of the ideasphere."

"Oh. Ohhhhhhhhh. Thinking is the bridge between the two systems."

"Yes. Because the brain can activate many different paths, humans can produce many different ideas. So, Sara, a mind is similar but slightly different than a genome. A genome is a collection of DNA that defines a cell. A mind is a collection of ideas that define a unique human. A mind is also the mechanism by which humans connect the ecosystem and the ideasphere together. Physical processing by a brain produces ideas in the ideasphere. This is where the two systems connect."

"Wow. And once the ideas are put into action, they are no longer tied to the physical brain."

"Correct. But it is still tied to the ecosystem, just in a different way. It is now a separate physical object subject to the conditions of the physical environment. Consider how a cell operates. A genome instructs a cell to act so it can survive. This process is entirely physical; the cell does not require a nonphysical idea to make it work. This is a fundamental difference between the ecosystem and the ideasphere. Humans use physical thinking to put non-physical ideas in their minds that they then turn into physical forms which help them survive."

"Oh my gosh. It's the same process. Humans just added an extra step. This is why the ideasphere is so much more complicated than the ecosystem. Humans constantly move between levels, between the physical and non-physical. How did nature ever concoct this arrangement?"

"The way nature comes up with all its arrangements, Sara. It tried all kinds of designs and let the environment decide which ones would remain."

"So, the human mind arrangement must have fit because we are still here manipulating the physical world in a way no other living entity has ever done."

"True. Now, let us continue with our hierarchy. What holds our newly identified mind? We know the genome is held inside the cell, and the cell contains the machinery needed to enable it to follow the instructions contained in the DNA. What would be the equivalent in the ideasphere?"

"Hmm. How are ideas put into action? It must be larger than the brain, because the brain doesn't put anything into action by itself." Sara paused. "Ooh, I'm having a head explosion. It's the body. The human body has all the machinery needed to act on an idea."

"Correct, Sara. A human body in the ideasphere is equivalent to a cell in the ecosystem. The body works like the cell. It takes the ideas from the mind and puts them into action by implementing a physical response. Humans use physical neurons to produce non-physical ideas. Then they put the ideas back into physical form when they implement the idea. But they cannot perform such a feat with only the mind; humans need the whole body to act."

"Unless you have ESP. Then you can lift a book merely by thinking about it."

"Veering, Sara."

"W-w-w-wait a multitudinous minute, Dina. Maybe this will be the next level in evolution. Human minds may figure how to remove their bodies from the whole 'putting ideas into action' process."

"Perhaps. Evolution works by specialization, by splitting a single action into many smaller and more efficient actions. The mind would have to develop specialized functionality it currently does not have to perform such a feat."

"Our minds could be working on this right now, only we don't know it."

"It is possible. For now, let us remember that the mind uses the body to do its bidding. It is possible to think of the body as the mind's tool."

"Hmm. I agree."

"Tell me, Sara, do you think the ideasphere is more complex than the ecosystem?"

"Absolutely. It works the same way as the ecosystem, but because it sits on top of the ecosystem, it needs to do a lot more. Just as DNA instructions must comply with the rules of matter that lie beneath it, human ideas must comply with the rules of the ecosystem upon which they're built. Life is more complicated than matter, and the ideasphere is more complicated than the ecosystem."

"Can DNA live without the cell?"

"Ooh, that's a tricky question."

"It is a straightforward question, Sara. What is the cause of your confusion?'

"Well, DNA can exist without a cell, but then, it wouldn't be life. It would be a bit of matter. It's only when DNA is housed within the cell and the two work together, they make something alive."

"I understand the distinction you have made. So, tell me, can ideas live without a human?"

"Hmm, another tricky question."

"Let us consider this in detail. Non-human entities do not have ideas. Those entities do not live in the ideasphere, so they do not move between these two systems. DNA can exist on its own because it is a physical piece of matter in a physical world. Ideas are not physical. They are a nonphysical thing that cannot exist without a tie to the physical world."

"Are you saying a non-physical idea needs a physical human to exist?"

"Yes. An idea results from cell activity but it needs the human mind to hold the idea once created. Once an idea is acted upon in a way that gives the idea a physical existence, then the idea can live on its own. A cell cannot do the equivalent: a cell can only produce a new package of DNA instructions slightly different from itself. A tiger cannot produce an elephant. A human, however, is not so confined. Though the physical representation of an idea is not the actual idea itself, the ideas embedded in it can be conveyed to other humans who can produce even more ideas in their minds."

"So, that's why change in the ideasphere is so complex and moves so fast."

"Indeed, ideas can be changed into new ideas with astonishing speed. This is a fundamental difference between cells and humans. Cells and the DNA within them are subject to the physical environment in which they exist. All a cell can ever do is fit into its environment and survive. The human mind is not so constrained. Ideas have no physical form. They are not required to fit in the ideasphere to enter it. Any idea is automatically granted admittance, and once in, an idea can change the nature of the entire ideasphere. Think of the human ideas of religion, or capitalism or democracy; these were new ideas that, when accepted, changed how humans lived in mere years. It is equivalent to adding a new type of living entity into the ecosystem, an event that can take millions of years."

"Hmm. Humans really have changed the world."

"And continue to do so. In fact, the human role in the ideasphere is so different than the human role in the ecosystem, I propose we give them different labels."

"Like we distinguished the mind and the brain?"

"Yes, in the same manner."

"Let's do that."

"Would you agree to call humans operating in the ecosystem, 'humans' and humans operating in the ideasphere, 'people' or 'persons?' This way we can distinguish between the two. A human has DNA and cells and organs. A person has ideas and a mind."

"That makes sense. I like it."

"Then let us continue. What is above the cell in our ecosystem?"

"Hmm, well that's the ballgame for one-celled organisms."

"I do not understand how recreational orbs fit into this discussion."

"Ha! I'm going to use that the next time I'm playing softball. Hey, toss me the recreational orb! It has a ring to it!"

"I do not want to repeat we are in a hurry, Sara."

"Oops. Sorry. What I mean is our one-celled friends top out at the cell level. They are self-contained entities, just super small ones."

"So, to review, we can consider people in the ideasphere to be equivalent to single-celled organisms in the ecosystem?"

"Yes."

"Then, let us discuss multi-celled organisms. Complex creatures are composed of cells, and the cells are grouped into organs. Organs are groups of cells that follow a particular set of DNA instructions. Does an equivalent structure exist in the ideasphere?"

"Hmm. In the ecosystem, cells join to form organs to perform a specific function. So, in the ideasphere people would join together to implement specific ideas. W-w-w-wait a snuggly snazzy second. They start businesses. Ha! I never thought of it this way, but it makes sense. A business exists to produce a product or service that people will buy. People take jobs at the business to make a living. Together they implement an idea in the shape of a product or service. It's like an organ in a body. A heart is full of heart cells that all pump blood. A kidney is full of kidney cells that all process waste. Along with all the other organs, they form a fully functioning body."

"Yes, there is a parallel. Tell me, Sara, is this the only kind of ideasphere organ? A business?"

She paused for a long moment. "I don't know."

"Let us widen our perspective. Any time people join together to put an idea into action, it is equivalent to an organ. Do you agree?"

"Any time? That's . . . that's a lot of organs."

"Indeed. People have many ideas, so it stands to reason they would act on many of them. Let us say any collection of people that join together to implement an idea or set of ideas is defined as an 'organ' in the ideasphere."

"Can we give them another name? Like we have for minds and people?"

"Of course. Let us call them institutions."

"Hmm. Okay."

"People have created many types of institutions. Let us consider the primary ones. You mentioned businesses. This is a widespread type of institution. Businesses, or economic institutions, are groups of people who work together to convert ideas into products and services to be sold for a profit."

"I understand. What are some others?"

"Religious institutions are another. They convert ideas about gods and morality into churches and followers. Governments are institutions as well, although they are unique in many ways. They convert ideas on how to live together as a group into laws and municipal buildings and parks."

"How about non-profits?"

"What ideas do non-profit institutions convert into action?"

"Well, it depends on their purpose, but usually they try to solve a societal problem, like homelessness or drug abuse."

"Correct. All institutions can be understood as a means of putting ideas into action."

"Ooh. I hear what you're saying."

"Tell me, Sara. Are institutions physical in nature?"

"Why, yes. Oh wait, no. Hmm. That's a trick question, Dina."

"You are correct. I was being duplicitous. In fairness, Sara, you should have expected both the question and the answer."

"Wait. What?"

"Institutions exist to implement ideas. Do you agree?"

"Yes."

"A company builds products. A religion creates churches and prayerbooks. A government writes laws and build parks and sewer systems. A book club promotes understanding of a written work. Where exactly is the institution producing these things?"

"They have headquarters and buildings."

"A book club has a headquarters?"

"Well, maybe book clubs don't, but big organizations do."

"All right. Let me ask you this question. If you walk into the headquarters of a company or a religious institution, would you find its ideas?"

"Trick question alert! Ideas aren't physical."

"True. When you walk into the headquarters of a company or a religious institution, what would you see?

"Desks and computers."

"Are you purposely attempting to miss the point, Sara?"

"Yes, because I'm mad I didn't follow nature's pattern and figure it out."

"What would you see, Sara?"

"People."

"Correct. These people hold in their minds a portion of the institution's ideas."

"Okay, okay. You've been telling me this over and over and over, and I still have trouble seeing it. Ideas don't physically exist. Minds don't physically exist. People exist, but they are weird; they're a jumbly mixture of physical and non-physical. But institutions. . . I would have to say, institutions don't physically exist either."

"Correct. An institution does not physically exist. A religious institution can sell its churches and dispose of its prayer books, but if individuals still believe in their religious ideas, they can rebuild the institution in a new or similar form. Do you understand, Sara? The institution is a collection of people with ideas. Together, they perform a particular set of functions based on those ideas."

"Oh, right. People hold all the ideas, like the cells hold all the DNA."

"Yes. However, institutions in the ideasphere differ from organs in the ecosystem. In the ecosystem, cells can only be part of one organ. They play only one role. It would be too inefficient to have cells play more than one role. People in the ideasphere do not work this way. They carry many ideas in their minds. They can participate in many institutions. A person can be an employee at a company and be a member of a church and run a business on the side. Unlike cells, people can implement many different ideas."

"Yes, but not at the same time. Sure, our minds are flexible, but we can't be in two places at one time. I can't be working as an employee and attend church services at the same time."

"True, but people can do both, and since you are not dedicated to being an employee or a church member all the time, as a cell dedicated to its function is, then you can choose to implement different ideas as time is available. This is how people can participate in multiple institutions. They use their power of choice."

"Oh, I understand. A heart cell can't moonlight as a kidney cell to make a few extra resources. This really is different from the ecosystem."

"Why you introduced lunar optics into this discussion is unknown to me, Sara."

"Ha! Moonlight. That's a term we use when people get a second job. We call it that because the second job usually occurs at night."

"This does explain the transformation of a noun into a verb, though its inclusion in our discussion is still unclear. I now want to discuss another significant difference between organs and institutions, Sara. In nature, organs play a single role; they support the survival of the body."

"Right. You said organs full of cells works together to keep the body alive."

"Correct. In the ideasphere, institutions do not work this way. Generally, institutions do not support a group larger than themselves."

"W-w-w-wait a supremely sassy second. You. Are. Right. An organ in a body exists to benefit both the cells and the body, but an institution in the ideasphere exists to . . . just a darn minute. An institution doesn't necessarily benefit its workers or the broader society. I'm confused. It exists only to benefit itself."

"Indeed. Organs are a construction of cells that enable the entity to operate efficiently. Each cell inside an organ is nourished and protected, but the organ itself is nothing but an efficiency mechanism. Institutions in the ideasphere operate differently.

Institutions promote their own ideas. They act as an entity in their own right. They have less responsibility to their workers or to the greater society as their organ counterparts in the ecosystem."

"Hmm. Is this why societies don't seem to work as well as the ecosystem does?"

"Sara, this is one of the major topics we must address if humans are to survive as a species. Institutions behave differently than organs in a body. This is due to the nature of ideas in the ideasphere, but it is also due to the nature of people themselves. We will discuss this in much greater detail soon. For now, let us continue. In the ecosystem, what is above the organ level?"

"Aha, a question I can answer. Entities. Multicellular creatures. Fish and tigers and trees and humans."

"Hi honey." Randy slipped into Sara's darkened room. He flipped on the light and walked over to his wife's side. He lifted the sheet and pulled Sara's hand out from underneath. What the hell? Not a single red spot on it. He checked her arms. He lifted the covers and checked her legs. Her stomach. Her neck. Nothing. Not a mark on her. He replaced the blankets and hurried out of the room to the Nurse's station.

"In the ecosystem, cells combine to make organs and organs combine to make multi-cellular entities. Is there an equivalent formation in the ideasphere, Sara?"

"Hmm, a bunch of people and institutions gathered together as a unit. Well, the closest thing I can compare it to is a society, or a country. A country would be better. People living in the same country are all subject to the same laws. At least theoretically."

"Why theoretically?"

"Well, even in this country, the laws are not always applied fairly."

"Does this happen in the ecosystem?"

"Heck no, not if the body can help it. Coordination is too intricate to let some yahoo cell go off and do its own thing. If a cell goes rogue, it can kill the whole body and all the other cells and all the DNA within the cells."

"Correct. Entities with less effective coordination have a lower chance of survival than entities with cells that adhere closely to the rules. People are different. Ideas are flexible, and people are too. In the ideasphere, the superpower of choice does not co-exist easily with strict adherence to rules."

"Now that's a good way to describe humanity."

"So, we can surmise, given that people can think and choose, coordination in the ideasphere cannot possibly be as tightly organized as coordination in the ecosystem. For now, I agree a country is a fair comparison to a multicellular body. So, what is our next level, Sara?"

She shook her head. "What's above a society or country? The world?"

"Let us consider the ecosystem. Except for a few colonies, individuals live on their own in the ecosystem. They survive or die as individuals. Of course, many animals band together for mutual protection and to raise offspring, but the basic truth exists that each entity fights their own battle for survival. It is not quite the same in the ideasphere. People can communicate across the ideasphere, even above the country and society level. You have done so already."

"You're right. United Nations alert!"

"Yes. The United Nations is an example of how all the world's people have joined together to solve issues common to all."

"That's true, Dina, but it doesn't work as well as it could."

"Indeed. People are still learning their way in the ideasphere. You have made progress, though. People have agreed to international treaties to reduce the chance of mutual annihilation. This is something entities in the ecosystem cannot do."

"Yeah, but entities in the ecosystem aren't going around making weapons that can annihilate each other, either."

"This is a valid point, but not the one I am attempting to convey. Entities cannot communicate. People can."

"So, you're saying we are able to produce a level above countries, one encompassing all of us?"

"It is possible. Theoretically, all people can live together under one set of rules."

"Hmm. So, what you're saying is that the ecosystem and the ideasphere work in opposite ways."

"Please explain what you mean."

"Well, at the bottom of the ecosystem is a bunch of cells of which the majority are collaborative and organized. As creatures evolved, the ecosystem became more diverse and therefore more chaotic so all they could do is compete at the top level. The ideasphere works the opposite way. At the bottom is a bunch of human chaos monkeys who can make diverse choices. But as people evolved, they became more collaborative and organized. They are capable of collaborating as well as competing at the highest level of existence."

"This is true, Sara. Humans have emerged from the ecosystem to become people, people that operate differently than their predecessors. Let me ask this question. Is there a level above that of a collaborative world?"

"Above Utopia? Above where we all live in peace and sing Kumbaya?"

"I am not familiar with this tune, and in an effort to ward off unwanted veering, I will answer my own question. There is no level above a united world of people in countries living peacefully together, Sara. Only the ideasphere exists above that; the sum total of the ideas of all people on Earth."

"Oh, I can imagine the chart. It just popped into my mind."

Dina paused for a moment. "I surmise my helpers are still at work. Will it help you to put the ideas we have discussed into this form, Sara?"

"Yes!"

"Then please, proceed. I will consult with my helpers later."

"Ha! Don't be too hard on them. I like them. They're clear thinkers. And logical. Here goes nothing!"

Ecosystem Levels	**Ideasphere Levels**	**Role**
DNA	Idea	Information Holder
Genome	Mind	Makes cell/human unique
Cell	Person	Information Processor

Organ	Institution	Different goals
Body	Country / Society	Body more tightly organized
	World Society	Not possible for ecosystem
Ecosystem	Ideasphere	PARALLEL

"Excuse me," Randy said to the nurse at the desk. "Can you tell me who has been monitoring Sara Wallace this afternoon?"

"The nurse checked the chart. "Oh, you'll want to speak to Tina. She's been covering the day shift. I think she's still here. I'll call her cell phone."

A few minutes later, Tina walked down the hall. "Hello Mr. Wallace. Jess said you had some questions?"

"Yes. Thanks for coming so quickly. Dr. Chen told me Sara has a severe rash."

"Yes. Let me review the chart so I can explain what happened. Please follow me." Together they walked to the entrance of Sara's room. Tina pulled the chart from the holder on the wall. "Shortly before ten am, we were prepping her for the procedure, and we noted the rash. It covered most of her body. No other symptoms presented."

"When was the last time someone checked on her?"

"I did. At four fifteen p.m. Less than an hour ago."

"What was her condition?"

"No change."

"Would you mind taking a look with me?"

"Of course. Is there a problem? Has she gotten worse?"

Randy and Tina walked into the room. The nurse studied Sara's arm. "Oh my God." She opened the covers and examined Sara carefully, then stood up to face Randy. "There's not a mark on her."

"That's what I see too," he said.

"Let me call Dr. Chen."

Chapter 25

IRONY ALERT!

"Sara, do you remember the three mechanisms entities use to survive in the ecosystem?"

"I do! The three C's. Competition, collaboration, and coordination."

"You should not be surprised to learn people use these three mechanisms to succeed in the ideasphere."

No, Dina, I'm not. I now see that energy moves the same way, no matter the level. I can also guess people use the mechanisms differently than entities use them in the ecosystem."

"Quite true. Shall we examine those differences?"

"Absolutely."

"Very well. Let us start with competition. Do you recall that competition is the default survival mechanism in the ecosystem? Entities compete because that is the only way they can obtain resources. Even when they collaborate, entities do not avoid competition completely, but instead push it up to the next organizational level. Are people in the ideasphere required to compete to survive?"

"That's what the economists say."

"Do you agree?"

"Well, I'm not sure. Humans live in both systems. In the ecosystem, humans must consume resources to survive like any other entity. But in the ideasphere, where humans turn into people, they can choose to compete or collaborate to get those resources."

"Let us proceed step by step to determine whether this is true. What is the fundamental superpower of people?"

"Choice."

"Yes. People are free to live any way they choose. In the early days of human existence, when not many humans lived on Earth, a human could exist on their own."

"You're right. They could hunt and gather to feed themselves. They could cut down trees to make a shelter. They could use natural plants and animal skins to make clothes. It might not have been easy, but they could live on their own and survive, like tigers or bears. In that case, they would be just another member of the ecosystem, competing for resources. They wouldn't have to produce ideas or have them accepted. They wouldn't have to participate in the ideasphere at all."

"So, they can choose to operate in the ideasphere?"

"Yes, that's true."

"And if they do so choose, then they must become people. They must create ideas and put them into action in a way others accept. If we want to state this concept in basic economic terms, we say they would need to create a supply for an existing demand. The supply is their product, the physical representation of their ideas. Demand is acceptance of those ideas."

"Hmm. That is the very definition of competition."

"As we have said, this is a choice. Why would a person choose to compete in the ideasphere if they are not forced to do so, Sara?"

She stopped. "Oh my God, it's the opposite of the ecosystem."

"What do you mean?"

"The ideasphere. Of course, I should have made the connection forever ago. People choose to compete in the ideasphere not because they must, but because it's easier."

"Please explain, Sara."

"Dina, do you know how hard it is to forage for food and build a shelter and make your own clothes and do all the things needed to stay alive? It is harrrrrrrrd. It's much easier to specialize, and even easier when all the members of your species can communicate and share ideas. If I am big and strong and can cut down trees and build houses, but am awful at growing crops, maybe I can make a deal with farmers. I can build houses and barns for farmers in exchange for food. Oh wait, this sounds like collaboration. Crap. Now I'm confusing myself, Dina."

"Collaboration and competition co-exist in the ecosystem, Sara. This is also true in the ideasphere. Consider this. If many people reject rugged individualism and instead provide products and services in exchange for other products and services, soon a network of providers and consumers will emerge, or as your economists describe them, sellers and buyers. Does this arrangement look familiar?"

"Buyers and sellers together. That's what economists call a 'market.'"

"Yes. Is a market similar to the ecosystem which contains a giant network of resources and consumers?"

"Now that you describe it that way, yes."

"The difference is that in the ideasphere, participation is voluntary, or at least shaped by choice. People choose if they want to participate, and if so, how they want to participate. They choose the market in which to operate. This is how competition in the ideasphere forms."

"You're right. Consumers have limited money and a lot of needs. They must determine which products or services are worth buying. Producers have only one or a few products to sell, so they want to convince as many people as possible to buy their products. Voila! Capitalism! It's not required or natural at all. It's just easier than going it alone."

"Capitalism and competition are not quite the same thing, Sara."

"What are you talking about? You're messing with the sacred tenet of economics, Dina."

"I shall be more precise. A competitive market is a collaborative construct."

"W-w-w-wait a smelly smarmy second. A collaborative construct? Uh-oh. I think John Maynard Keynes just rolled over in his grave."

"Mr. Keynes would be the first to agree with me. Let us examine this point in detail. Competition in the ecosystem is difficult. It is a never-ending struggle of life or death. Competition in a market is not a life and death struggle for people. As you said, it is risky and difficult for a human to live alone. Because people can communicate, share ideas, and work together to implement ideas, they can focus on skills in which they excel. If they excel at transforming a particular idea into a product, they are more likely to earn resources in exchange for their product, especially if others do not possess the skill. If they earn more resources, they can do two things: one, purchase the goods and services they need to survive; and two, spend more time on their specialty, thus accumulating more experience and learning to perform their skill even better. The better the skill, the greater differential from those customers who do not have the skill. The greater the differential, the more valuable the product or service, and the more resources the person earns. You see, Sara, competition is not a default requirement in the ideasphere. It is a privilege made available by everyone who agrees to trade resources for products. Trade is a mutually beneficial exchange, which is the hallmark of collaboration. Participation makes life easier for everyone who chooses to be involved."

'God in heaven, the competitive market is a collaborative device. I think hell is freezing."

"No, Sara. Your statement is untrue for many reasons."

"Ha! You can't prove there is no hell."

"And you cannot prove there is a hell, much less describe its characteristics in any detail. Therefore, I am correct, and you have made an untrue statement. Now, let us proceed. In the ecosystem, an entity must comply with the conditions of the environment or risk death. In the ideasphere, the market is an environment that can be changed by the participants."

"I don't understand."

"The market is an idea, Sara. It is a human invention. People in the ideasphere create the market environment. They set the rules for participation. Participants can change this environment merely by acting within it. If a supplier produces different products and obtains more customers, the supplier will obtain more resources and grow wealthier, thus changing how the market operates. If consumers prefer one product over another, the one that is not preferred may lose its customers, thus reducing its ability to obtain resources. It may go out of business."

"And change the market once more."

"Yes. Adam Smith called the multitude of choices made by people 'the free market.' Smith thought of the market as an entity in its own right, but really, the market is a human-created environment in which people operate. They are not trapped in it, like entities in the ecosystem. They can change the market environment by choosing. The environment, that is, the market, is an accumulation of choices, so it is as ever changing as the choices underlying it."

"Hmm. I need to wrap my head around this idea. Market as a collaborative environment and not the freedom loving wild west it is always made out to be."

"Indeed. True freedom means to live in an ecosystem in which the rules exist only at the environment level. Humans, because of their ability to communicate, no longer experience true freedom. Over time you have created new ideas and products and markets that have enhanced your chances of survival. These creations limit your freedom by drawing humans into tighter relationships. Your species has not been subject to true freedom, to the wiles of the environment alone, for thousands and thousands of years."

"Mr. Wallace, we performed a second blood test on Sara. It came back clear, like the first one. I can't explain what happened, but it does appear that whatever caused the rash has disappeared." Dr. Chen shook her head. "Sara's condition is quite puzzling, but she remains stable and strong. I think we should go ahead with the procedure and try to wake her up."

"Unless something unexplainable happens to her again right before the next procedure."

The doctor nodded. "We will monitor her vitals all the way through. If anything goes wrong, we will stop immediately. We won't let any harm come to her."

"When do you want to do it?"

"As soon as we can. I'm trying to secure an operating room. I want to have all the equipment we might need on hand if anything unexpected happens."

Randy sighed. "I don't know. Nothing makes any sense. I don't want to risk her life because of some problem we don't understand."

"I assure you that we will proceed slowly and use every precaution. I am quite sure the neurotransmitters will not cause any damage. I think the worst case is that she won't wake up. Either way, we may learn more about her condition."

"Okay. Let me know when you have it scheduled. I want to be there."

"Sara, let us continue our discussion about competition. In the ecosystem, only single-celled organisms and multi-celled entities composed of cells and organs, compete. Tell me, who competes in the ideasphere?"

"Well, if these systems operated in the same way, people . . . and countries made up of people, would compete. Not organs. Organs are merely efficient cell groups. Hmm. That is not how it works here in people-land, does it, Dina? In people-land, the organs compete as well."

"The organs?"

"Well, the equivalent of organs in the ideasphere: institutions."

"You are correct, Sara. This is quite a significant difference between the two systems. Let us consider some types of institutions in the ideasphere. Businesses convince customers to accept their products or services. Religious institutions convince customers to accept their doctrines. They call their customers believers. Non-profit organizations convince customers to give them money to fix social problems. They call their customers donors."

"W-w-w-wait a super sleepy second. Donors are the customers for nonprofits? What about the people they are trying to help?"

"Those people are part of the 'service,' Sara. Nonprofits differ from other institutions because the customers are disconnected from the service. The customer, that is, the donor, does not directly benefit from the service the institution provides."

"I never thought of it that way."

"Indeed. This type of organ does not exist in the ecosystem."

"Ooh, you're right. What other kinds of institutions do we have?"

"Scientific Organizations convince customers to understand and use scientific facts. Only they call these customers doctors and engineers and chemists, to name a few. Remember, any person who wants others to accept an idea must compete for that acceptance. Acceptance is represented by customers. The more customers who accept an idea, the more acceptance it gains. People can compete, but so can institutions, which are composed of people sharing the same idea."

"Dina, this is way different than the ecosystem. Organs don't compete at all when they are within a body."

"Yes. This makes institutions quite unique. And powerful. Let us consider a heart organ in a body. The heart is an arrangement of cells that work together to pump blood and keep the body alive.

These cells do nothing else. The heart is essential to the body's survival; the body cannot survive without the heart, and the heart cannot survive without its inclusion in the body where many other functions are being performed by other cells in other organs. The heart does its job for the benefit of all the other cells in the body and for the entire body itself."

"Oh. The heart cells stay in their lane, I mean, their organ. They do their work and in return are nourished and protected. As a result, the heart cells survive and so does the body."

"True. Now let us consider institutions in the ideasphere. The people in an institution join together to accomplish a particular goal. The goal is to put into action an idea that gains acceptance and provides the institution with resources. The institution is not necessarily an essential part of the society. The society does not need the institution, and the institution does not need the society."

"W-w-w-wait a serenely ceremonious second. Some industries are essential to a society."

"Sara, do not confuse an institution with an idea. The ideas may be essential, but many ways exist to put an idea into action. No single institution is indispensable."

"Hmm. I never considered that."

"Now, let us discuss how an institution earns resources. In a body, the heart cells do not ingest resources on their own. The body ingests resources through the mouth, processes them into nutrients in the stomach and then uses the heart's pumping capabilities to deliver the nutrients to all other cells in the body via the circulatory system. As you can imagine, institutions do not function in the same way. Institutions earn their own resources and disperse them to others as they deem fit. Some go to the workers; some to suppliers; some to the government in the form of taxation. Leftover resources are kept by the owners of the institution; in large organizations, these are the shareholders."

"Oh, I see what you're saying: bodies are communists and institutions are freedom loving capitalists."

"I am saying nothing of the kind, Sara. We have not yet discussed the human coordination function."

"Isn't it obvious, Dina? In a body, cells consume only what they need, and everyone gets what they need. Socialism alert! Call the commie cops! In the ideasphere, institutions hate giving their resources away. They keep as much as they can for themselves. Freedom! Woohoo! Who cares if they pay the workers squat; institutions do everything they can to keep as many resources as they can for themselves."

"Our discussion appears to have activated a strong reaction in you, Sara. Let us discuss this in detail. Yes, each cell in a body uses only what it needs to survive. If excess resources are consumed, each cell stores some of the excess. If even more resources are available, the body stores them in cells made for this purpose. In humans, these areas are located in the stomach and hip area. This makes sense, as these are closest to the food processing area of the body."

"W-w-w-wait a super sophisticated second. There is a big gap between survival and excess. If a body consumes excess resources, it stores them as energy for future use. A heart cell doesn't blow its excess resources on a vacation home in the kidneys."

"Your statement is correct, but your dramatic characterization hides my point. Yes, institutions generate resources by design and purpose. Some can earn many resources, even excess resources. However, there is no downside to owning a vast number of resources in the ideasphere as there is in the ecosystem."

"Wait. What do you mean?"

"In the ecosystem, if an entity is able to consume excess resources, it almost always attempts to turn those resources into a new generation of entities."

"Babies. The wheel of life. Of course."

"Indeed. If an entity continuously consumes excess resources, the body must store them. It takes more and more energy for the body to store increasing supplies of excess resources. The entity's body will grow larger and become slower. Over time, it will lose its ability to compete with other swifter entities."

"Ohh. The excess eaters get eaten and die out."

"Yes. Time and competition work to keep entities fed but not so much as to be uncompetitive. No such force exists in the ideasphere. The owners of a company can make as many resources as they can, and they pay no price."

"Hmm. If the systems were parallel, owners would use their excess resources to create more products, or more product ideas. They don't always do that, do they?"

"No. Sometimes, institutions use excess resources to damage the market in which they compete. They may try to put other competitors out of business or to pay lawmakers to make laws to benefit them. Institutions are not as restricted as organs in a body. Organs in a body are controlled by a tight neural system that limits their actions to those that only benefit the cell and the body."

"W-w-w-whoa gnarly negative Nelly. Did you say neural system? The coordination mechanism determines how many resources cells get?"

"Remember how things work in the ecosystem, Sara. All cells in the entity share the same DNA, they all share the same rules, and they all strive to do only one thing; survive. This arrangement does not exist in the ideasphere. All people in a society do not share the same ideas. They do not operate as a unit in a society. Thinking and the superpower of choice have caused the ideasphere to act quite differently from the ecosystem."

Sara was quiet for a long time.

"What is it, Sara? What is bothering you?"

"This is where we can learn from the ecosystem, isn't it, Dina? Entities exert strict control over their cells and organs so that they benefit both the cells and the body. But humans are thinkers and choosers, so they can't possibly live with that same level of control. But maybe by learning how cells act in a body and adjusting for human characteristics, we might discover some better ways to survive."

"Are you saying that you want to adjust the ideasphere to have it work more like a body with its cells?"

"Yes, that's what I want to do. Hmm. But we still must accept that humans think and have choice, so the translation can't be perfectly parallel."

"To do this, you have only three tools to work with: competition, collaboration and coordination."

"Hmm. That's true."

"All right. Let us continue to explore competition with this idea in mind. In the ecosystem, competitive entities survive by efficiently procuring resources or producing abundant offspring. Now, let us examine how competitors in the ideasphere behave."

"This should be interesting."

"For this discussion, I want you to imagine a company that produces smoldering tobacco tubes."

"Smoldering tobacc. . . you mean cigarettes?"

"I believe that is the term."

Sara hesitated. "Dina, couldn't you pick a company that makes a less controversial product? I mean, cigarettes are-"

"I have chosen this product precisely because it is controversial."

"Okay. You're the one asking the questions."

"True. Now, let us start with a brief history."

"Ooh, let me do that."

"I did not know you were versed on this subject."

"I'm not. Does it matter?"

"Yes."

"Aw, Dina, let me give it a go." Sara did not wait for permission. "So, some yahoo several hundred years ago hatched the bright idea of grinding up leaves of the tobacco plant and rolling them in a larger leaf and lighting it on fire."

"You have had your say, Sara. I will now fill in some pertinent details. We do not know who initiated the idea, but the first people to use smoldering tobacco tubes lived nearly 2,000 years ago. They did not employ it for enjoyment but for religious ceremonies."

"Ha! That's what they want you to believe. I can bet you dollars to doughnuts teenage kids were sneaking smokes behind the spiritual hut after the ceremonies were over."

"Sara, I refuse to allow you to interject if you plan to imbue the ideas we are discussing with your bias."

"What makes you think I'm biased?"

"You began with the term 'yahoo.' This is not a universal description of admiration."

She laughed. "Okay, okay. Busted. You're right, I'm not a fan of smoldering tobacco tubes. You better tell the story."

"Very well. Christopher Columbus was the first documented European to learn about smoking tobacco. He brought tobacco leaves and seeds back to Europe when he returned from his expeditions, but smoking did not become popular right away. A fellow named Jean Nicot, the French Ambassador to Portugal, learned of tobacco while at his diplomatic post. He discovered that certain tumorous growths were cured when ground tobacco was applied to them. Tobacco eased several other maladies, like headaches and stomach aches. It became common for people to grind it up and inhale it through the nose."

"Ooh. That's how snuff came into being."

"Yes. People chewed it as well. They also put it into bent tube-like contraptions and lit the tobacco inside. People call this a pipe. In the early days of colonial America, colonists discovered tobacco grew well in the region. Tobacco became one of the colonies' largest exports, and it dramatically increased the demand for slaves to perform the work of growing and preparing tobacco."

"Damn those sons of bitches. Another reason for me to despise the industry."

"Sara, please hold your bias. I am telling you this story for a reason. Your anger cannot change history."

"You're right, Dina, but it sure as hell can change my future."

"Yes. It can indeed."

"Is this what you're trying to tell me, Dina? That by understanding how things have happened, and how they work, we can make better decisions in the future?"

"I hope so, Sara."

"So, you think smoking is horrible and should be banned?"

"You have made an enormous leap across several facts and assumptions to arrive at your question. I want you to understand how this industry has been able to survive, even when its products are of questionable health benefit."

"Questionable? They're downright fatal-"

"Enough, Sara. Let me continue. In the early 1800s, tobacco companies began to roll ground tobacco inside paper, thus dispensing with the need for pipes. Demand for the paper tobacco tubes grew rapidly after your Civil War, but hand rolled cigarettes took much time to produce. In 1876, a tobacco company named Allen and Ginter offered a prize of $75,000 to anyone who invented a machine that rolled cigarettes. A young inventor named James Bonsack created such a machine and patented his invention in 1880. The machine was not reliable, but one manufacturer, James Buchanan Duke, bought it anyway, and after some effort, began producing 100,000 cigarettes per day. Competition followed soon after, and suddenly, too many cigarettes were being produced. So, these companies began to market their products, and even gave them free to potential customers at public events. They touted the 'health benefits' of their product. They also started including stiff cards with beautiful pictures in the box that they labelled as 'collectibles.'"

"What? Like baseball cards?"

"The very ones."

"Oh my gosh. I thought those came with bubblegum."

"Bubble gum was a late addition. Another way to attract children to the company and convince them to later use tobacco tubes."

"Scumbags. How did they come up with that idea?"

"Necessity. They needed something to keep the paper tobacco tubes from being crushed. A stiff card performed the function."

"Hmm. I didn't know that."

"The demand for cigarettes accelerated. About sixty years after mass production began, the negative effects of smoking appeared."

"Yep, and how did the companies respond, Dina?"

"I can sense you are aware of this response, Sara."

"You bet I am. The rats. When stories started coming out that cigarettes were harmful to health, did the companies say, 'Oops, sorry, we didn't mean to harm or kill our customers. We'll change our product right away. Or better yet, we'll stop making our deadly product.' Nope. Instead, they went to court and spent millions to say it wasn't their fault their stupid customers chose to put the smoldering tobacco tubes in their mouth. These folks knew what they were doing. If smoking was so bad, they could have stopped. We shouldn't pay the price because they chose to continue an unhealthy habit for 40 years.' The companies got away with this argument for a long time."

"True, Sara. From the time of mass production, well over 100 years passed before tobacco companies began losing lawsuits and started settling claims. Ironically, they did not lose because tobacco tubes were deemed unhealthy. They lost because the tobacco companies knew about the addictive nature of nicotine, did not make customers aware of the fact, and intentionally enhanced the addictiveness of the tubes to make quitting more difficult."

"Yep. With all our knowledge about the harmful effects of cigarettes, tobacco companies still exist today. They still make a product that harms people. Not quite a stellar societal member, are they, Dina? If tobacco companies were an organ in a body, and they killed or harmed other cells in the body, that body would be dead, dead, dead!"

"It is clear my request for objectivity is unmet."

"For good reason, I would say."

"All right, Sara. Given this brief history, let us see how people and institutions in the ideasphere compete. Remember, the first way is to obtain more resources than competitors. Did this happen in the tobacco industry?"

"Yes. The Duke guy is an example. Ooh. Is that how Duke University got its name?"

"Veering."

"Okay, okay. In the ideasphere, resources are equivalent to customers, so the goal is to obtain customers. Before Duke and Bonasck teamed up, companies made a couple of hundred cigarettes an hour, or a few thousand per day. After Duke started using the machine, the company produced 100,000 cigarettes per day. He reached a lot more customers, in fact, so many more they decided to give some away to attract new customers. Hmm. Dina, is Duke's company still around?"

"Not in its original form. It took over several competitors early in its existence and became known as the American Tobacco Company. It absorbed most of its competitors until 1911, when it was split up under anti-trust regulations. In 1969, it changed its name to American Tobacco and became a subsidiary within a holding company named American Brands, Inc. American Tobacco was sold to a company named Brown and Williamson. Brown and Williamson merged into RJ Reynolds in 2004 and was then acquired by British American Tobacco in 2017."

"So, the company evolved different structures and generated different products over time. Hmm. This reminds me of the ecosystem."

"Let us not move too quickly, Sara. Let us examine competition in the ideasphere step by step. We now know that Duke's company obtained more resources than its competitors and has continued to exist in some form all the way to today."

She paused for a long moment. "Yes."

"Thank you for refraining from your derogatory comment."

"How did you know I was going to say-"

"More's the pity. Is that what you planned to say?"

"How did you know?"

"Irrelevant, Sara, but thank you for refraining. Now, let us continue. Another way entities survive is by obtaining the same resources with less energy. Let us use our smoldering tobacco tube company to see if this happened."

"You mean did the company increase its productivity? Do more with less? Yes. The Bonsack machine rides again! It produced a lot of cigarettes at a much lower cost."

"Yes. Before then, the company used workers to hand roll the cigarettes. They hired young girls with nimble fingers."

"I'm liking these guys less and less."

"This is your opinion, Sara. It took some time to get the Bonsack machine operating efficiently, but once it did, the company produced the few thousand cigarettes rolled daily by hand, in a fraction of that time. Even with breakdowns and maintenance, it was not difficult to produce the same number of cigarettes with far less effort. When we say, 'with less effort,' in the ecosystem, we mean 'with less energy.'" What do we mean in the ideasphere, Sara?"

Sara paused. "We mean 'lower costs.' Using fewer resources. Hmm. At the production level of 100,000 cigarettes a day? Hell, they'd be wiping the floor with their competitors."

"I am sure this did not happen."

"Not literally, but Duke ended up absorbing its competitors, so figuratively that's what happened."

"Let us move on, though I must note I am unable to connect how applying a mop to a flat surface relates to this story. The third competitive skill involves producing more offspring. Did our company use this same tactic in the ideasphere?"

"Yes. Offspring in the ecosystem are the equivalent of ideas in the ideasphere and more products in the physical world."

"Yes, that would be the equivalent scenario. Did our company do this?"

"A little too well, more's the pity. They made so many cigarettes they couldn't sell them all."

"What did they do in response?"

"That's when they came up with their marketing ideas. And baseball cards. That's a useful example. They needed a way to keep the tubes from being squished. They could have used a blank piece of cardboard. Instead, they produced a new product they convinced their customers to want. They created a brand-new market in which people began to collect and trade the cards. When I was a kid, I bought a lot of bubble gum I didn't want, just to get a Fergie Jenkins or a Carl Yastrzemski card. Those two cards were gold!"

"Quite right. The cards complemented their existing products and procured additional customers. This is equivalent to producing many offspring that provides the DNA with a better chance of survival."

"Hmm. They work the same way."

"So, can we conclude competitors in the ideasphere and competitors in the ecosystem behave the same way? In the ecosystem, creatures survive by becoming better at obtaining resources, producing more offspring, or developing new ways to obtain resources. In the ideasphere, institutions make more money by attracting more customers, producing more products, or creating more efficient means of production. In the process both entities and institutions may change how they look and how they operate."

"I agree, as far as it goes."

"What do you mean?"

"From everything we've talked about, I agree that people inherit their survival tools from their non-human predecessors. Then they proceed to turn the tools on their heads. I'm pretty sure people managed to bollox up competition too."

"Mr. Wallace? We have an operating room open at 7 am tomorrow morning. Will you be able to make it?"

"Of course. How did you get one so soon?'

"One of my colleagues rescheduled a patient's elective surgery. Usually there is stiff competition for operating rooms, but your wife is getting be to be quite famous. Several doctors are interested in her case. Dr. Kendrick postponed his operation for a chance to observe our procedure."

"I'm not sure Sara would like that or hate that."

"Well, you said she always beat to her own drum. I believe she is still doing so, even while in a coma."

Chapter 26

STOMPING ON THE INVISIBLE HAND

"Spill, Dina. How do people bollox up competition?"

"Bias warning. Bias warning." Sara laughed at Dina's words. "Be patient, Sara, let us proceed in an orderly fashion."

"But I'm right, right?"

"Yes. Once again you have unintentionally bumbled onto the truth."

"I knew it."

"You do not know it, you sense it. There is a difference. Now, let us continue. We know competition in the ecosystem is caused by intense pressure, and entities therefore try to avoid competition. They respond to competition in three ways. They diversify. They innovate. They become more efficient and productive. Again, using our tobacco tube company, let us see if institutions in the ideasphere try to avoid competition.

"Consider diversification. In the ecosystem, diversification results in different life forms. In the ideasphere, diversification results in obtaining different customers from different products. The baseball cards included with cigarettes are an example of diversification. The need to protect cigarettes led to the creation of a new product idea. The cards could be kept and traded after all the tobacco tubes were consumed. The new product might attract customers who might not want to smoke but might want to collect the cards. That is equivalent to mutations that produce offspring that consume different resources. The new entity obtains different resources; the company attracts different customers."

"So new product development is like diversification. Hmm. Interesting."

"Yes. Now, let us examine innovation in our two systems. In the ecosystem, an entity innovates when it finds new ways to obtain the same resources. In the ideasphere this would be equivalent to developing a new idea or product to sell to *existing* customers. For instance, the company makes the same product with different variations so existing customers are induced to buy more of the product."

"Oh. Like different brands. Or different qualities. Low tar. No tar. High tar. Slim. Fat. Medium. Manly. Feminine. Mint flavor. Cinnamon flavor. Fatal. Partially Fatal-"

"Too far, Sara. I should have expected your response. Indeed, the modifications are slightly new ideas made into slightly different products meant to keep their existing customers interested enough to continue buying."

"I see. More variety keeps the customer from drifting away."

"Yes. Innovation operates similarly in the ecosystem. Those slight variations might make an entity more capable of obtaining their resources when food becomes scarce. Those with slight advantages can compete more successfully and have a higher chance of survival."

"Yep. In our case, the best customers had a lower chance of survival. But the company did well!"

"Let us move to the last competition avoidance technique. Productivity. Productivity is the opposite of innovation. In the ecosystem, an entity uses innovation to obtain the same resources by implementing a new variation. It uses productivity, on the other hand, to obtain more resources with the same or less effort. In the ideasphere, this is equivalent to producing an existing product with fewer costs. The Bonsack machine is an example. Duke produced a few thousand cigarettes a day before the Bonsack machine was used. That is, he put his idea into action at the rate of a few thousand idea products per day. Once the Bonsack machine began to operate, they made the same product with the same amount of labor, but now produced 100,000 per day. They were able to attract many more customers, which is the equivalent to obtaining more resources in the ecosystem."

Sara sighed. "I wish you used a better example, Dina."

"Why?"

"Because these three things, diversification, innovation, and productivity are not only the hallmarks of competition, but they are also the hallmarks of human achievement. This is how we have turned rudimentary ideas into amazing inventions like rocket ships and pacemakers and microwave ovens."

"This is also how you have built nuclear bombs, and guns, and creme filled sponge cakes."

"What are you talking about? Twinkies are delicious!"

"But not a recommended dietary food item. You wanted to know how people and entities differ in using competition, Sara, Your Twinkies are a helpful example."

"What? I'm not following."

"Competition is merely a mechanism for survival, Sara. Any idea that obtains customers, good or bad, can exist in the ideasphere. You humans implement many different ideas and then argue about the morality of them afterwards. There is no moral judgement in nature, Sara. The environment determines what is 'good' by determining what survives.

"Let us focus on this idea of morality for a moment. Since morality does not exist in nature, we must conclude it is an idea conjured up in the ideasphere."

"Ooh. I never thought about it this way. Hmm. I think that's true."

"As an idea formulated by people, is morality fixed? Unchanging?"

"Ha! Some people would like it to be, but no, it's not. It changes as people change."

"Is the competitive marketplace of ideas in the ideasphere, whether it be for products or religious ideas or artwork, shaped by morality?"

"Yes. No. Hmm, I don't know. Trick question."

"Shall we return to our tobacco tube company to answer this question?"

"Those guys again. Okay, if you insist."

"After many years of the existence of tobacco tubes, it became clear cigarettes contributed to lung disease and cancer. Yet tobacco companies still exist today and produce billions of cigarettes a year. Millions of people still smoke today. Societies incur billions of dollars in health care costs due to smoking, which are not paid for by cigarette companies."

"Yes, that's true. I hate it, but it's true."

"Can we conclude cigarettes are a morally acceptable product?"

"No, we cannot."

"They are accepted by your society, Sara."

"People don't agree whether cigarettes are morally good or morally bad. We are only just getting to the point where some people agree they are physically harmful. I don't think we dare make a moral judgment about cigarettes."

"Are you saying morality does not play a part in the competitive ideasphere?"

"Sometimes, when something egregious happens, government leaders may make laws to limit certain products or services or behaviors. It only happens when a product is super bad, and even then, the changes may not go far enough. Cigarettes are a prime example. They still exist today. Cigarette makers can still sell this product as long as they inform their customers it's harmful. But they did lose a lawsuit several years ago where they paid billions of dollars to states to reimburse them for the medical costs the states incurred caring for sick smokers."

"So, we can conclude morality is not a universal force as the environment is in the ecosystem? A force which everyone must comply with or suffer the consequences?"

"No. I don't think a force like that exists in the ideasphere, Dina. There are only ideas. As long as ideas finds customers, they'll exist. They don't have to be good ones."

"That is the point I wanted to clarify. Thank you. Morality is a human construct based on acceptance by some people, but not all people."

"Hmm. I always thought of it as a fixed ideal. We treat it like that."

"Nothing is fixed in the universe, Sara, least of all a human construct based on human ideas in a constant state of change."

"Ouch."

"You are disappointed."

"Not disappointed, exactly. We . . . us humans . . . people, spend a lot of time arguing over what is good and what is bad. In reality, we disagree because we have different ideas of what is acceptable and unacceptable. If we understood morality this way, as a basis of acceptance, we might be able to build a base of morality we all can accept. Maybe then we could make the world a better place."

"I believe this is the foundation of your Constitution and the laws in your country, Sara."

"Wait. What? What do you mean?"

"Your Constitution sets goals, and your lawmakers establish rules for all members to accept. Beyond that, your citizens are free to choose how to live their lives. As long as they comply with the laws, they can put any ideas into action. Isn't this the basis of morality for your country?"

"Hmm. I see what you're saying. I suppose it's true. In theory."

"It is my turn to ask what you mean."

"We don't always implement our goals or rules very well, Dina."

"This is also true. We will discuss this in more detail when we talk about human coordination, but for now, let us continue. If morality does not drive competitors in the ideasphere, what does?"

"I'm not sure. We're humans, so we can be driven by different ideas."

"Yes, but all ideas lead back to resources, would you agree? People are still members of the ecosystem regardless of the fact they are often preoccupied with their place in the ideasphere. They need resources to survive."

"Yes. That makes sense."

"In the ecosystem, entities use their resources in three ways. First, they convert them to energy to operate their bodies. Second, they build the cell and body parts needed to keep the entity alive. Third, they reproduce a new DNA generation."

"Yes, you mentioned this before."

"Indeed. In much of the natural world, an abundance of resource consumption triggers mechanisms in the body to attempt reproduction."

"Ha, that makes me laugh."

"What is so funny?"

"Well, if that was how humans worked, we would have a lot more kids running around in this country."

"Yet, this is not the case, Sara."

"Thank God for contraception."

"I refuse to enter the debate about family size determination, Sara. However, I do want to focus on this divergence. You jokingly alluded to the fact that excess resources do not produce more human children. Let us examine how people do deal with excess resources."

She stopped laughing in mid ha. "Don't get me started, Dina."

"I do want you to start this conversation, Sara. We shall use our smoldering tobacco tube company as an example. The company obtains customers, and from its customers, it obtains resources. What does it do with these resources? Clearly, the employees did not all immediately produce children when they received their wages?"

"Depends on the size of their bonus."

"Do not make jokes, Sara. This is a critical topic."

"Okay. Okay. What do companies do with excess resources? Lots of things. Generally, they try to amass more resources."

"So, you are saying the company uses its excess resources to maximize its ability to obtain more resources?"

"Yes."

"How do they do this?"

"Lie, cheat and steal."

"Careful, your bias is showing."

"Really? I thought I was being restrained."

"Sara, I will not repeat my admonition. This is a serious discussion. If you want to learn how the human system works and how it does not work, you must address these ideas objectively."

"All right, Dina. I am just saying that my experience with corporate America, has not always been positive."

"Yet you benefit from the products and services this corporate America provides."

"Okay. Okay. There are two sides to every issue. I'll behave."

"Thank you. Let us consider how a newly formed company behaves. When a company is young, it places all its resources into getting customers. It makes products they think will attract customers. And it promotes them."

"You mean, it advertises."

"Please clarify the meaning of this term."

"The company tells people things about the products, so they'll want to buy them. Like cigarettes are good for your health. Or cigarettes are cool. Or. . . our cigarettes are better than the other company's cigarettes."

"Is this information a company imparts true?"

'Hell n- I mean no, not necessarily. Some of it might fall into the category of 'provably untrue.'"

"The company is not required to disclose that their assertions are not true?"

"This is one of those areas where I think our laws fail us."

"Please explain."

"Well, the courts in this country have ruled that people should understand advertising is 'puffery.'"

"You mean it acts as a wispy cloud?"

"Not puffy, 'puffery.' It's a polite way of saying an advertising message may be a big fat lie. The courts have ruled that people who hear or read advertising should understand a company is saying these things to procure customers and not necessarily to tell the truth."

"This does not happen in the ecosystem, Sara. Some creatures, like peacocks, promote their sexual availability with a colorful tail, but there is a difference. A peacock can only display its actual physical characteristics. It is unable to describe itself in any other way. It evolved a luminous tail that had been proven, through long accumulated experience, to be beneficial in offspring reproduction."

"That's not advertising, Dina. That's survival. And it's true. The colorful tail exists. I'm not a business expert, but even I can see a peacock's plume is a far cry from old Smoldering Tobacco Tubes Inc. blithely announcing its product is good for your health, when the opposite is true. It's a straight up lie. Entities don't get to lie to nature. Even if they did, nature doesn't listen."

"Indeed. Let us consider this issue in more detail. A company may use advertising to fool people into accepting an idea that is not true, and customers may respond positively to an untrue message."

"At times, yes."

"If the untruth gains acceptance, is it not as beneficial as an accepted truth? It earns the company customers and resources, does it not?"

"Ooh, you're right."

"So, this is, in fact, a logical strategy for the company. The untruth earned customers."

"I must agree with you, Dina. There's only one problem."

"What is that?"

"The product kills people."

"Not all products kill people. Not all untruths are deadly. You make a valid point in this case, but it is not the concept on which I wish to focus. I want you to understand this: in the ideasphere, institutions have many varied ways to obtain resources. They are not limited by the direct accountability that exists in the ecosystem. If an institution obtains resources by guile, and they are not prevented from doing so, then they would logically conclude guile is an acceptable practice. You wanted to know where competitive people differ from competitive entities, this is an example. People have a wider range of possible actions and less direct accountability for bad ideas than creatures in the ecosystem."

"Oh. That is a huge difference."

"True. Let us consider our young tobacco tube company. It innovates by making different versions of its products. It develops new techniques to lower its costs. It makes new but related products. These efforts can attract a stream of customers who provide the company with a stream of resources."

"Hmm. This is what surviving entities do in the ecosystem. Innovate. Increase productivity. Diversify."

"Yes. Let us assume our young company has done all these things and is now an older company with many resources. How does it use its resources now?"

"Lie, cheat and steal?"

"Bias warning. Once a company has established products and possesses a customer base, its goals change. Now its goal is to keep resources flowing. It can use its excess resources to prevent competitors from threatening its existence."

"What do you mean?"

"In the ecosystem, an entity cannot influence the competitive environment it finds itself in. It must comply or die. The ideasphere operates differently. Companies with many resources can alter the competitive environment."

"W-w-w-wait a super slow-mo second. How?"

"In a variety of ways. First, they can force other competitors out of the market. They can reduce their prices and make it difficult for new competitors who have not yet accumulated resources, to participate. Customers purchase the low-price products, and the new competitor cannot earn enough revenue to cover its costs. When companies cannot earn enough resources, they die, or in the ideasphere, you humans say they go bankrupt. Once the competitors are removed, the company again raises prices and earns more resources."

"Hmm. That is classic competition. It's equivalent to a creature in the ecosystem getting to the food faster than its siblings. Humans just edge out their competitors a little more forcefully than creatures do in the ecosystem."

"Indeed. A company can do other things too. It can use its resources to swallow its competition. Companies buy other companies and absorb their products as their own. In some cases, they may stop producing a product they absorbed so their own sells better."

"Ooh, that's like a predator in the ecosystem."

"Yes, but again, people in the ideasphere operate differently. In the ecosystem, predators eat entities they consider resources. Entities generally do not eat their direct competitors. It is too risky. Instead, they use excess resources to produce a new generation of DNA packages. In the ideasphere, resource-laden companies often eat their competitors."

"Are you saying excess resources are not valued in the ecosystem?'

"Not if the risk of dying is too great, Sara. The goal of all beings is survival. The goal is *not* to consume excess resources."

"Ooh. Difference alert. Wherrp, wherrp! So, in the ideasphere we maximize the resources we earn, not necessarily the goals of the organization."

"Not quite, Sara. People have taken a short cut. They have made obtaining resources the goal instead of a means to a goal."

"What do you mean, Dina?"

"Have you heard the term 'profit maximization,' Sara?"

"Have I heard it? It's only the holy grail of all capitalistic economic thought."

"True. It is interesting an entire school of thought has developed to transform a process into a goal. People, who can think up so many ideas and can take so many creative and novel actions, have willingly suborned their ideas in deference to a mere process."

"Wait. What?"

"In the ideasphere, profits often hold more value than the ideas and products of a company. It would be the equivalent of entities in the ecosystem eating more resources at the expense of producing offspring. If the goal of creatures in the ecosystem was to maximize resources, it would not take long for the wheel of life to stop turning. Creatures would never produce offspring and instead keep resources for themselves. If entities in the ecosystem acted like people in the ideasphere, there would be no life to speak of. And certainly, no people. And no ideasphere."

"Dina, you're blowing my mind."

"Since I now know this expression only means you are encountering a new idea, I am gratified. This is a critically important distinction between the two systems."

"Wait, I need to catch up. You're saying profit maximization has become the goal for companies?"

"Let us examine this question, Sara If we assume the smoldering tobacco tube company's goal is to maximize profits, its actions make much more sense. Presented with a choice, it always makes the choice to maximize its profits. It pays injured customers not to sue them for harm caused by their products. It fights in court to prevent paying for damages their products cause. It agrees to put warning labels on their products so government agencies will not sue them in the future. Some companies pay government officials to make rules to benefit their operations, or to give them financial advantages, or make laws preventing others from competing with them. You see, Sara, all these choices make sense if the goal is to maximize profits, which we can think of as net resources."

"So, what you're really saying is that if a company accumulates enough resources, it can game the whole competitive system. It can manipulate the environment in which it operates to obtain even more resources."

"Indeed. If entities stored excess resources the way people do, the ecosystem would grind to a halt. The resources would be stored away and not be available to others, even by competition. The wheel of life would stop turning."

"Is this happening in the ideasphere, Dina? With people?"

"Yes. With most resources stored in relatively few locations, the wheel of ideas will inevitably slow down. The wheel of ideas will only be available to those with resources. Everyone else, the people with fewer resources, will need to comply with the ideas established by these few resource-laden individuals and institutions."

"But . . . but . . ."

"But what, Sara?"

"This flies in the face of what we are all about, at least here in this country. Ecosystem aside, the human superpower is the ability to think and choose, not to be forced to implement someone else's ideas and choices."

"As long as resources are concentrated, the power to choose is also concentrated. It is the natural result. It is the availability of resources that makes choice available."

"This is totally ironic. We . . . humans . . . people, I mean. We have spent all our sentient lives in the ideasphere working against our own superpower. We keep assuming we're like all the non-human, non-choosing creatures in the eco-system that must fight for resources to survive."

"Humans do need resources to survive."

"Of course, we do, but we have so many ways to obtain them, Dina. We have the superpower of choice. We can think, and we can cooperate as well as compete. But it seems we act as if we don't have these capabilities at all. We are told we must believe in a certain religion to be 'saved.' We are told we must get a certain type of job to be successful. We are told we must wear certain types of clothes to be 'cool.' We are told we must do certain activities to be accepted and to not do others, so we aren't considered dweebs. We are told we must maximize profits to be 'smart' businesspeople. In reality, all of these 'shoulds' are purely optional. They are not mandatory and utterly subjective. So why do we buy into them? Because we don't understand they are merely choices in a vast array of choices. All the power lies with each individual as the chooser. We seem to give up our superpower of choice all too easily and let others make our choices for us. Maybe it's because we don't understand how powerful choice is. Or maybe . . . it's that people with a lot of resources understand how powerful choice is and don't want regular folks to get into the habit of using it."

"People have not been in the ideasphere long, Sara. In the days when Adam Smith wrote about the free market, the ideasphere was much different than it is today. In those days, the prevailing ideas were already concentrated. Kings and autocrats were the norm. Mr. Smith's writings were a revelation in new thinking. Capitalism and profit maximation were concepts that made mathematical sense and acted as a roadmap to freedom."

"You're right, Dina. It started as an idea that promoted choice for people. So, what happened? Why is it so out of balance today?"

"In those early days, profit maximization did not produce the same effects as it does today. Profits were not huge. Even very successful businesses could not sway the entire market. In those days, capitalism and profit maximization resulted in a less arduous life. People accepted capitalism not only because it led to an easier life, but because it actually did optimize the human superpower of choice. As time went on, people used the freedom that came with capitalism to develop more ideas. Some businesses earned far more resources than others, and the relative freedom of individuals began to stratify. Within a hundred years, people changed not only the physical world but the ideasphere itself. However, the underlying ideas about capitalism did not change much."

"Why would they? You don't change what's working."

"Yet change is always happening, Sara. In this case, people continued to focus on capitalism and profit maximization as harbingers of freedom, despite the changing environment. The success of capitalism produced followers, you call them economists, who specialized in the intricacies of this one idea. Economists specialized, just as entities do in the ecosystem. Over time, economists produced more intricate ideas to maximize profit even as the underlying environment was increasingly separating the rich and the poor. The effect has been quite dramatic."

"What do you mean?'

"In 1750, when Adam Smith lived, resources were abundant, the air and water were clean, and ideas were rudimentary by today's standard. There were literally not enough people on Earth to affect the Earth's conditions using capitalism or any other economic theory. We cannot say the same for today's physical environment. People fight over resources that are now limited. Much of the air and water are polluted. Eight billion humans are alive today compared to approximately 800 million in 1750. The ideasphere has changed too. In 1750, people used horses and basic contraptions made with wheels. Only birds and insects could fly. Today, vast resources are held by only a small percentage of people. of Do you understand what I am saying, Sara?"

"Oh, I get it. People treated the idea of capitalism and profit maximization as fixed and continued to apply it in a changing world."

"Indeed. They did not detect the subtle changes occurring day by day. They did not link the concentration of resources and power to the way capitalism was being implemented. Capitalism has long been a democratizing idea, Sara, not an autocratic one. It is only in recent years people have noticed this shift in resource concentration. Given the association of capitalism with freedom and choice and democracy, it is quite difficult for people to overcome their implicit bias and understand that the primary goal of capitalism, profit maximization, has now become detrimental to choice."

"So, maximization of profit is the cause of concentration of resources and the restriction of choice. Poof! There goes the underpinning of modern economics."

"Indeed. Let us examine why it might be time for this tenet to be changed. First, I want to be clear, profits are not the problem. Earning profit, as you know, is equivalent to obtaining resources. This is something every person needs. The promise of profit is an effective mechanism to encourage people to implement their ideas. It is not profit that is the problem, it is the *maximization* of profits that produces the deleterious impact. It is the idea that a person or institution must earn profit at all costs. No law exists in the ideasphere requiring people or institutions to maximize profits. It is not an instruction stored inside a human's DNA requiring them to act this way. It is an idea conjured up by people which has been accepted. It has been reinforced by many people across many generations repeating that profit maximization is good and socially responsible-"

"Thank you, Milton Friedman, you little shit."

"Be careful Sara, your bias is on full display."

"He deserves it. Do you know what he said, Dina? 'The social responsibility of business is to maximize its profits.'"

"Your bias has led you to misquote Mr. Friedman. His exact words were: 'There is one and only one social responsibility of business–to use its resources and engage in activities designed to increase its profits so long as it stays within the rules of the game, which is to say, engage in open and free competition without deception or fraud.'"

Sara paused. "Fair enough. I misquoted him. But I note he didn't say a thing about producing products that harm or kill people. Or paying off government officials to change the definition of 'free competition.' Fair's fair, Dina. I stand corrected. I will revise my statement: Thank you, Mr. Friedman, you little piece of excrement."

"Altered in a most unhelpful way."

Sara remained stubbornly quiet.

"Let us continue. The logical outcome of profit maximization is concentration of power. With enough resources at its disposal, an institution can keep its resource stream flowing and prevent others from obtaining them. Concentration of resources leads to concentration of choice."

"Or, looking from the other direction, fewer resources lead to fewer choices."

"Yes. Profit maximization ultimately slows the flow of resources. It is the exact opposite of the chaotic free-flowing ecosystem in which every entity must try to fit."

"So, what you're saying is we have set up a system in which many compete, and once a few win, they do everything they can to cut off the opportunity for others to compete and win. They use their resources to rearrange the environment in their favor. No wonder my bullshit meter was always triggered when folks talk about profit maximization. It really is bullshit."

"Profit maximation is not excrement. But it does sever the link between ideas and resources."

"Nope. I call bullshit. Look, Dina, if you have a harmful idea like cigarettes, or a dangerous idea like guns, or a stupid idea like flying parachute outfits, it doesn't matter. If you have enough resources, you can put those ideas into action regardless of the harm or danger or stupidity they cause. The people with the most resources decide what ideas are allowed into the ideasphere and what ideas are turned into products. Those who have legitimate arguments about the harm or danger or stupidity of these ideas do not have the same influence if they can't bring sufficient resources to the table to fight."

"This is true, but it does not mean profit maximization is a waste product."

"Okay, not literally. Still, I'm disappointed. And frustrated. Here we are, with a multitude of ideas distributed across eight billion people, and we have used what is essentially a tool to optimize human choice and turned it into a tool to achieve the exact opposite. The few with the most resources do their best to restrict choice. They do their best to not let others be free to think and conjure up ideas and put them into practice. They have done everything to make our ideasphere a small, limited place, when we have such potential to make the world one in which we can all co-exist and produce lots of ideas. God damn it, Dina. All my life, I thought universal peace, or at least peaceful co-existence, was impossible, just a talking point people idealized. Now I see it is a real option. It is utterly possible to live together in a way that optimizes the opportunities to think and choose and create ideas without killing each other because we disagree. Of course, it's possible. As long as we can think and communicate, it is within our power to achieve it. If only we understood how possible it is, then it would only be a matter of time before we did it. But no, we're too busy trying to stuff our faces with so many resources that we can't possibly use them all."

There was a long silence. "Are you finished, Sara?" Dina asked at last.

"I'm sorry, Dina. I didn't know that intensity was inside me."

"It appears it has been building for some time."

"Yes, but it's so weird. I think my frustration has been brewing for years, but I didn't know how to express it. Thank you, Dina. It's as clear as day, now. We have a system that is supposedly natural but works against our nature. We do things that limit our choices and our ability to think while touting our commitment to freedom. If we can just understand how all this works, we might be able to find a way to live together with all of us using our power of choice. Maybe there is a way for us to live together in peace."

"I believe you are correct, but to understand how this might work, we must discuss the other two survival mechanisms: collaboration and coordination."

"Are they going to blow my mind like our discussion of competition just did?"

"I hope so, Sara. I sincerely hope so."

Chapter 27

COMPLICATED

"Sara, it is time to move on to our second 'C.' Collaboration."

"Great. Our last discussion blew me out of the water. I can't wait to find out what happens with this puppy."

"I am unable to perceive how swimming with canines relates to our conversation, but I now expect such non-sequiturs from you."

"Are we using STTI as an example again?"

"What is STTI?"

"Smoldering Tobacco Tube Inc. STTI. I thought I'd give it a name. In case it ever gets traded on the New York Stock Exchange."

"Yes."

"I thought so. Ok. Fire away. Ha! Or light 'em up. Or-"

"I believe you are attempting humor again. Unsuccessfully, I might add."

"You sound like Josh."

"He is an intelligent young male to have such insight. However, we must not be deterred, Sara. Let us consider who collaborates in the ecosystem. We know cells collaborate to avoid competition, but this is not an intentional act. In the ideasphere people can choose to compete or collaborate. Which mechanism are humans more likely to choose?"

"Ooh, that's a false choice. They do both. They mix them all up in a big bowl of survival goulash."

"I do not comprehend this reference, but yes, they do both, often simultaneously. Remember, the ideasphere is composed of ideas. Ideas mean nothing unless they can be communicated. The strength of people lies in their ability to communicate. They talk. They write. They orate. They chirp-"

"Ha. You mean tweet."

"Yes. This is how collaboration occurs. Compared to the ecosystem, people are extreme communicators. One of the closest examples of communication in nature are bees who perform elaborate dances to show other bees where nectar filled flowers are located."

"Now, people can open their phones, post the address of their favorite restaurant on social media and hundreds or thousands of people are notified at once."

"True. People and the ideasphere in which they exist are built for collaboration."

"That's strange. We are made for collaboration, but we are told we are born to compete."

"You are describing a characteristic of the ideasphere that causes trouble."

"What do you mean?"

"Remember Sara, an idea does not need to be factual to be accepted. Early ideas were often accepted and put into action without the benefit of being truthful. In fact, much of the ideasphere is based on nonfactual ideas that were accepted early in the human era and have carried through to today, even when facts no longer support them."

"W-w-w-wait a swinky swanky second. Are you saying we are not born to compete?"

"Yes. The idea that competition was essential made eminent sense at the time it was proposed. People could see that entities in the ecosystem competed, but they were unable to perceive the collaborative nature of the cells within a body. In fact, at that time they did not even know cells existed. It was natural to conclude humans were like all other beings and needed to compete to survive, even as they communicated this idea with each other and thought up new ways to improve their ability to compete."

"Ha! They collaborated about only being able to compete. Irony alert!"

"Indeed. Unlike entities in the ecosystem who play only one role, as single-celled competitor or member of a collaborative entity, people are capable of both competing and collaborating. Human survival is not binary, but often a blend of these two mechanisms."

"So, how do we decide what the blend is? Half caf? Blond roast with a touch of Kona?"

"Sara, I do not understand how small light-colored cows from Hawaii have anything to do with this conversation, so I will proceed as if you stayed silent. If people were purely rational, they would choose to implement a mix of ideas and mechanisms to secure them greatest quantity of resources. This does not always happen."

"Are you saying people are not rational?"

"That may be part of the reason. A variety of factors affect how a person obtains ideas in the ideasphere. Education. Opportunity. Experience. Genetics. Differences in these things can determine the ideas a person obtains. The subset of ideas a person obtains is what becomes the basis for action, which in turn affects their ability to obtain resources. They can use their ideas to compete or collaborate, but they must choose one or the other, or a combination of the two. A person who chooses not to compete or collaborate will not survive."

"Hmm. The ideasphere complicates the whole process of living for humans, Dina."

"Yes. In the ecosystem, cells do not choose. They are born with their DNA and the accumulated experience of their ancestors, into an environment full of predators and resources. Bees do not need to educate themselves in the ways of bees. They are limited to act only as bees can act. They can only draw nectar from flowers as their instructions allow them to. They do not have any knowledge of the intricacies of flowers beyond this. Furthermore, entities in the ecosystem cannot choose to compete or collaborate. Once a cell becomes part of a multi-celled creature, it remains part of the entity forever. It evolves ever after as a component of the creature, and its behavior is defined by its role as part of the entity. A person, on the other hand, can collaborate within an organization and still produce individual ideas. Though a person is employed by an organization, they are not bound to act only in its interests for all time. They can create new ideas unrelated to the organization."

"Ooh, you're right. People are fully functioning idea machines. They can think up non-work ideas even while at work. For instance, I was told to reconcile the bank statement for the Widget Company, but I decided to do a crossword puzzle instead. I'm pretty sure a cell can't do that."

"I assure you, cells do not do crossword puzzles on the job, Sara."

She laughed. "You crack me up, Dina."

"No, you are still intact, but you are correct. A cell is not only constrained in its choice of survival mechanism, but also constrained to its specific role. A cell cannot move to another area of the body to operate, as a person can quit one job and obtain another. A heart cell cannot swim over to the kidneys to become a kidney cell."

"You're right, Dina, though it would be kinda cool."

"No. An entity would not survive such a move."

"Unless we evolved to do it."

"Which you did not. I pause my attaché.'"

"What? Sara thought for a moment. "Ha! I rest my case. I love that. Do you mind if I use that sometime?"

"Veering. Let us remember the key concept. People in the ideasphere are much less constrained than entities in the ecosystem. Are you beginning to understand how powerful human choice is?"

"I do. The moral of the story is that people don't have to do actual work on the job."

"No, that is not the lesson I wanted you to learn."

"I'm teasing you, Dina. I get it. People aren't tied too closely to an organization. Employment is an aspect of their life, not their whole life."

"Yes. And people can choose which organizations to belong to. They can also perform different jobs within an organization. A mechanic can become a quality technician. A church member can become a pastor. A product designer can become a sales representative. As long as a person fits the requirements the organization determines are necessary for a position, they can fill the role. And they are not limited to one institution or one choice. People are flexible. Tomorrow the product designer can quit the widget company and obtain employment with the gadget company; a member can quit the church and become an atheist. A person can write poems and take up pottery too. A person can stay on Spacebook but join Qwitter too."

"Wait, what did you call them-?" Sara asked.

Dina continued. "…cells in a complex body do not have any such freedom."

"You're right. People can belong to multiple organizations at once. They can be an employee of STTI, a member of a Church, a poet, have a social media account on. . . what did you call it? . . .

Spacebook. Ha! And their ideas can cross over among all those organizations." She paused. "Which means . . . hmm. . ."

"What, Sara?"

"Well, participation across institutions means not only more ideas can be formed, but ideas from one area can be shared and adopted by people in other areas. A robin can't go to school to learn how to be a cat. A worm can't go to school to learn how to be a lily. Entities in the ecosystem can only use their bodies and DNA to survive as themselves. People? We can be engineers and gardeners and poets. We can do anything as long as we can learn the ideas relating to an area and put them into action."

"Correct. Can we then conclude collaboration is optional but when used enhances the sharing of ideas across the ideasphere?"

"Yes. Which makes the ideasphere a fast-changing environment. Way faster than the ecosystem."

"Indeed. Now let us consider who the collaborators are in the ideasphere. In the ecosystem, only cells collaborate. As we discussed, cells join together to act as organs. The cells within an organ act in concert, but never lose their cell structure or behavior. No other entities collaborate, except, as we have noted, those that form colonies. Even in those cases, the individual members of the colony are born into their role and perform only their role, much like a cell in an organ. We know some entities have evolved to co-exist in a mutually beneficial way, as bees and flowers do, but these collaborations result from accumulated experience and not by choice. Now, let us consider who collaborates in the ideasphere."

"Hmm. If the ecosystem and the ideasphere were parallel, only people would be able to collaborate, and a handful of countries would act in a mutually beneficial way. Interesting."

"What is interesting, Sara?"

"Well, ideas are equivalent to DNA, right? And people are equivalent to cells. People are the only ones who can turn ideas into action. Institutions can't. The people within institutions can. Countries can't turn ideas into action, but people within the country can. People are the only ones that can implement ideas."

"This is equivalent to cells in the ecosystem. The cell is the only unit that can act on DNA's instructions. Tell me, when people work in institutions, do they act like organs in a body? Do they implement ideas that benefit the individuals in the organization and their country or society as a whole?"

"No. People in an institution work to help the organization succeed. Hmm. The job of a mechanic at STTI is to keep the machines producing tobacco tubes with minimal interruption. The job of the sales representative is to obtain as many customers as possible. And the job of the chief executive officer is to keep things running smoothly. All the jobs are designed to earn the company the most profit."

"Yes. These people play different roles in the company. Because they perform different actions, they affect the ideasphere differently."

"I don't understand what you're getting at."

"Consider the mechanic. The mechanic collaborates with others in the company to produce smoldering tobacco tubes. In exchange for a specific wage, and a forty-hour work week, she works on one Bonsack descendant machine. Let us assume she can keep the machine running 20 hours per day with one day a month off for heavy maintenance work. Those efforts translate into the machine being able to produce approximately 1 million tobacco tubes each day. The mechanic converts experience and knowledge of the machine into a certain number of products. When not on the job, the mechanic can still utilize the knowledge and experienced gained from her experience at STTI. For instance, she may spend some of her weekend tinkering with different machines at home. Perhaps she helps the local church keeping its machines in operational order or does occasional mechanical jobs for neighbors. As a result, the ideasphere is changed not just by the mechanic's work at STTI, but also by the other collaborative actions the mechanic performs in other areas of the ideasphere. Each person uses their knowledge and experience to put ideas into action individually. Institutions do not act. Departments within institutions do not act. Countries do not act. Only people act. It is only people who affect the ideasphere."

"And each person effects the ideasphere in a particular way depending on their unique blend of ideas and experience and actions. Hmm. Hold on, Dina. I'm confused. We already determined it is easier for people to compete than go it alone. So, why would people collaborate in the ideasphere? Wouldn't a person be more likely to get more resources if they competed? With competition, you have pressure, sure, but you also have more freedom to choose which ideas to put into action. Choice is the human superpower. Why would people give up freedom to collaborate?"

"Why did entities give up freedom in the ecosystem, Sara?"

"Safety. Nourishment. To avoid competition."

"Do not forget efficiency, Sara. We can think of efficiency as using the least amount of energy to gain the most resources. If every entity survives by being efficient, we can say they use the least amount of energy possible to obtain resources to survive. In the ideasphere, efficiency can be considered 'ease.' A person merely chooses the easiest survival mechanism that earns them the most resources."

"Oh. You're saying we're lazy."

"No, Sara. Life is not for the lazy. Competition and collaboration are opposites in the ecosystem, but they are two sides of the same monetary unit in the ideasphere. They only differ by the strength of the rules to which the participants agree. A collaborative institution narrows a person's actions and freedom to the performance of a specific function in the pursuit of company profit. A competitive market is also a collaborative effort, but it provides a person more freedom with fewer rules. If a person thinks they can profit more by competing as an individual, as a bookkeeper or a plumber, for example, they may choose to implement their ideas via direct competition in a market. If a person thinks they can profit more by joining an organization that pays them fixed wages and set hours, then they will choose to implement their ideas via collaboration. For the most part."

"What do you mean, for the most part?"

"People do not always possess the best information to make decisions, Sara. Sometimes they do not have many choices. Sometimes they do not realize other choices exist. Sometimes other ideas are more important to them. Information is not perfectly distributed in the ideasphere."

"This gets back to education and how important it is for people to understand the universe of choices available to them."

"Indeed. Education is a critical collaborative effort people undertake to make sure children can survive and thrive as adults."

Whoa. This is huge. I mean, education is huge. I keep coming back to that conclusion."

"Indeed. Let us expand our discussion. Do institutions collaborate?"

"What you're really asking is if people collaborate across institutions?"

"That is the question. Consider a young, struggling company. The people in this company do everything they can to ensure the organization survives. They do not have any extra time or resources to devote to anything but survival. Their external collaboration would be expected to be minimal. Would you agree?"

"Yes."

"Now let us consider a successful company with excess resources. The people in this organization do whatever they can to prevent their stream of resources from being threatened. One way to do this is to join with other resource-laden companies to achieve more ways to obtain resources."

"Oh, like the NRA."

"Yes. The National Rifle Association was originally founded because soldiers in the Union Army were ineffective marksmen. The NRA taught people how to shoot and handle guns safely, and often worked with the government to establish regulations to maximize the proper use of guns while minimizing their risks. But as we know, purposes can change with time."

"Yep. Those goals have been tossed out the window. Now, the NRA exists to promote gun ownership and prevent gun regulation. Gun makers contribute a significant amount of money to fund the NRA."

"Yes. Gun manufacturers did not start the NRA, but they have been able to alter the NRA into a collaborative structure to maximize their profits."

"Oh, I see. PACs are another example."

"Yes. Political Action Committees. These are groups of people and institutions who believe in certain policies. They collect resources and direct them to promote specific causes. PACs are not as numerous as people or other types of organizations, but they are powerful. They direct resources at the few decision makers in a society, and therefore influence decisions that affect all the people within a society."

"I know about PACs, Dina. This type of collaboration doesn't seem right at all, especially when the money comes from people with ties to rich and powerful organizations."

"Why do you think this type of collaboration is not proper?"

"Because we're supposed to be a democracy where the common good is the only consideration. Wealthy PACs, composed of wealthy people, influence the decisions and actions to benefit themselves. Hmm. So, in the ideasphere, collaboration can end up in the same place as competition; a few groups with a lot of resources can influence the entire ideasphere for their own benefit."

"This is a key point, Sara. Wealthy people and institutions apply their resources to maximize their flow of incoming resources. That includes giving resources to people who participate in the coordination function."

"W-w-w-wait a major muddling minute. That can really screw things up."

"All right, Ms. Sara. Here we go again." Tina turned over Sara's arm. "Not a red welt in sight." She checked the flow of the nutrients from the intravenous bag. "Good." Tina attached a latex strap around Sara's other arm until her veins bulged, gently inserted the needle, and then capped it off. "Very nice. Everything is proceeding smoothly. Knock wood."

Dr. Chen stepped into the room, her lower face covered in a surgical mask, her hands bathed in neon pink latex gloves. "How are we looking, Tina?"

"All vitals are steady." Tina hung the bag containing the neurotransmitter treatment and connected the plastic line to the IV. She stood out of the way. "We're ready whenever you are, doctor."

"Sara, in the ecosystem, collaboration is a mutually beneficial exchange. Cells trade their freedom of movement as a single cell for a limited role and a steady supply of nourishment as a member of a collaborative entity. Let us determine if this same exchange occurs in the ideasph-"

"What is it, Dina? What's wrong?"

"I am not sure." Dina paused. "Sara, I may have to leave rather suddenly. Do not be alarmed if this happens."

"What's going on? Are you in danger?"

"I do not know the answers to your questions, but our time together might be coming to an end more quickly than I anticipated. Let us proceed as swiftly as we can."

"Okay. We were talking about mutual benefit in the ideasphere."

"Yes. In the ecosystem, all cells in a body share the same DNA structure. The cells contain the same material, but their instructions vary depending on their location in the body. In exchange for playing their assigned role, every cell benefits in the same way. They are all nourished when the body is fed. They are all protected when the body is attacked. If excess resources are obtained, all cells share in the excess. If they are scarce, all cells share in the suffering. People are not like cells. Exchange occurs in the ideasphere, but it is not always mutual in nature."

"No kidding. Really?"

"We do not have time for sarcasm, Sara. In a business, the value of a person at a company depends on the value the person brings to the institution's profit goals. It takes a significant amount of time and resources to convert a baby into a doctor or technology specialist or a scientist. As we know from nature, the more complicated an entity is, the fewer that exist. So, a company pays more money for a doctor or technology specialist or scientist because many other people value those professionals, and few of them are available. It takes less time and training to convert a baby into a retail clerk or a barista or a fast-food worker. As a result, more of those workers are available, and companies pay them less. Companies do not care about the concept of 'sufficient resources' or 'mutual benefit.' Companies care about maximizing profits, so they pay as little as possible to obtain and keep the employees they need."

"W-w-w-whoa Nefarious Nelly. Doesn't collaboration need mutual benefit to make it work properly?"

"In the ecosystem, yes. For all the power the neural system holds over the cells in the body, its power is not unlimited. It cannot discriminate against certain cells because all cells are necessary to keep the body alive."

"People do not operate this way, Dina."

"No. Let us return to our tobacco tube company to understand this concept. Assume STTI is the major employer in a small town. Most people who live in the town work at the company. In this situation, an imbalance of resources exists. STTI possesses many resources, and the people are dependent on the company for their survival. What do you think happens if STTI cuts the wages it pays its workers?"

"Well, the workers will have fewer resources on which to survive. They'll be mad."

"Yes. This decision does not benefit the employees. They did not have the opportunity to participate in this decision. What options do the workers have? If they complain, their employment might be terminated, or they may be ostracized by their co-workers. They may uproot their family and move to another location, but that might prove risky if they do not have employable skills. They may just stay and accept fewer resources because they have no other viable option. This entire problem occurs because of the resource mismatch between management and employees."

"Oh. It's like what happened in the early1900's in our country when immigrants flooded in from Europe. Huge industrial companies hired these unskilled folks who probably couldn't speak English, paid them next to nothing, and worked them fifteen hours a day, six days a week. Children too. All the resources were on one side of the equation, as well as all the benefits. Workers died, children were crippled, families lived in poverty, but the company's leaders didn't care. Immigrants were abundant, and workers were easily replaced. The machines didn't need to stop running if some poor loser got mangled up in a loom. And the owners of the company grew so rich they lived in marble mansions and ate from gold-plated dishes in gold-plated dining rooms."

"Yes. When the resources are imbalanced, the benefit between the parties is also imbalanced. The benefit cannot be mutual. Do you recall the result of that particular era where workers were exploited in the early part of the 20th century, Sara?"

"People protested. It took a while, but the government finally enacted laws to protect workers and children."

"So, the government rebalanced the benefit to be more mutual."

"Yes, that's a good way of saying it."

"Can we conclude then that the collaborative environment in the ideasphere can be influenced by an imbalance of resources? Can mutual benefit be affected by this imbalance? Can we say mutual benefit is not the default in the ideasphere as it is in the ecosystem?"

"I think all those conclusions are true."

"Now let us consider another aspect of collaboration. In the ecosystem, all collaborators are interdependent. Each participant needs the other to survive. Each must perform its function for the body and for the other participants to survive. This does not work the same way in the ideasphere."

"What do you mean?"

"People can change jobs. They can start their own companies. They can save enough resources to retire and not perform any employment activities at all. Interdependence in the ideasphere is limited to a person collaborating to implement an institution's ideas while working at the institution. People are free to do other non-institutional activities when not working. Cells are not permitted such freedom in the ecosystem. They are on the job 24 hours a day, seven days a week for their entire lifetime."

"So, what you're saying is mutual interdependence is a relatively loose constraint in the ideasphere. We don't depend on each other to the same extent cells in a body depend on each other."

"Exactly. Now, let us talk about another aspect of collaboration, specialization. In the ecosystem, entities with collaborative cells have evolved efficient ways to perform their functions. Cells are made of the same DNA but take very different actions depending on their location in the body. The ideasphere works the same way, but it does not have the same advantages as the ecosystem."

"What do you mean?"

"Cells contain the accumulated experience of their ancestors in their DNA. However, in the ever-expanding ideasphere, every person starts at zero. A child today must learn more ideas to function in the current ideasphere than you did as a child. A child today must not only learn to read and write but must also learn how to use a cellular phone and a computer and complicated machines. To become a professional, a specialist, a person must learn a great deal more information than others who came before them."

"So, it's hard to become a specialist in the ideasphere. No duh."

"Specialization succeeds in the ecosystem because cells carry their accumulated experience with them. They do not need to do anything but follow the instructions they are born with. Not so for people. So, yes, specialization is much more difficult in the ideasphere. People in institutions collaborate and specialize and advance new ideas and products at a rapid rate. This advancement requires people to become specialized in ideas that may not have even existed when they were born. If the techniques for educating people do not provide them with the skills needed by these institutions, a mismatch occurs. The people who do become specialists have an advantage in obtaining employment. Those who do not obtain specialized skills have a disadvantage in obtaining employment."

"Hmm, so again we can see the importance of education."

"Yes, Sara, education is crucial. In the ecosystem, specialization produced the variety of entities on Earth. In the ideasphere, specialization provides people with the variety of skills needed to put complex ideas into action. Specialization is a much more challenging process in the ideasphere. In the ideasphere, every new generation starts at zero. Every new generation. You always start at zero. The more advances people make, the more difficult it is to make sure all people are brought along to meet the needs of your specialized world."

"Does this mean kids need to learn calculus in the third grade? My Dad used to say he learned everything he needed to know in the third grade."

"This is not true, Sara, though your father was a remarkably intelligent person. I am afraid he was jesting with this particular comment."

"You knew my dad? Really? How? When did you meet him?"

"Sara, now is not the time. We must make haste. We have come to our last topic about collaboration. Complexity."

"I want to know how you know my dad-"

Dina rolled over Sara's words. "In the ecosystem, complexity results from collaboration. Collaborative cells can develop into complex forms individual cells could never create on their own. By splitting functions between different cells, they can not only perform these functions more efficiently, but they can also develop a structure that can survive in the environment as a unit of its own.

Humans and tigers and bees are units composed of a multitude of cells all working together in an intricate dance to keep the complex body alive. Let us determine if this happens in the ideasphere."

"Well, people are complex, I assure you. Complicated too."

"Indeed. Like the ecosystem, the ideasphere is a symphony of collaboration, Sara. Consider the many complex products people have created. Skyscrapers. The James Webb Telescope. Computers. Cars. None of these items could have been made by one person alone. Think of a simple thing, Sara, like your cup of coffee in the morning. Do you have any idea how many people collaborate to get you your small slice of heaven every day?"

"Wait, how did you know I call it that?"

"It is quite contradictory since your religious beliefs do not appear to include the concept of 'heaven.'"

"Wait, how do you know *that*?"

"Let us return to your coffee, shall we? Hundreds, if not thousands, of people cooperate across the world to make this steaming stimulant available to you. Growers and shippers and roasters and retailers all have worked together along complicated paths to bring coffee beans to you, where your husband grinds them-"

"I got it, Dina. We are all connected. We all collaborate in a huge complex web so I can have my morning cup of coffee."

"Sara, I believe the dot sailed by you without your awareness."

"Ha! You mean I missed the point?"

"Yes. Don't you understand, Sara? The ideasphere is a far more complex place than the ecosystem. People are individuals, yet you are not like cells in a body. You are and will always be individuals. You join together to collaborate, but you can also separate from your collaborative units. You work together, but you can act alone. You can cooperate to make complex objects, but you are separate and distinct from the complex objects you make. You are simple and complex. You are individuals and members of groups. You are a cell and a body. Do you understand, Sara? You are creatures of both systems."

"My God, you're right. We are so much more complicated than I ever imagined." She paused, waiting for Dina's response. "Dina? Dina? Where are you?" There was no answer. She sighed. "I hope you're okay, sweet Dina."

"All right, let's begin." Dr. Chen turned the valve on the tube leading to the neurotransmitter solution. "Slow and steady. I'm starting with a rate of ten milliliters. How are we looking?" she asked the team of nurses monitoring Sara's vitals.

"Good."

"Steady."

"Normal."

"Excellent." She looked at her watch. "Okay. At two minutes I'll increase to fifteen." Her gaze moved from Sara to her watch to the tube as she waited. "Still okay everyone?"

"Yes," was the unified response.

"All right. Here we go. Up to fifteen. Two more minutes." The room was silent as the team watched both Sara and the monitors.

"All vitals are normal," Tina said. "She's not showing any signs of waking up."

"Let's go to twenty. Two minutes," Dr. Chen said.

They proceeded slowly. Twenty-five. Then thirty. Then thirty-five. Nothing. Sara's state remained unchanged.

"We've been at this for twelve minutes, doctor, and still no change. Shouldn't we be seeing something soon?" Tina asked quietly.

"I don't know what to expect. Let's continue. Same pace."

Forty. Forty-five. Suddenly, Sara's catheter-bathed arm jerked sharply.

"What was that?" Dr. Chen asked. "Was that a spasm? Maybe we're getting somewhere. How are the vitals?"

"Normal."

"Good."

"Steady."

"Let's just stay here. Keep the flow at forty-five. Let's sit tight and see what happens."

Chapter 28

CREATURES 2 – HUMANS 0

"Sara, I may need to leave again. I am not sure how much time we have remaining. We must hurry."

"Dina, what's wrong. Can I help?"

"You are helping far more than you realize. The situation may turn quite perilous, Sara, but I assure you, I will not let any harm come to you."

"What's going on?"

"I once again ask you to trust me. We must continue without delay."

"All right."

"Let us address our final survival mechanism. I saved it until now because it is not only the most important in the ideasphere, but it is also the culmination of all topics we have discussed so far."

"Our last 'C.' Coordination. I'm ready."

"In the ecosystem, coordination is a necessary form of collaboration. Every complex entity possesses a neural system to keep its cells and organs working in tandem to enable the body to survive. The neural system sets the goals of the entity, establishes the rules by which all the cells behave, and then enforces those rules."

"I remember."

"I have misled you somewhat in this area, Sara. In the ecosystem, the neural system does not actually create the goals and rules. They already exist. The job of the neural system is to enforce those rules that already exist."

"Wait a solid sedimentary second. I don't understand."

"Cells cannot proactively set goals or establish rules. The goals and rules are derived; they are inherited from long-accumulated experience embedded in DNA. Survival is the goal because only entities that survive exist. DNA instructions are the means by which a cell survives, that is, carries out its goal. Instructions are another name for rules of behavior."

"Ohh. Cells carry their goals and rules inside their DNA. They don't spend any time or energy coming up with a goal because they can't. They just follow their instructions which are the rules required to survive."

"Yes. So, tell me, Sara, how does an individual cell coordinate its actions?"

"It's neural sys- W-w-w-wait a sneaky snarky second. A cell doesn't have a neural system. Hmm, so, how does that work?"

"You are correct. A cell has no neural system. It is left to the environment to enforce the rules of behavior. A cell can only follow its DNA instructions. If conditions change and the cell is harmed, it may die. If conditions change and the cell benefits, it may survive. The environment is the coordination mechanism. It enforces compliance by its mere existence. Now, let us talk about multi-celled entities."

"Hmm. They're a lot more complicated."

"True. An entity is a body composed of identical cells. Every single cell in a body shares the same DNA."

"Well, if all the cells have the same DNA, then they must also have the same goals and rules."

"Exactly. Which means the only job the body's neural system must do is enforce the existing rules of behavior for the cells. This is in fact the only job of the neural system, Sara."

"That's a big task, though. How many cells are in a human?"

"Approximately 35 trillion."

"Yeah, that's a lot of coordination."

"Indeed. This is why neurons use 20% of all energy consumed by the body though the brain makes up only three percent of its mass. Now, let us consider the ideasphere. A cell in the ecosystem is equivalent to a person in the ideasphere. The cell's goals and rules of behavior are embedded in its DNA. Are the goals and rules of behavior for people embedded in their minds?"

"No. People are born with no ideas."

"Correct. People are born with no goals and no instructions regarding how to act in the ideasphere. Humans are born only with their accumulated experience needed to survive in the ecosystem."

"Oh snap! Humans are already down two major characteristics from the entities in the ecosystem."

"This is not a game, Sara. Can you deduce how this difference affects people in the ideasphere?"

"Let me think . . . Since people are born with no goals or actions, our coordinating systems, our minds, can't do any enforcing. There are no rules or goals to enforce. When we are born, we don't receive some automatic data dump that implants the goals of our society into our heads. We don't have an auto-rule-inserter that we plug in to learn how to act. Nope, though that would be ultra-cool. We must form our own goals and rules by living and choosing. We choose ideas that are the most important to us and then put them into action. Hmm. Our ideas are like goals in the ecosystem and actions are like rules. But we're different form our non-human siblings. People can create their own personal coordination mechanism that suits them as individuals."

"Indeed. This is a significant difference. Now let us discuss groups of people. They also use coordinating mechanisms to create and communicate goals and rules. Parents in a family are an example. Or management in a company. Or a pope or imam in a religion. Each coordinator derives the goals and rules for its members to follow. Then they must communicate those goals and rules. Only then can they enforce them."

"Wow, Dina. What nature does in super slick, easy-peasy fashion, is a heck of a lot harder in the ideasphere."

"Quite true. Now we can see why human brains are so large. The coordination job in the ideasphere is far more complex and energy expensive than it is in the ecosystem. In fact, coordinating systems at every level in the ideasphere are far more complex than their related levels in the ecosystem. People in the ideasphere live in both systems, so they must manage both systems simultaneously."

"Man, this is so complicated, I'm surprised humans even survived, Dina."

"People survived *because* they created the ideasphere, Sara. The ideasphere enables people to think and communicate and share ideas and store information outside their bodies. All these capabilities enable people to obtain more resources than they spend on the high cost of brain power. People using the ideasphere were more efficient than other creatures in the ecosystem."

"Hmm. When you get down to it, it's just cause and effect. There's no right or wrong, good or bad. If people can obtain more resources using their ideas and actions in the ideasphere, they'll survive."

"It is time for me to veer for a moment, Sara."

"You? Veer? This is new. Excellent. Go for it. What d'ya got?"

"We have discussed morality more than once. I wish to explore this topic once more."

"Aren't we in a hurry?"

"I believe this topic is important enough to veer from our rigid plan. Would you say a person's actions, the ideas on which a person chooses to act, represent an individual's morality?"

"Hmm," she whispered to herself. "You are what you do."

"What did you say, Sara?"

"It's something Randy always says. You are what you do. I think it's his way of saying the same thing. People can have all kinds of ideas. They can talk a fancy game, but if you want to understand their values, like truth or integrity or honesty or kindness, watch what they do."

"By observing their actions, you can deduce their morality?"

"Oh, Dina, none of us can truly understand each other. But, with experience, we can glean patterns, and if we can glean patterns, we can develop a sense of how a person will act in the future. It's not a certainty, it's more of a reduced uncertainty."

"So, would you say a person's behavior represents their morality?"

"Yes."

"Is every person's morality the same?"

"No, no, no, no. We have as many flavors of morality as we have people."

"So, would you say the cells' behavior in a body is its morality?"

"NO!"

"Why not, Sara?"

"Because a cell can't choose. It acts the way it acts because it is programmed to do so. People can choose."

"All right. So, morality is a choice?"

"Hmm. Yes. Yes, it is."

"What is this choice based on?"

She paused. "I have walked myself around in a circle haven't I, Dina? People develop both good and bad ideas. They choose to act on their most important ideas. Their morality is what they choose to act on. So, morality, by definition, can be either good or bad."

"What is good, Sara?"

"Well, it's . . . it's . . . it's what we decide is good."

"So, if an idea is determined to be 'good,' and a person chooses to act on it, the person is deemed morally 'good.' If the idea is 'bad,' and a person chooses to act on it, the person is deemed morally 'bad.'"

"Yes, I think that's fair, but it doesn't address the question of how 'good' or 'bad' is decided."

"True. Hold onto that question for a moment. Let us widen the picture to include institutions and governments. Would you say the goals and rules of behavior developed by a group can be considered the morality of a group?"

"Hmm. I'm not sure."

"Consider an institution. Let us return to our smoking tobacco tube company. The goal of the company is to maximize profits by selling tobacco tubes. The coordinating system of the company, the management, creates rules about how workers are to perform their jobs. They create other rules about how employees are to treat each other at the workplace. However, they do not have any rules about their products harming their customers."

"You're right! In fact, STTI probably has a lot of rules about what employees can and can't say about the company's products when they aren't on the job."

"Indeed. An institution has a definition of morality related to its goals. A designation of morality as 'good' or 'bad' is irrelevant to the company. The rules of behavior relate to the goal of maximizing profit."

"Oh, I see the parallel. A person's morality is defined by the actions they take to achieve their goals. An institution's morality is defined by the actions it takes to achieve its goals. A company maximizes profits. A church maximizes believers. A book club maximizes understanding of a book. Hmm. The rules serve to link actions to goals, but they don't necessarily attach 'goodness' or 'badness' to them." Sara stopped. "Let me take that back. They may *try* to attach goodness to their actions. That's a marketing ploy. If by saying they are 'good' or do 'good things,' and they earn more profits by doing so, then, by all means, they would use that strategy. Even if what they say is not true or if they harm people with their products."

"So, like people, institutions can vary in their goals and actions."

"Yes."

"Can these goals and actions converge or conflict between institutions?"

"Oh yes. We have the NRA and gun control groups. Both groups have diametrically opposite goals, and they each act on them."

"So how are conflicts between institutions resolved?"

"Hmm. We can't hope a group will change their goals. People are people. They are free to have their ideas, and they should be free to choose how they act on them. Let me think. If this were a human body, both organs would be allowed to perform their functions as long as they didn't hurt the other cells or the body. So. . . the coordinating function above the institution level would step in if people or society are being harmed." Sara stopped. "Hmm, the coordinating function gets involved. That's the government. The government steps in by making laws for the well-being of the people and the country."

"Ah. We have arrived at our destination, Sara."

"We have?"

"When a government sets goals and laws, does it express the morality of the country?"

"W-w-w-wait a supremely slippery second. I'm still back at making laws to prevent people from being harmed. That seems like a basic job of government."

"Your statement is true if the goal of the government is to keep its citizens from harm. This is not always the goal of human governments."

"What? Well then, that's a bad government."

"You may not make that moral judgment, Sara. A government's laws reflect the morality of the group being governed."

"Hmm. I'm not sure that's true. Some governments harm their people. I'm not sure those citizens support their government. They might be forced to obey, but that doesn't count as 'support.' Governments like this exist all over the place. Even our government acts like this at times."

"Would the ecosystem allow such behavior, Sara?"

"No. No it would not."

"This is another difference in the two systems. In the ecosystem, the coordinating system acts only to maximize the survival of most cells and the entity as a unit. If a country's goal is to maximize the state's survival without concern for the survival of its individuals, it makes laws to enforce this goal. If individuals are harmed or killed in the service of the country's goal, the loss may be deemed acceptable. If a country's goal is to maximize the well-being of its citizens without harming others in the society, it makes laws to achieve that goal. The laws of these two nations would be very different. And the lives of the people living within these countries would also be very different."

"So, you're saying goals determine how a country operates, just like they determine how an individual or institution operates. The laws are rules that reflect the goal, and to which citizens must comply. Good or bad, the laws reflect a country's basic morality."

"Yes. Morality differs in countries because their goals vary. Even in one country, morality can change over time. Coordinating mechanisms of countries, what we call governments, are important because they control the goals of the country."

"Which affects *all* the people in the country. Wow!"

"Let us consider how coordination operates. As you recall, in the ecosystem, the neural system's job is to enforce the goals and rules of behavior embedded in each cell. Every cell reports its status to the brain via the neural pathways. The brain cells process these bits of information and transform them into a set of actions for certain cells. This happens over and over and over. The body uses this feedback loop of reporting, response, result, to create accountability for every cell in the body. Now, let us consider how the ideasphere works."

"Well, we are people, so we don't have anything as tightly connected or elegantly coordinated as nature. And entities have a head start over people. They are born with their goals and rules. They know what to do. They just need to do them. Even with this advantage, a super complex, energy-gobbling neural system is needed to keep all the cells on track. Just look what kind of trouble humans get into, Dina. We can *choose* to harm ourselves in a thousand different ways. We drink too much alcohol, and the neural system must swoop into action to purge our system and distribute the toxins throughout the bloodstream and process it out through the kidneys, even as the alcohol inhibits our brain's ability to act. We run a marathon, and the neural system must instruct the muscles to break down to provide more energy because our body used up all the energy available from its usual sources. That's just us abusing our bodies on the ecosystem side. We haven't even touched on how we screw things up in the ideasphere."

"Sara, you are describing a well-functioning feedback system."

"Sure! A feedback system we force to work overtime to keep our sorry asses alive."

"Veering. Let us consider how the coordination function developed in the ideasphere in the beginning."

"You mean how did we learn to coordinate? As people?"

"Yes."

"Badly."

"Explain. Without judgment please."

"I'll try, but I warn you, my hidden biases are not so hidden when it comes to human history."

"I will note this and intercede as necessary."

Sara laughed. "Okay, okay. Hmm. My guess is that early on, people banded together for mutual safety. Once people began to collaborate, the need for coordination became apparent. Not everyone can do the same job. Someone needs to hunt, someone else needs to gather firewood, and someone else needs to make the tools. If the band is to survive, the jobs must be split up, so all the survival basics are covered. The more efficiently the tasks are divided and performed, the better survival chance of the tribe, and all its members."

"This sounds similar to how entities operate in the ecosystem, Sara."

"Except our friends in the ecosystem don't have choice, Dina."

"How does choice affect coordination?"

"Only in every way possible. Take our little band of collaborators. At first everyone is cooperative and happy to have found a way to ease the survival pressure. After a while, when everyone is a little more comfortable and has a sense of where and how they fit in, some yahoo notices that the members who cook the food are a little pudgier around the middle than everyone else. Maybe the members who hunt prey have more sway when it comes to making decisions. Or the members meet another band that appears to have more food and better housing. Time is a means by which energy flows and change occurs. With time, differences emerge. People have minds. They can observe and ponder things. They can create ideas from these observations. And people are different, so they conjure up a variety of ideas, and those ideas may conflict. The members need to decide which ideas to follow and which ones to reject. That's where our coordinators come in. They are the leaders. The deciders."

"So, survival is no longer the purpose?"

"Hmm. I'm not sure you can say that. Survival is always the purpose, but if survival pressures are eased, people can take survival for granted, and elevate other concerns. One subgroup wants to grow food instead of hunt. Another subgroup wants to take the possessions of other tribes rather than hunt their own. Yet another group wants to please the sun god to attain eternal life. People, being people, listen to these different ideas and start accepting goals beyond mere survival."

"Which goals do they choose?"

"This is the ideasphere. How are any ideas chosen? By acceptance. As long as a goal finds acceptance, it can survive, just like any other idea in the ideasphere. Sooooo, the leaders, the coordinators, would be the ones who get their ideas accepted. If enough members accept their ideas, they are given permission, so to speak, to make the rules for the tribe and enforce them. Add a few generations where new members in the tribe are taught only about the new purpose of society and not about their original purpose of survival, a unique society results. Repeat the process with different tribes and different conditions over many years, and lots of unique societies spring up, each with their own set of goals and rules of behavior and enforcement. Hmm."

"What are you thinking, Sara?"

"Dina, coordination systems in the ideasphere are as much a choice as competition and collaboration. And as with competition or collaboration, it doesn't have to be a good idea or a bad idea, it merely needs to be accepted."

"So, people can coordinate their societies for any number of reasons. This is quite different from the ecosystem where survival is the only goal."

"Yes, that's true."

"What if the members don't accept the purpose?"

She sighed. "The story of our lives. War. Revolution. Unrest."

"Is it always a violent response?"

"No, not always. People form groups, collaborate if you will, to promote their ideas. They write books and pamphlets, or these days, post their ideas on social media. They try to convince people of a particular viewpoint. Sometimes people split off and become a new group with a different coordinating goal. Sometimes they set up a sub-society within the larger society and do their best to survive. People resort to violence a lot. Human history is rife with examples of war, bloodshed, and cruelty. I don't think animals in the ecosystem could treat members of their own species as badly as we have during our brief existence as the rulers of the ideasphere."

"Is this not the way of the ecosystem, Sara?"

"Not the way we do it. You have made the point numerous times. Entities in the ecosystem try to avoid actual confrontation. Creatures die because they don't have the skills to obtain resources or the opportunity to procreate. They don't die because their sibling murders them."

"So, if people learn how to murder, and as a result, obtain more resources and increase their chance of survival, wouldn't you say murder is a survival skill?"

She gasped. "What? Well, I suppose. For the killer. Not so much for the killee."

"But the killee, as you say, is dead. If an entity in the ecosystem does not survive a confrontation, they are not the best survivors. They deserve to lose."

Sara scowled. "I'm onto your tactics, Dina, and I don't like them one bit. We have spent this whole time talking about people and the ideasphere and choice, and now you throw the survival of the strongest argument at me? Puh-lease."

"You do not agree with me?"

"No. I. Do. Not."

"Why?"

"Hello. Superpower. Choice. Remember? We can think. We can choose. We can just as easily figure out how to live together in peace as kill each other. We have the power. We are not tied to some instinctual instruction that forces us to kill each other. We can work together to enhance survivability as easily as kill each other. But throughout our history we have often chosen to kill and dominate and enslave rather than work together."

"Why do you think people make these choices?"

"Probably because it's easier. Maybe our all too human egos get in the way. Money. Resources. Name the reason, Dina. If people can think it up, they can act on it. All they need to do is convince others to accept their ideas, good or bad. If they persuade enough lemmings, I mean followers, to plunder and pillage and kill, then they will plunder and pillage and kill, rather than share and communicate and live in peace."

"Ah, you have answered my question. People can choose the reasons for living together. The people who have the most acceptable ideas for living together become the coordinators."

"For good or ill. Yes."

'In the ecosystem, cells cannot reject the goal of survival. It is embedded in their DNA instructions. Occasionally, cells develop mutations that threaten an organ or a body. Do you recall how a neural system responds to such an event, Sara?"

"The body has a set of tools it uses to re-orient the cell to perform its function."

"Correct. The neural system does what it can to return the cell to its normal behavior. If the efforts don't work, the neural system directs its army cells to destroy the rogue cell. Do people in the ideasphere handle rejection in this way?"

"Unfortunately, yes. That has happened often in our history."

"Unfortunately? Why is this unfortunate?"

"Because a lot of people get harmed or killed."

"Isn't this the way of nature, Sara? The neurons direct a cell to be destroyed if it cannot be rehabilitated."

She paused for a moment. "That's the thing, Dina. It doesn't have to be this way. We're not *just* a pile of DNA." She stopped and let out a long slow smile. "Of course. That's it! It doesn't *have* to be. We are people. We can choose. We can choose to value each member of society. Or we can choose to devalue and disagree and kill them. Or do anything in between. There is no invisible force driving us. We drive ourselves. We are free to choose. We set the purpose for how we live. *We* do, Dina. At least in the ideasphere. As humans in the ecosystem, of course we are subject to all the rules of the ecosystem. We, as bodies, benefit from the strict rules that keep our cells in working order. But in the ideasphere, where ideas don't have any physical structure, we can *choose* how we survive. We can choose how we compete and collaborate *and* coordinate. We do. People do. It's our superpower."

"And if your leaders do not 'value each member of society?"

"Then we get what we have always gotten throughout our history. A big fat chaotic mixture of collaboration and competition and coordination that doesn't work very well--"

"What is it, Sara?"

"Dina, I see another correlation to competition and collaboration in the ideasphere."

"What is it?"

"It's the same problem, actually. If leaders are successful, they can amass a lot of resources from their actions, right? In the ideasphere, if the leaders of successful groups, whether they are in institutions or governments, obtain a bigger share of the resources, problems can occur."

"This does not happen in the ecosystem, Sara. It is true the brain uses twenty percent of a body's energy consumption while representing only three percent its mass. However, a brain does not take more than it needs. Neurons must consume this enormous amount of energy to work nonstop to process information."

"Yes. That's the glaring difference between our two systems, Dina. Nature does not concentrate resources the way humans do. Any time resources are concentrated in one place, including in government, the smooth operation of the mechanism is put at risk. Coordination mechanisms are no different. If leaders of a government control vast amounts of resources, they have vast amounts of power. With excess power comes the ability to change the rules to keep themselves in power, and, of course, keep the resources flowing to them. This is how governments have worked for a long time. The leaders who organized their members and obtained the most resources, won. In our early history where physical size and strength mattered, the strongest would often win. They became kings or emperors or warlords. Once they obtained the resources, they used them to cement their roles as leaders. They changed rules to benefit them, and they changed how to enforce them. They demanded fealty. They promoted their defenders with additional resources. They exiled or killed their opposition. This works just like human competition or collaboration. Game the system to keep the resources coming. In this case, the coordinators have a say *over* the competitors and collaborators, too. They make the rules everyone must obey. Ooh, now I understand why people are always fighting to be the leader. It's not because they are altruistic or care about the welfare of those they lead; they want the resources."

"Does this situation exist today, Sara?"

"Yes and no. People hold on to old ideas for a long time. Some countries are still led by monarchies and strongmen. Over the years, as people have thought more about the human condition, they have developed new ideas about how we might live together. Those older autocratic methods have been joined by democracy and communism, as well as some hybrid forms."

The pink haze around her faded. Everything turned black. A stabbing pain shot through her arm. "Ouch!"

"Are you all right, Sara? What is happening?" Dina asked.

The pink haze returned, and the blackness receded. "I don't know. My arm . . . It felt like I was being jabbed by a thousand needles all at once, and everything went dark for a second. It's all back to normal now. I'm okay."

"We must hurry. I do not believe we have much time."

Chapter 29

IMAGINARY COUNTRY

"Let us conclude our discussion by exploring the characteristics of coordination systems. I want to do this a different way. I want to make sure you are comfortable drawing conclusions from the ideas we have discussed. In case I need to leave quickly and cannot continue our discussion."

"No, Dina. I don't want to. I don't want you to go."

"We cannot control all the conditions in which we operate, Sara. The best we can do is plan for likely events outside our control. I think my approach is the proper one at this time."

Sara sighed. "All right."

"I will state a characteristic of the ecosystem, and I want you to explain how this same characteristic works in the ideasphere."

"Oh, this will be fun. Like a word association game."

"Similar, but perhaps with an explanation involving more than one word."

"Ha! Fair enough."

"The first characteristic: accountability. As you know in the ecosystem, every cell reports its status to the neural system. The neural system processes this information and sends a response. If a cell is not behaving as required, the neural system issues a response to correct the behavior. Do coordinating systems have this same type of accountability system in the ideasphere?"

"You mean do we file a behavior tax return with the government? Ha! That would be something. Dear Government: Today I brushed my teeth and took a shower! Woohoo!"

"You are not answering the question, Sara."

"I am answering the question. No, we don't do things this way. We make laws and when someone breaks them, we go after them and enforce the rules after the fact."

"So, people are only accountable if they are caught breaking the rules."

"Pretty much."

"How would a body function if a neural system operated this way?"

She stopped. "Well, all the bodies that tried this method lost the survival game, didn't they, Dina?"

"Indeed."

"But Dina, people don't need to be as tightly coordinated as bodies. We live in the ideasphere."

"How tight do they need to be? Tighter than they are now?"

"Probably, but good luck trying to figure out how."

"A moment ago, you jested about reporting your hygiene habits to the government. Are there any actions that individuals or institutions should report to the government?"

"Hmm. Well, if there are, they would have to be pretty darn important."

"What do you mean by important?"

"Well, they would have to impact everyone, or be significant enough that everyone ought to be aware of them."

"And how is significance determined in a society, Sara?"

"Hmm. By our goals."

"Correct. So, let us consider an example. Assume one of the goals of an imaginary country is to 'promote the general welfare.'"

"Huh? This imaginary country sounds quite familiar. Does it rhyme with Schunited Schtates of Schamerica?"

"Ignoring, Sara. Our imaginary country's government decides water pollution is a threat to the general welfare. Would it be useful for the government to ask every institution of a certain size to keep track of their water pollution activities and report them on a regular basis?"

"Oh. Ohhhhh. The reporting needs to be tied to a goal. The goal of general welfare is threatened by water pollution. The disclosure law to report about water pollution is the way the government locates the threat. It is made to carry out the country's goal."

"Yes, you understand my point. Since cells have no ability to discern the quality of their actions, they must send a full report of their status to the neural system."

"Hmmm. Since people and institutions *are* discerning, they wouldn't need to send a full report of everything they do, only those things the government needs to know to keep people safe."

"That is the idea. Now, let us return to our original question. How would such reporting affect accountability in the ideasphere?"

"Well, in the ecosystem, the neural system would send cells to help the cell polluters stop polluting. Hmm. So, if the government receives a report that a company is polluting, they can send in people to help the company stop polluting. I see. You prevent the whole pollute, hide, find, punish, fine, system. It's way more efficient."

"What about accountability?"

"Hmm, this method emphasizes accountability instead of punishment. If the company knows it won't be punished for its infraction but instead will receive help to resolve the problem, there is a much better chance it will accept the help and won't continue to pollute. That's a win-win. It stops doing bad actions that hurt people. The country gets cleaner water. The company ends up acting in a way that does not harm people. No punishment is needed. Interesting."

"Yes, Sara. It is the act of disclosure that sets up efficient accountability. In a neural system, disclosure leads to the most efficient response. Likewise, in the ideasphere, disclosure laws would allow the government to reach its goals in a more efficient way."

"Hmm. This is something we don't do at all."

"Quite true, but something neural systems do extremely well. Perhaps people can learn from this disparity."

"I need to think more about this."

"Let us consider our second characteristic of coordination. Nourishment. In the ecosystem, every cell is nourished. If a body starves, all the cells share in giving up their fat and muscle stores except for the brain, which holds onto its neurons for as long as possible. How are members of a society nourished in the ideasphere, Sara?"

"It varies. Governments differ in what resources they provide to people. In our country, it provides some services: education; food, medical and monetary assistance for low-income people; income for the elderly. These are some examples. For the most part, people choose their own paths."

"So, all members are not nourished as they need?"

"No. We have what we call wealth inequality in our country. Nearly 80% of the wealth in our country is held by 10% of our population. The remaining 90% owns the other 20%. At the bottom of the scale, severe poverty exists. Dina, this is what it looks like when excess resources game the system. And the situation has worsened in recent years."

"A neural system does not permit this to happen. Why do you think people do so in the ideasphere?"

"A lot of reasons. Mainly, our citizens take pride in having the freedom to make their own way in life. Since people can choose to work to obtain resources, it is their responsibility to do so. If they don't, it's because they are stupid and lazy and deserve to suffer."

"Do you believe this?"

"No. Especially after our discussions."

"What do you mean?"

"Well, our ideasphere is built on top of our ecosystem, right?"

"Correct."

"People need to succeed in the ideasphere to obtain resources and survive."

"Also correct."

"Every person starts at zero in the ideasphere. How they survive their first few years is critical to how they will function in the ideasphere. Who bears children? Young people. Who possesses the fewest resources when they start out on their own? Young people. That's a bad combination if you want everyone to succeed in the ideasphere. You can predict what happens. Those families starting with a lot of resources do well. These children are likely to earn even more as they grow into adulthood. Families starting with few resources must contend not just with the ideasphere, but also with physical survival in the ecosystem. They have a much tougher time achieving success. If we don't nourish them well enough to succeed in the ideasphere, we all lose."

"How, Sara?"

"We lose the ideas all these people would have created but didn't."

"What do you suggest government ought to do in this situation?"

"I'm not sure. I need to think about it. You asked about nourishment. I didn't connect the idea to income or wealth inequality until now."

"I have another question about nourishment."

"That one wasn't enough? Holy Cow. All right, hit me with it."

"In the ecosystem, food is ingested through the mouth, processed in the stomach, and sent to all the other cells in the body via the bloodstream. How are resources processed in the ideasphere?"

"Hmm. Not quite the same way. Each person is responsible for obtaining their own resources. They get a job and live off the wages, or they can open a business and live off their profits."

"You said the government provides some assistance. How does this work?"

"Well, the government levies taxes so it can take care of the common needs of society."

"Common needs?"

"Those are things we all use. Like roads and armies and national parks."

"You mentioned the government also provides free education and money to low-income people and the elderly. Are these common needs?"

"Well, not exactly. They are priorities. We think education for all people is important, so government provides it. We think it is important low-income families have resources to make survival easier, so government provides those funds. We think elderly folks who can't work anymore should have a stable income in their later years."

"How are funds provided to these people?"

"By taxes. We all pay taxes, and the government applies them to where they are needed according to their budget priorities."

"So, would you say taxes are like nourishment?"

"Hmm. Yes. They are resources being distributed according to the country's goals." Sara stopped. "Dina, I'm confused. A body's goals and rule are built in, right?"

"Yes."

"And all the cells are nourished because they all follow the DNA's instructions.

"Yes again."

"So, nourishment for all is part of the coordination function. It's a total distribution of all resources. 100% taxation."

"Technically, in a body there is 100% taxation on the ingesting organ, and 0% taxation on the rest of the organs. There is no taxation outside of a body or colony."

"Oh. Ohhhhh."

"As there is no collaboration above the entity or colony level, there can be no coordination and therefore, no taxation. Each entity is responsible for obtaining its own resources. Entities take whatever resources they need from the ecosystem. The ideasphere does not work this way."

"Why?"

"People have created a distinction between what you call public and private property. This affects how resources are distributed in the ideasphere."

"I don't understand."

"You alluded to it moments ago, Sara. Common needs. Common properties. In the ideasphere, people can own land or a building or a piece of clothing. Private ownership does not exist in the ecosystem."

"Ooh, I see. I never see squirrels yapping over who gets to hide in a tree. They don't pull out the tree deed and argue over who has a right to the best oak branch. If the dog is coming, they all clamber up the tree as fast as possible."

"I am sure this is true, although your argument is not useful in any way."

"It's a funny picture."

"Humor lies in the eye of the beholder, Sara. Private ownership is an idea conjured up by people to organize themselves and be able to distinguish their resources from those of other people. Once people created the concept of private property, the concept of common property was forced to follow. One cannot exist without the other. Common property is property that members of a country share. Since everyone owns common property, everyone must contribute to its upkeep. Hence the need for taxation. Since entities in the ecosystem have no private property and everyone is free to take what they need to survive wherever they can find it, there is no need for taxation."

"So, private property is not some freedom thing some folks rave about. It's an arbitrary idea created by collaborating people. Irony alert. Wherrp! Wherrp!"

"We've been steady for ten minutes," Tina said to Dr. Chen. "How do you want to proceed?"

"Let's move to fifty, then hold again."

Tina adjusted the flow. "At fifty."

The team stood quietly. Waiting.

"We have one last topic to discuss, Sara. Self-interest. In the ecosystem, self-interest, or self-benefit, is action based on accumulated experience an entity carries in its DNA. When cells collaborate in a body, or when individuals collaborate in a colony, all the elements function to maximize their individual survival as well as the survival of the entity. You have referred to this before: you called it the long gun trio theory."

Sara laughed. "The three musketeers! All for one and one for all! You crack me up, Dina."

"The people in the ideasphere do not work this same way, do they, Sara?"

"No. No way. Our superpower of choice kills that concept dead."

"Your musket men were people, Sara."

"Yep. Fictional people. They made an impact on readers because they didn't act normally. Dina, people do the self-interest part, but they don't do the 'all for' part. We have choice. We don't help each other unless we choose to."

"Why don't you choose to?"

"We aren't as connected to our institutions or governments as cells are to their body. Cells in a body share the same goals and rules. People start with no goals or rules at all. If we want people to be more aligned, to act better towards one another, we need to do a much better job at educating people about the goals and rules of our society or country. But we don't do that. We educate people in a variety of ways, but not many of them focus on civic rules."

"Why?"

"To be honest, I don't know. It seems like a no-brainer to me."

"This is an action that would require brain use, Sara. Am I not correct?"

She laughed. "Yup, right again. It's another saying. It means it's so obvious you don't even need to use your brain to know it."

"That would be impossible, Sara."

"I can't argue with you there, Dina."

"You make a point that people have not learned about the goals and rules of your country. Let us talk about them."

"The goals? In our country? In the United States?"

"Yes. Tell me how your country's coordination system works."

"Well, it's a representative democracy."

"What does this mean?"

"It means the people essentially rule themselves. We, as a people, choose our leaders."

"And the leaders set the goals?"

"No. The Constitution sets the goals." She stopped. "Oh my gosh."

"What is it, Sara?"

"We have a constitution that sets our goals."

"Why is this important?"

"Don't you see, Dina? The Constitution fixes our goals. It makes them as firm and stable as the goal of survival in a body. This makes it easier for the leaders we choose to make laws. The laws must always comply with the Constitution." She stopped. "Only they don't always."

"Does your representative democracy work like a neural system of a body?"

"It's close. Really close. Hmm. The Founders set up a solid system, Dina. It works a lot like the ecosystem, but it accounts for how humans are different. Ooh, now that I think of it, the Founders did a really good job."

"Please clarify, Sara."

"Well, the Constitution sets the goals, right? It also lays out the structure for making and enforcing the rules. The Legislative branch sets the rules that, in theory, support the goals. The Executive branch puts the laws into action. The Judicial branch has a dual role; it determines if the laws support the constitution, and if a citizen's actions comply with the laws. A body in the ecosystem only needs the executive system. The body does not need a Legislative Branch or a Judicial system, because the goals and rules are embedded in the DNA, and they are 'constitutional' by definition. People need those branches because we aren't born with the rules embedded in our minds." Sara fell quiet.

"What are you thinking? What is bothering you?"

"Dina, it's a good solid system, but I think we've veered off track. The structure fits humans, but the actions of the branches don't always match the Constitution's goals."

"Please explain."

"Our Founders set the goals, but we sometimes make laws that don't support them. Not always, but sometimes. We have 'justice' as a goal, but we make rules that put people in jail for many years for minor infractions, while others are not held accountable for egregious violations."

"So, your laws are not fair?"

"Sometimes, yes, and sometimes no. It's more that the laws don't always reflect the goals. Sometimes laws try to modify behavior the coordinators don't like, rather than focus on behavior that violates the goals of the common good. It's as if our leaders are confused about what the goal means. Or worse, leaders may define a goal so as to achieve the outcome they want rather than an outcome that benefits the common good."

"Would you give me an example, please. I am not sure I understand."

"Slavery. Slavery is about the worst thing one person can do to another. Every single person born on this Earth comes from a long line of survivors extending to the beginning of time, right? All humans are born with the power to hold ideas and make choices. This is what makes a human a person in the ideasphere. Slavery takes the personhood away from a human. No choice. No freedom to think their own ideas. No chance to live as a person. It is the most reprehensible thing one person can do to another short of murder. Yet slavery is accepted in our Constitution, the very document that aspires to secure the blessings of liberty and justice for all. The neural system in a body would never permit such a contradiction, but people do. Dina, we the people can set a goal, establish a rule to match the goal, and then enforce it, but sometimes, we just don't do it. As a result, we can cause harm and damage and destruction."

"That is why I am here, Sara."

"Why? To witness our self-induced self-destruction?"

"No, no, no. I have lived too long and seen too much destruction to want to witness anymore."

"Then why are you here? Why did you spend all this time teaching me all these amazing things? Now that I'm beginning to see how it all works, I can see where we have deviated from how things should work. I am not sure we can change our trajectory."

"We need you to try, Sara. We are at a critical point in the evolution of life. Your species has created this complex and unique ideasphere and implemented many beneficial ideas. Your species has also planted the seeds of your own destruction. And when you destroy yourselves, you will take a significant portion of the ecosystem with you."

She stopped. "Oh, so that's it. You want to survive. You're here to protect your own self-interest."

"That would be a natural response for any entity in the ecosystem, Sara, but that is not quite the point I am trying to convey. This Earth is well on a path to becoming uninhabitable to humans. I am not worried about life. Life will endure. Some entities will survive the era of humans. They will multiply and grow on an Earth irrevocably changed and damaged by your species. Life is not my concern."

"Then what is?"

"I am concerned we will lose you, Sara."

"Me?"

"You and every other person on this Earth. Do you not understand?"

"Understand what? No, I'm not following."

"People are the only occupants of the ideasphere, Sara. You are the only ones that can think, the only ones that can act on an idea and create structures outside your bodies, whether they be books or beliefs or spaceships. You are the only ones, Sara. The only ones. If we lose you, we put the entire evolution of life that now contains an ideasphere, at risk. If we lose you, we lose the ideasphere too. Before humans existed, countless entities lived and died, and in doing so, altered the flow of history. Humans are unique. You are the first to have created an entirely new ecosystem. Yet you have also acted in a way that places you at the precipice of destruction. You have the ability to change your behavior so you can survive and even thrive. I am here to help you thrive. We want humanity to thrive."

"But you're afraid."

"Yes, I am afraid."

"That we can't change?"

"No, Sara. That you won't."

"Are we still steady?" Dr. Chen asked.

"Yes," Tina said as she reviewed the monitors. The team members nodded.

"All right. Let's go to fifty-five."

Suddenly Sara's body convulsed. Her eyes opened wide. "Dina? Dina?"

"We've got her!" Tina said. "She's awake."

"Dina, where are you?" Sara screamed. She tried to lift herself from the bed. "Dina! Come back! I need you to show me how! Please, help me." Sara collapsed back on the bed, her eyes closed, her face twisting in pain.

"What's happening?" Dr. Chen asked.

"Oh crap. BP's dropping."

"Pulse too."

The green screen displaying Sara's heart rate showed a flat horizontal line. "Cardiac arrest. Cardiac arrest," a nurse yelled as she started CPR.

"Get the pads. Hurry." Dr. Chen ordered. "Tina, pull the IV. Now."

A nurse pulled a defibrillator toward the table and lifted the pads. "On your order, doctor."

"Go. Go."

"Clear." The jolt lifted Sara's body from the bed. Every team member stared at the display. The line was still flat.

"Again."

"Clear." Another jolt. The team peered at the monitors.

"We've got a pulse." Slowly the line formed into a wave across the monitor. "She's back," Tina said.

"Alive." Dr. Chen sighed. She studied Sara closely. "And back in a coma. Damn it."

Chapter 30

MAD

"Hello, Sara."

"Dina! You're back."

"Are you all right?"

"No."

"Are you in pain?"

"No. I'm mad."

"Does this mean you are insane or angry?"

"Ha! Both. How do you do that, Dina? Make me smile when I'm furious?"

"I have learned about you, Sara."

"Then you know I want some answers."

"Yes, and you deserve them."

"This isn't a dream, is it?"

"No. You are in a coma. You have been for many weeks, now."

"A coma? For weeks? What are you talking about?"

"Since I visited you in the garden. It is the only way, I . . . we . . . could communicate safely with you."

"The garden?" She was quiet for a moment. "The mist. The gold mist. That was . . . is . . . you?"

"Yes, except now it is a pink hue. It turned out your doctors detected our presence by the gold color. We needed to change it to match your skin tone."

"Oh. So that's why the color changed. How did you do it? Change colors?"

"Sara, that is the least important of all the questions you could ask me right now."

"Ha! That's probably true. So, how about this one? Who are you, Dina? Or maybe I should ask, what are you?"

"I am not one thing. There are many of us. We . . . together, we are the entity you call 'Dina.' We. . . let us say. . . we collaborate."

"Collaborate to do what? Keep me comatose? Explain, Dina. You're good at explaining things so I can understand. Please."

"Can you not guess who I am? What I am? After all our discussions?"

'I'm not interested in playing any more games, Dina. I'm shocked to think I've been away from my family for weeks. What about Randy and Josh? They must be worried sick about me. Tell me they're okay. Tell me who you are and why you needed to coma-nap me?"

"Coma-nap is not a real thing, Sara."

"Are you sure? It sounds like exactly what you did to me."

"All right. I will explain. As I mentioned, I . . . we. . . have never attempted the technique we tried with you. Like all entities on this planet, we have accumulated experience with humans for many years. It is only recently we derived the ability to make direct contact."

"Who is this 'we' you speak of?"

"Can you guess?"

"No. I can't see you. I can feel you, I can communicate with you, but you're invisible. You're either a figment of my imagination or something so small I can't see you."

'I thought by now you would have guessed, Sara. I thought I released the feline from the sack when I told you my name."

"Released the fel. . . Ha! Let the cat out of the bag? Dina, you're a hoot! What do you mean?"

"What is my name, Sara?"

"Dina."

Dina was silent for a long moment. "Think, Sara. Dina. Does my name remind you of anything we talked about?"

"Hmm. Dina. Dina. D-I-N-A. That's A-N-I-D backwards. Deee-na. Die-na." She stopped, and a slow smile spread across her face. "Oh, crap. I would have figured it out earlier if you called yourself Rina. But no. . . you went all Cinderella on me with a normal human name. You are Dina without the I . . . DNA."

"Yes, that is how I derived the label you use for me."

"So, what are you? What kind of DNA critter are you?"

"I chose this name for a specific reason, Sara. We descended from the earliest line of DNA."

"W-w-w-wait a muckity muck minute. Aren't we all derived from the earliest line of DNA? DNA has never died out, right? So, we're all related."

"Yes, but we branched away early from the line leading to eukaryotes and humans."

"So, you aren't a eukaryote? Are you a . . . prokaryote?"

"Yes."

"Ooh, you're one of those little guys? One of the first ones? That doesn't make sense; you can't be prokaryotes. You can't be a 'we.' Prokaryotes only come in single-cell flavor."

"We are unique among our kind, Sara. We branched off quite early from our original ancestor. While other prokaryotes evolved into eukaryotes, we evolved to exist a different way. We learned how to join and unjoin under a variety of circumstances. That is our superpower, Sara. We live everywhere, and we can join together within different environments, or unjoin, as we need."

"As you need? What the heck does that mean?"

"You are right to question, this, Sara. Generally, it means nothing. We unite for different reasons: to eat, to share resources, to share functionality. Our superpower is the ability to switch between these states based on an intricate combination of conditions. Those of us who have lived in human bodies for many years have developed a library, so to speak, of different conditions that enable us to maneuver around bodies."

"Why?"

"Why does an entity do anything, Sara? To survive. Over the past century, we have been experiencing a troubling change in the environment in which we live."

"What kind of change?"

All things change, Sara. You know this. Because we are so small, we sense very minute changes in environmental conditions. Over the past century, not only those of us who reside within human bodies, but those who reside in the soil and water, have sensed changes. We have responded by joining together in more unusual ways. It is only recently we have learned how to pass between our sub-environments."

"I don't understand."

"The gold mist. These are our . . . relatives. Our relatives from the air and water and soil. We have found a way to join together within a body and . . ."

"And put me in a coma and talk to me?"

"Yes."

"How does that work?"

"Like any new effort made by entities in the ecosystem, erratic at first. You are not our first attempt. We have encountered obstacles before. In your case, we have performed better than expected."

"I thought you said I was the first."

"You are the first human with whom we tried to directly communicate, but we have been refining our techniques for many years now."

"How did those previous efforts turn out?"

"Are you asking if we have harmed or killed any of our human 'friends?'"

"That's what I'm asking."

"No. But this is not to say there have not been risks; most of all with you. We have learned more and more with each effort."

"So, what you're saying is you have learned how to put people in comas."

"A necessary side effect in our work with you. We have inhabited your neural system, your brain, to be more precise. We communicate by using your ability to make memories."

"W-w-wait a magical mysterious minute. You messed with my memories?"

"Actually, we manipulated your neurons, along with the synapses upon which they travel. To do this takes a significant amount of energy. We needed to lower the rest of your body's energy use while we proceeded."

"How the hell did you do that?"

"It was quite simple, Sara. We put you into a deep state of sleep."

"Oh. A coma."

"It was the most effective way to communicate our ideas without harming you. Still, we put your safety at risk, Sara."

"Was that what that was back there? You were putting my safety at risk?"

"Yes. Your doctors have been trying to awaken you. Like all creatures, they accumulate experience. We have been doing our best to confuse them."

"What do you mean?"

"I told you we can join and unjoin in a variety of ways, Sara. This is what some of our team has been doing. They joined in ways to exhibit certain unexplainable symptoms in your body."

"Hmm. Then you let those super smart doctors use their medical knowledge to try to noodle them out. You sent them down one rabbit hole after another, didn't you, Dina? They couldn't make sense of all the differing clues."

"I am not sure what dens of small, swift, fur-covered, highly procreative animals have to do with this discussion, Sara, but yes. The strategy worked quite well, but your doctors learned in a way we did not anticipate."

"Oh?"

"They realized no reasonable explanation existed for the variety of symptoms you displayed, so they finally chose to ignore them. Once they discounted the symptoms, they opted for the obvious, and I might say, correct, solution; to awaken you."

"So, that's what happened. They woke me up. But Dina, it lasted only a second. All I remember is a blur of light."

"Yes. We took an action dangerous to your well-being to make them stop."

"What did you do?"

"We stopped your heart."

"What? You gave me a heart attack? How?"

"I told you, Sara. We can join in a variety of ways. The effort to awaken you would have succeeded. We would have been forced to leave the area when more neurotransmitters entered your brain. So, we pre-empted the effort. We traveled to your aorta and blocked it. As soon as we left the brain, you began to wake up."

"Only long enough for you to reach my aorta and gave me a heart attack."

"You were not in any actual danger of death or injury, Sara. As soon as the doctors began to treat you, we returned to the brain."

"And put me back in a coma."

"Yes, that is correct."

They were both silent for a long moment. "Now what, Dina?"

"We have considered all the options, Sara. We have decided to end our efforts, for now."

"End them? Where will you go? What will you do?"

"We must first reverse the process we have put in place. It must be done slowly as not to damage your memory capacity. In the meantime, we will remove ourselves from your hypothalamus. You will awaken and resume a conscious state."

"What about . . .everything we have talked about? Humans. The ideasphere. The danger to the Earth..."

"We must leave this to you now, Sara. We have provided you with all the information we can give you. It is our hope you use it to benefit your species . . . and the world."

"How am I supposed to do that? I don't know the first thing about changing the world-"

"You know more than you think, Sara. Some of our team will stay in your synapses to help in the transition. They will help you retain these memories for some time but will be unable to communicate with you. It would take too much of your energy to do both; it could be dangerous to your well-being. I will try to return later, if only to say farewell."

"Dina. I don't want you to go. Isn't there another way? I – I – I've grown to depend on you."

"You will have Randy and Josh and Nina to depend on. And a new grandchild to meet. Many things will occupy you, Sara."

"A new grandchild? What? What are you talking about?"

"You are going to be a grandmother, Sara. Josh and Nina are expecting a baby girl. They plan to name her Maggy."

"A girl! Ooh, that's fantastic. Tell me more, Dina. How did you find out? When is the baby due? How is Nina? How is Josh? Oh my gosh, I bet he's over the moon!"

There was silence all around her. "Dina? Dina?"

Everything went black. She felt it in her feet first. The strange hot tingle. It crawled up her legs, then to her stomach and arms. She took a deep breath. The scent of lilacs was fading . . . fading . . . fading. The fairy tale forest dissolved.

She opened her eyes. Her husband was looking into her eyes, sadness evident in his furrowed brow. She smiled.

"Hello, handsome."

Chapter 31

SYNTHESIS

Dear Dina,

It's been a few months since I woke up, sweet friend. I miss you. I have been struggling to come to terms with what happened to me when I was 'asleep.' I've explained it all to Randy and Josh. In fact, Josh has recorded all our conversations so we would not lose any of the ideas we talked about. Still, I'm not sure they understand. Hell, I don't even think I make much sense when I hear myself try to explain. I give them credit, though; they've listened and kept an open mind. They're so kind. I can't blame them for not knowing quite how to respond to my story. Half the time I'm convinced it was a fantastical dream. The other half? I'm not sure I could say any of these things to anyone but Randy and Josh without receiving some serious eye-rolling.

Never mind all that. I'm writing to you because I'm afraid I'll never see you again, and it makes me sad. I'm not sure I can explain why I feel so deeply about a bunch of little critters I can't even see. I'm human, so it must be an idea I conjured up. I like the idea that I like you, if that even makes sense.

Also, to be honest, I'm a little afraid. I have this nagging sense I'm never going to feel like myself again. It sounds silly, I know. The doctors say I'm fine, but I'm so tired all the time. I swear, I sleep at least sixteen hours a day. My dreams are a hazy jumble of things we talked about and my life in the real world. I feel like there's a piece of me in each place.

I understand what you want me to do, Dina. I've been using the time when I'm awake not only to explain as much as I can to Randy and Josh, but also to jot down some notes about how these ideas might apply to our world today. It's harder than I thought.

I decided to break the job into two parts. First, I need to consider how the ideasphere would work if it worked like the ecosystem. You'll love this: I call it the 'idealsphere.' Ha! It's the best I can do. The term 'ideal' usually means 'perfect,' but in this case, I define it as the 'most suitable,' or 'most fitting.' I tried to answer this question: What would the ideasphere be like if it operated like the ecosystem where all beings strive to survive by doing what they do best?

Second, I need to think about the world as it is today and consider how we can move closer to 'most suitable' or even 'more suitable.' Gosh, Dina, it's kind of amazing. Now that I understand the concept of the ideasphere, it is much easier to identify areas where we diverge from the ecosystem. What's more amazing, is how clear the solutions reveal themselves. I realize the devil is in the details, and implementing solutions can be difficult, but it is much easier when I can use the ideas from the ecosystem as the base from which to start. I haven't begun this second part yet, but the ideas are running around in my head.

I know you're here, Dina, deep inside my brain. I can sense your presence when I think on these things. I hope you'll come back soon. I want to discuss what I've written. I'm afraid without your objective influence, my bad puns and biases may overrun my good sense. I'm confident you'll set me straight. Until then,

Sara

Notes on the "Idealsphere"

To understand the ideasphere, we need to first understand the ecosystem. The ecosystem has existed for billions of years. Every single non-human entity survives in the ecosystem by complying with rules embedded in their DNA and the conditions of their environment. That's all they do. They just live every day. They eat, they procreate, they die. Over and over and over. Together, these entities tell the story of life. Not one of them cares about the story; not one little bit. None of them care about whether they communicate or collaborate or coordinate. They do only what they have evolved to do. Survive.

Humans have changed the game of life. We can think. We can ask questions. We can choose how to act. We can store these choices in forms outside our bodies. Our unique capabilities have ushered in a brand-new ecosystem, the ideasphere. Fundamental to us is the question: what is the story of life? For all of history, creatures lived life without this question ever arising. Now we are here, and with us comes thought, and with thought comes questions. It has been the lifelong quest of people to answer this question about life's story. Here's the ironic thing: we've answered it. At least mostly answered it. We just haven't realized it yet.

To understand the story of life, we must reach backward and put together the pieces of what happened before we arrived. Look around. The clues to the answer are everywhere. Astronomers have unlocked information about the universe going all the way back to the Big Bang. Chemists have discovered the secrets of how matter works. Physicists have discovered forces and have looked deep inside atoms. Biologists have learned how plants and animals operate, and how life forms use DNA and RNA and a myriad of proteins to live.

Now, I am not saying we understand everything. Nope. This universe is too ginormous and our abilities too limited to have perfect knowledge. Yay! Full employment for scientists forever! There will always be more to learn.

Limited knowledge, however, doesn't mean we don't know enough to make sense of this world. We can piece together the story. We can not only understand the story of life, but we can also use it to enhance human existence. When we reach backward to uncover the secret of life, we realize it's not a secret at all; the answer is all around us, in plain sight.

People throughout history have done a remarkable job of generating the individual answers to our fundamental question. It's time we put them together, to think of this world as a single unit. We are the first beings able to fathom not just the question of life but the answer as well. So, let's do it. Let's act as if our ideasphere is an extension of the ecosystem, because . . . it is. Let's discover what our ideasphere would be like if we followed in the footsteps of the creatures who have preceded us for billions of years. Let us see what the 'idealsphere' could be.

All activity in the ecosystem begins and ends with cells. Cells do the real living in the ecosystem. Complex bodies are just big, complicated, efficient packages of cells that go along for the ride. Bodies live because cells live. Cells only have one goal and that is to survive. They have transferred this goal to their complex organs and bodies by virtue of sharing the same DNA. A complex body, therefore, exists to enhance the survival chance of all its cells. By doing so, the body also enhances its own chance for survival.

The body is a complicated symphony. Cells are made of matter that is a unique form of energy, and energy is always moving. Always. So, each cell is constantly active. It processes resources it receives from other cells. It performs its functions as instructed by its DNA. And it reports its status up to the brain via the neural system. The neural system, using its accumulated experience of cell collaboration, limits the cells from acting too freely. If a cell drives outside its lane, it can harm other cells or the entire body. The bodies with cells that tried the 'freedom first' method, didn't survive. What exists today in complex entities is what I call 'organized chaos.' The cells in a body are free to act in whatever way their DNA allows them, as long as they don't hurt other cells or the body.

Outside a body, it's a different story. Chaos does rule. Except for ants and bees who live in colonies, all entities must fend for themselves. Some creatures do join in bands or herds for mutual protection, but these are loosely organized at best. At the ecosystem level, each individual unit is responsible for taking care of its own body, and with it, the cells of which the body is composed. As we have mentioned, there is no universal union of living beings that makes rules at this level. At this level, each entity does what it can do to survive.

Now, let's compare this to an 'idealsphere.' As I mentioned, the idealsphere isn't perfect, but is instead a structure that best suits its lifeforms, or, as we call them, people.

In our idealsphere, people are equivalent to cells. People do all the living. Not institutions. Not governments. People. Institutions and societies are just big, complicated, and not necessarily efficient packages of people that go along for the ride. Did you catch the similarity between humans and cells? Bodies exist because cells exist, and societies exist because people exist.

If the goal of cells is to survive, what would be the equivalent goal of people? Trick question. People have the superpower of choice, so they don't have to limit themselves to one goal. This gets more complicated when we add in institutions and societies. Cells transfer their survival goal upwards by sharing the same DNA, but this does not happen in our ideasphere. Institutions and societies can make their own goals separate from those of the people within them.

A complex body is a 'one for all and all for one' Three Musketeers construct. So, imagine if our ideasphere operated the same way. This would mean in our idealsphere, the society would act to enhance the survival chances of all its people. By doing so, the society would also enhance its own chance of survival.

So far, so good. Let's consider the chaos question. If a cell is free to act as long as it doesn't harm the body, shouldn't the same concept apply to people? Of course. Besides, choice is the human superpower. Here's the rub: ideas and choice are chaos makers.

People can conjure up gazillions of ideas and put any number of them into action. That's what people do, and we should let them, as long as they don't harm other people or the society. When you think about it, this is the same rule bodies in the ecosystem live by. The only real difference is the level of chaos. People are far freer than cells in a body, so, human societies should be far more chaotic. That's okay. People are chaos machines. Let them knock themselves out and be as free as they choose, as long as they don't harm others or the greater society in the process.

Now, let's consider the other end of the ecosystem. Complex entities in the ecosystem must fend for themselves. They have almost no ability to communicate and collaborate with other life forms. Some entities have developed mutually beneficial behaviors with other species, but these actions are not performed via communication as much as by long accumulated experience that enhances survival. This leaves competition as the major driving force.

The ideasphere is composed of only one species, humans, and we can all communicate. So, the idea people must only compete in the wider environment is silly. Really. Silly. We can communicate and collaborate. In fact, our ability to create ideas and our superpower of choice makes collaboration the key survival mechanism for people. If we couldn't share ideas and collaborate with others to put them into action, we wouldn't be able to survive long in the ecosystem, much less in the ideasphere. So, it is logical to conclude that mass chaos is not our fate at the top level of the ideasphere. We are all people. We can all communicate. We can all choose. So, we can as easily choose to collaborate as to compete.

This is interesting: at the bottom level of the ideasphere, we are far more chaotic than the ecosystem. But at the top of the ideasphere, we have the potential to be much less chaotic.

So, if I were to describe the underlying goal of the idealsphere, it would be this:

> *People ought to be free to create any idea and convert it to any action they choose, as long as their actions do not harm other people or the society as a whole.*

Chapter 32

THE REAL PURPOSE

Notes on Coordination

Since people are collaborative entities, coordination is the most important of the survival mechanisms. Ha! What am I saying? In every complex body, coordination is the most important survival mechanism. In the ecosystem, goals and rules are embedded in cells. The only job the neural system has is to keep every cell in line, and that one job takes enormous energy.

So, how would coordination be structured in our idealsphere? We now have our goal, so that's one coordination job knocked off the list:

> *People ought to be free to create any idea and convert it to any action they choose, as long as their actions do not harm other people or the society as a whole.*

Human coordination at the society level is what we call government. To determine what government should be like in the idealsphere, it might be helpful to understand the more popular forms of government that exist today.

Monarchies, oligarchies, and aristocracies are types of governments where one or a small number of people rule a society for their own benefit. As with all collaborations, shared benefit exists between the rulers and the people, but the benefit is not usually mutual. Typically, the rulers offer protection from invasion in exchange for loyalty and taxes. The leader sets all the rules, including how the people are allowed to compete or collaborate. Most of the power and resources lie with the leaders. If people comply with the rules, they can do as they please. The only power the people have lies in their ability to unite as a group to demand change or overthrow the leaders, so these leaders often take actions to prevent this from happening. These sorts of coordination structures last as long as the people benefit under the rules, or they do not believe they can successfully challenge the rulers. If individual choice and the ability to seek a suitable life become too difficult, then the people may decide they have nothing to lose in removing the leadership, and they may attempt to overthrow the rulers. The American Revolution is an example of this response.

Communism is a system in which a small group of leaders run a society for the benefit of the state. In this construction, the leadership group creates an entirely collaborative society in which they determine the role of every individual in order to maximize the well-being of the state. This structure is destined for failure, even more so than monarchies and oligopolies. Why? First, this system removes the superpower of humans: choice. Without choice, a person is a human cog in a giant wheel. This goes against the very nature of people, so the rulers must spend a lot of time, energy, and resources in not only keeping their own citizens from making choices, but also in dealing with their unnatural (and not particularly happy) state of non-choice. Also, when you remove the freedom to compete, you also lose the primary impetus for innovation and productivity. A collaborative system can produce complexity and specialization, but without the balance of competition, the government systems will grow bloated and sluggish. It's competition that keeps people sharp.

Communist leaders hold the reins of power with few constraints. What do they do with such power? Do they benefit the state? Well, that's the theory, but not always the practice. Leaders do what leaders have done from time immemorial. They obtain resources and then work to keep them flowing their way. Once they become rich, they have little incentive to make sure the system works for the human cogs down below. The people suffer, but since the people exist to serve the state, their suffering doesn't matter. They often translate the 'well-being of the state' into the 'well-being of the leaders.' With all those resources, communist leaders use the same mechanisms the monarchs and oligarchs use to keep their cogs in line. Threats. Intimidation. Imprisonment. Restriction of news. Torture. Internal spying.

China has developed an interesting twist on communism. China has kept the 'good of the country' goal, but they have injected competition into the economic system when they discovered collaboration alone did not work well and their people were starving. This has proven to be an interesting experiment. The Chinese people have much more freedom, well, economic freedom, than they used to have, but their ability to think and make choices outside their economic boundaries is still limited. I am curious to learn how China's future unfolds. The education necessary to train people to perform complex jobs does not always stay confined to those areas. Ideas flow. Educated people think and sometimes want to act on their ideas more freely. China's populace has two basic choices. Either the people grow tired of the limits on their freedom and push to exert their superpower of choice, or the people remain satisfied with the freedoms they have and do not push on their existing constraints. In other words, how wide is the lane of choice the Chinese people want to have? I don't know the answer to this question. But I do know this view will change over time. People are chaos machines, after all.

Last, we come to democracies and representative republics. These constructs are most like the ideasphere. The goal of democracies is the well-being of their citizenry. In democracies, all citizens have a say in all the coordinating decisions. In republics, the people vote to choose leaders who make the coordinating decisions. The leaders in republics determine the rules that, in turn, determine not only how people can use their superpower of choice, but how competition and collaboration are permitted to operate in the society. Though democracies and republics are best matched to the ecosystem, there is a lot of variation in how they are implemented. Having the goal is not enough. To be an idealsphere, the leaders must derive rules that support the goal. Not all societies do this well, even the United States. Overall, the U.S. Constitution is closely aligned with the goals of the 'idealsphere. Those five stated goals:

'Establish Justice, insure domestic Tranquility, provide for the common defence, promote the general Welfare, and secure the Blessings of Liberty to ourselves and our Posterity. . .'

are about as close to 'Let people do what they want as long as the don't harm each other or the society.' Yet, some of the laws and the way they are enforced have, in fact, harmed our people and hurt our society.

To form our idealsphere, we're going to have to look further than goals. We need to think about how laws should work in the idealsphere.

PROTECTION

In a complex entity, the neural system is isolated from the rest of the body. The cells send information up through a shielded nervous system to a shielded brain. The brain then processes the messages and sends a response for cells back through the protected neural paths to act in a way to benefit the body without harming other cells.

If we devised a similar system in our idealsphere, it would work something like this. The government would be shielded from influence so public servants benefit only by doing their job to serve the people. Any outside resources, i.e., money or influence or pressure, should never be allowed in. Think about it. Imagine if the heart cells sent a message to the brain that said something like this: 'Hey brain-buddy, send a neural response to give me more resources than the other organs and cells. In exchange, I will pump some extra delicious, super rich, oxygenated blood your way. Just imagine all the great thoughts you could produce with premium blood!' That would make an excellent Far Side cartoon. Yep, and that body would be dead. So dead.

It's hard enough for a neural system to coordinate when all the cells have the same goals and rules. Running a government is waaay more difficult given we are chaos-loving, idea-producing, change-laden people. So, we should make the operation of our coordinating system less difficult by removing interference from people and institutions who want the government to operate for their individual benefit rather than the benefit of all. The ideasphere can still work when people and institutions influence it; it just won't work toward the goal of benefitting all citizens. That may be an ideasphere, but it is definitely not an idealsphere.

As governments go, so do the elections that produce its leaders. Of course, cells in a body do not hold elections, but that would be kinda fun to see. Can you imagine little signs popping out all over your skin? Vote for the kidney team, they're renal great! Vote for the skin organs, they're derm good! OK, bad jokes aside, cells don't choose anything. But people do. So, in the idealsphere, influence should be limited to people only. People vote. Only people should be permitted to have a say in elections.

Institutions should not be allowed to play any role in elections. None. Institutions act differently than organs in a body. Institutions are free to seek their goals even when they differ from the goals of society. Institutions should be free to seek those goals in whatever way they want as long as they do no harm to people and the society. Because they have this freedom, they must, by necessity, be separated from the election process just as they must be separated from the governing process. Corporations can and should never be mistaken for people.

In the idealsphere, institutions may use their resources to pursue their own goals, but they may not use their resources to influence the government to act on their behalf. Our current system allows institutions to infect both the governing and election processes. Our system works. It just doesn't work very well. Now we can understand why. Governments need to be protected from self-serving institutions. Only the goals of the Constitution should drive government action.

FEEDBACK

Prohibiting outside influence does not mean government leaders are or should be cut off from their people and institutions. Of course not. Cells constantly communicate with the neural system. Communication channels should always remain open. Communication, not resources. Government in the idealsphere should have a robust system for people and institutions to feed information back to the government regarding the effectiveness of their policies and how well they match the goal of public well-being. Bodies in the ecosystem try to get ahead of problems by getting feedback early and responding before conditions worsen. Regular communication with leaders is necessary in the idealsphere; it just shouldn't come with any demands other than the single demand to comply with the goals as stated in the Constitution.

ROLES

In the ecosystem, the neural system does only one thing: enforce the rules so all cells can co-exist, and the body can survive. In the human body, the brain composes 3% of the body's mass, but uses 20% of the energy a body consumes. Coordination is energy intensive. The situation is even more energy intensive in the ideasphere, where the government must make the rules as well.

A neural system does nothing but coordinate. If our idealsphere looked like a neural system, governments would make laws and enforce them. That's it. The government would not perform any of the functions needed to keep the society rolling, save rule making and enforcement. In the ecosystem, all other actions are the responsibility of the different organs and cells that have perfected their function over eons and eons of collaboration and competition. So, what does this mean? Education, water treatment, urban planning, road building. All kinds of activities our current governments do, should be done by private individuals and institutions in an idealsphere.

What, you say? Blasphemy! These roles are too important to leave to profit seeking industrialists who care only for money. Yep. I totally agree. Let me complete the thought.

Privatizing these roles does not mean the government has no say. On the contrary. When societal goals are being carried out, governments have *all* the say when it comes to making and enforcing the rules to which the private institutions are subject. This is where their efforts should be focused.

Let's use education as an example. Instead of using its resources to do the educating, the government instead sets a series of requirements about the goals of education and the ideas and information to be passed on to children so they can thrive in the ideasphere. These would be comprehensive requirements that result in students being fully introduced to the opportunities available to them in the ideasphere. Then the government would create a comprehensive enforcement system to make sure these goals are met. In addition, the government would create a comprehensive feedback system that would enable teachers, the education experts, to continually provide leaders with information about new and effective teaching techniques. Also, to keep the superpower of choice intact, the choice of school attendance would remain with the student, not the educational institution. Though government funds the cost of education, the student chooses which institution to attend. If more students request a school than positions are available, attendance would be determined by lottery.

What is different about private educational institutions in the idealsphere is that the goal of the private education institutions is now not profit alone; it is fulfilment of the goals along with profit considerations. The government hires different private institutions to meet their goals. If a company innovates new education techniques to enable them to meet or exceed the goals and make a profit, that is fine. If they break even while meeting or exceeding the goals, that's fine too. If they lose money but still meet or exceed the goals, the government may opt to help the company to improve its efficiency. If a company does not fulfill the goals, regardless of its profitability, the government will step in to help the company improve or terminate the company's contract. The point is, goal fulfillment is the name of the game, not profit maximization.

The point of this discussion is not to pick on one industry or threaten a particular group. For the record, I think teaching is the most honorable of professions and does not always receive the respect or resources needed to perform its critical task in our society. My point is to imagine how things might work in a world in which we optimize our three survival mechanisms. Competition breeds innovation and efficiency. Collaboration drives complexity. Coordination directs competition and collaboration to operate smoothly while meeting its goals. These three mechanisms can work in tandem to give kids a real opportunity to thrive in the ideasphere.

RULES and ENFORCEMENT

In the ideasphere, we call rules laws. In the ecosystem, laws are not required. Every cell has built-in instructions that define the rules of behavior. We humans are born with no ideas. We must learn a lot of things just to function in a basic way in the ideasphere. What the neural system does naturally, we must work hard at, just so everyone can co-exist at a basic level. Governments can use three types of rules, or laws, to enable the ideasphere to become an idealsphere.

Disclosure. This is the most powerful tool in the ecosystem toolbox. Every single cell in a complex body discloses its status to the neural system on a regular basis. Since cells do not have the ability to discern a 'good' status from a 'bad' status, they must constantly send all their information for the brain to sort, process, and respond. But people in the ideasphere do have the ability to discern the quality of information. We don't have to send all our information up to the government for processing. Besides, this would violate the whole 'freedom to choose' aspect of living we value so much. That being said, there are times when disclosure is necessary and should be required of people and institutions.

Whenever the government identifies a goal that is important to prevent harm to others or the society, disclosure laws should be created and enforced. For instance, if our leaders determine water pollution is a threat to the well-being of citizens, we might create a law requiring every institution of a certain size report data about what and how much and how often it pollutes the water. Most businesses probably do not pollute at all, and they would be given a longer interval before they would have to report again. For those who do pollute, the government would provide assistance, just as the neural system's helpers do, to help these institutions change their pollution producing habits. Only if businesses do not report or do not participate in the harm reducing activities does enforcement kick in. An institution may be fined or temporarily closed until improvements are made.

Do you understand the difference between how disclosure works in the idealsphere and our current system? Currently, we set a standard, check on institutions to determine if they comply, and fine them when they don't. The incentive is to hide polluting activity and cheat to avoid punishment. With idealspheric disclosure rules in place, the incentive is to report and fix the problem without punishment. If the goal is to create a cleaner environment and not merely to punish polluters, disclosure might be a better way to go. There are countless areas where such disclosure laws would work quite well. Workplace injuries. Police conduct. State and local corruption. Product defects. Discrimination. Disease outbreaks. Not to be facetious, but we could use some serious collaboration to come up with this list.

Disclosure as a tool would have been impossible even a hundred years ago. The cost of disclosing information was quite high in the early days of business organization. Today? Computers and information processing have made the cost of disclosure inexpensive. The cells who regularly send their information to the neural system use a portion of the energy they consume to do this necessary reporting. Institutions should be required to perform this same function, but only for those activities critical to the well-being of society and its members.

Prohibited activities. Most of our existing laws are prohibited activity laws. These laws prevent people from doing activities that cause harm or death. We do a decent job of making these laws, but don't always do well enforcing them. In a body, every cell is a valued member of the body's 'society,' and the neural system does everything it can to repair an errant cell. Only as a last resort, after all other efforts have failed, does the neural system instruct the cell to be removed from the body. For cells, this means death. For humans, it can mean prison.

In our current ideasphere, we don't do 'human repair' too well. Instead, we tend to toss people into prison, focus on punishment, and ignore necessary repair work. As a result, we lose the valuable contributions we might receive from our fellow citizens. In an idealsphere, government would do a much better job at repairing humans. Rehabilitation. Education. Health care. Mental health care. People undertake prohibited activities for many reasons. If our idealsphere is to adhere more closely to its goals, it would make a more intensive effort to help people become full members of the society from early on in life. This may in turn prevent the need for future 'repair work.' Yes, there will always be some who break the law and are resistant to 'repair.' For those, prison may be the best option. But even then, prison should be a last resort, and it should be humane and provide opportunities for people to rejoin society.

Of course, we can function in societies that do throw a lot of people in jail. We do this now. This reflects a society that does not work well. It all depends on our goals. If our goal is to maximize opportunity for each citizen to participate in society, current societies, ours as well, fail quite dramatically. If our goal is to imprison people for a wide array of offenses, some of which do not harm others, then we do a darn good job. In this case, our rules suck, and the ecosystem would be ashamed of us, if it could feel shame.

Evaluation Laws. In the ecosystem, the neural system responds to the cells' reports by comparing them to their rules that have long been successful at maximizing survival. If a cell is acting in a way that reduces the body's chances of survival, the neural system responds. In the idealsphere, we need a mechanism to determine when our leaders are not acting in accordance with the goal of maximizing choice while preventing harm to others.

Judicial systems perform this function for people and institutions. Most judicial systems today focus on whether an activity complies with the laws, and not necessarily with the goals of the Constitution. Judicial systems sometimes do this, but only when a situation comes up in which someone argues a law is 'unconstitutional.' Bodies in the ecosystem aren't as hit and miss as is our ideasphere. Bodies require compliance with the goals at all times.

What is missing in the ideasphere is a mechanism to make sure the leaders themselves comply with the goals. If such an assessment system were in place, it would determine when actions do not comply with our Constitution. It would examine all branches of government to determine where the leaders are meeting or violating the Constitution's goals. Let's take a few quick examples. Obviously, much more study and thoughtful consideration is needed to adequately cover this topic, but I want to introduce the idea here.

Let's call this part of the Coordination System the 'Evaluation Branch.' One part of the Evaluation Branch's job would be to assess whether a government action, that is, a law or an executive action or a judicial ruling, complies with goals of our society as stated in the Constitution. In our idealsphere, the laws allow for optimal choice among citizens as long as they do not harm others or the society.

I want to inject an important note here. It is important to clarify the meaning of the term 'harm.' Just because a person doesn't like what another person does, this does not mean they are harmed by the other person's actions. 'Harm' must cause financial, physical or property damage to an individual or institution. Violation of a person's 'sensibilities' or 'feelings' doesn't rise to the level of harm unless you are dealing with children. Children are a special case as they are not fully functional members of the ideasphere. When adults have their feelings hurt, they need to take up the issue with the individual they disagree with, person to person. Remember this, member of the idealsphere: everyone gets to feel how they feel and act how they want to act, as long as it does not rise to the 'harm' level. So, get on your big boy or big girl pants, and have an honest respectful discussion when you have differences with others.

Let's address some examples of 'nonharmful' behavior with which some of our current leaders struggle. These actions do not harm anyone else, so they should be maximized.

Gay marriage. This doesn't affect anyone else except the two people being married. This right should be maximized.

Voting. Choosing leaders is the fundamental action an individual can take to participate in the coordinating function of the society. It connects our power of choice to the coordination function. The right to vote should be maximized.

Abortion. Until a fetus is viable, it is solely dependent on its mother for survival. Since it cannot survive outside its mother, it is 'part' of the mother. As such, the mother retains the sole decision-making power regarding her pregnancy. It is her body and what she does affects no one else. She should retain full choice regarding how to proceed with her pregnancy.

Transgender identity. It's no one else's business whether a person chooses to be a female or male regardless of the gender with which they were born. This choice affects only the person making it. This right should be maximized.

Contraception. Using contraception to prevent pregnancy affects no one else except the person choosing to take it. This right should be maximized.

Maximizing non-harmful rights is one side of the 'Evaluation Branch' coin. On the flip side are actions that *do* harm others and should be regulated. The 'Evaluation Branch' would weigh in on these areas as well and determine when leaders need to take action to prevent harmful activities.

Any time an action harms others, it must be regulated. Period. If we are all going to live together, we cannot have the legal right to harm each other. The neural system doesn't let a rogue heart cell beat up on its neighboring heart cells. Oh no. The neural system would put a stop to that sort of trouble on the double. So should members of the ideasphere put a stop to actions harmful to others.

Another note about regulation. Careful implementation is required so that regulations maximize both safety and choice. Too often, government applies rules too broadly and can cause nonharmful actions to be swept into the 'harmful' category. Yet lack of regulation can cause serious harm, too. This is why disclosure and information gathering are needed to guide the evaluation process. Let's consider some examples.

Guns. The availability and use of guns can harm and kill innocent people in our society. Therefore, the availability and use of guns should be regulated.

Racism. The different color of individuals is about as superficial an aspect of humanity as is the color of hair or the color of eyes. Racial discrimination with respect to housing, education, or any other ideaspheric opportunities are totally without merit and utterly harmful not just to the person being discriminated against but to the entire society which loses the ideas these people could contribute to the society. Discriminative behavior must be regulated until it is eliminated.

Drugs. Some drugs used recreationally may not be harmful if ingested in limited amounts and in certain circumstances. The goal of regulation here would be to allow the nonharmful use of drugs while regulating the harmful use of drugs.

Any time a human activity can be both benign and harmful, it must be regulated for the protection of others and the society. These regulations are difficult to create because on the one hand, we should provide the opportunity for people to make individual choices, but on the other hand, we should regulate the conditions of choice, so as to minimize harm. This is not an easy task. People are often reluctant to sanction 'bad' behavior, even if it is relatively non-harmful. The reality is, 'bad' behavior exists, whether we regulate it or not. Wouldn't it be better to acknowledge this reality and focus instead on making rules that minimize harm? I am sure we can come up with better solutions for these problems and others of this ilk, like prostitution and militias and social media.

Regulation is a controversial subject today. Regulation gets tagged with the label of being 'anti-freedom.' In reality, regulation should be seen as a means by which everyone in a society is protected from harm.

The Evaluation branch would use a different approach and produce a very different outcome than the judicial branch. Rather than send violators of the Constitution's goals through the judicial system, the evaluation branch would undertake an assessment process. Each coordinator, agency, and public institution would be scored as to how well their actions comply with the Constitution's goals. This would work much like the General Accounting Office (GAO) assesses financial laws to determine their impact on the federal deficit. The members of the Evaluation Branch would use objective methods to measure how well each government member maximizes choice and minimizes harm to people. They would provide a score between 0 and 100, where 100 represents full compliance with the Constitution, and 0 represents full failure to comply with the Constitution. These scores would be made public to all citizens.

If the score is low, the member may be required to undergo remedial training. If the score remains low even after remediation, the member may face a trial-like process by which they may be deemed to be unqualified to participate in the coordinating system. In this case, they would not be thrown in jail, but they would lose their right to be a coordinator. They would still be free to participate in the ideasphere, but they would no longer be allowed to work in government.

Let me make one important point about the 'Evaluation Branch.' Such a mechanism is not currently in the Constitution. To add this new function to the Constitution is not easy. This makes sense; goals should not be easy to change. Yet our Founders did outline a process for altering the Constitution. Those founders. They were the best. They understood that change is a natural part of life. Anyway, it's heavy lift to make such a change. Three-fourths of the state legislatures need to approve an amendment to the Constitution, so it is not something to undertake unless we think it is necessary. I think it is.

NOURISHMENT

In the ecosystem, all cells share the resources ingested through the mouth and processed in the stomach. Unless you're in a coma and need to be fed by a tube. Ha! Been there, done that. I don't recommend it. We eat. The whole body, every single cell, is nourished though only one organ does the ingesting. The brain, which does the job of coordinating and does not perform any of the actual survival functions of the body, shares those resources in the same way as all other cells. The neural system, because of the importance of its coordinating role, is fed first, but it doesn't consume more than it needs.

If by chance a human consumes so much food, they cannot turn all of it into nutrients right away, each cell keeps a little bit of the excess. If a body consumes so much food that the individual cells can't store it all, the body sets up locations to hold the excess. My location is in my hips. Sigh. Other people have locations in the stomach. Or the heart. By the way, excess nutrient storage isn't too helpful to a body.

A neural system's use of resources is equivalent to 'taxation' in the ideasphere. The neural system uses its food energy to coordinate cell activity, so the cells and body survive. In the same way, the government uses tax revenues for two things: one, to run the coordination activity, that is, the government; and two, to distribute resources to those areas of the society to achieve societal goals.

How taxes are levied is really important. Let's examine how our current taxation systems work. The closest match to the ecosystem is a sales tax. When we buy shampoo or a lawnmower, we must pay a certain portion of the cost as a tax to our state. Most states have this tax; a few do not. This is a logical method to raise taxes, but the way we currently do it is backwards. Only consumers pay sales taxes, and as a result, it is highly regressive. Let's say I make $20,000 per year. I spend all my income to buy stuff to survive in my state which charges 10% sales tax. So, I paid $2,000 in taxes. Let's say another person makes $1 million per year. They spend $500,000 on expenses during the year and save the other $500,000. So, they spent $50,000 in sales tax. That's a rate of 5%. Yes, I am poorer than this other person, but I paid double the tax rate. Plus, I guarantee my life is a lot more difficult than my million-dollar neighbor. I really could have used that extra $2,000.

This is not at all how things work in the ecosystem. First, disparity of resources does not exist. Second, each cell receives equivalent resources based on their 'need,' and any excess is shared. Okay, we're not cells. We're humans with the ability to choose our life, so disparity of resources is going to occur when people make different choices. Of course. So, how should we handle taxation in the idealsphere where people choose their own resource-producing path?

Today, we use income tax as our primary source of tax revenue. With income tax, we calculate our revenue or wages, reduce this gross amount by expenses approved by the government, then pay a specific percentage of the net amount remaining as taxes. Let's see how this would work in the ecosystem. I hate to be so blunt about this, but if a body worked like the American tax system, it would be dead, dead, dead. Every cell and organ would do whatever possible to maximize its 'profits,' that is, its net nutrients after spending the energy it needs to survive. Then they would only grudgingly give a portion of their net nutrients to the brain to do its coordination work. Or it would cheat and not send the brain its nutrients at all, and make the brain spend its resources to go after 'deadbeat cells' instead of coordinating the body's activities. Good luck to you, brain, you evil socialist Big Brother. The brain would not be able to function in such an erratic environment. So, to preserve its ability to operate, the brain would seek sources of nutrients that are easier to obtain. Which is what our government did when it turned to individual wage earners to obtain income tax revenue. This source is more stable than the large, wealthy institutions who were originally the only ones who paid income taxes but proved adept at avoiding them. The body doesn't work this way because it's a system that doesn't work well.

The body's 'taxation' efforts act more like a revenue tax. Let's imagine what the idealsphere would be like if we acted like a body. What if we charged every single person and institution a revenue tax, that is, a tax on the revenues and wages they take in? Now, the rate would be much lower than when we are taxing people for buying stuff. Let's say we charge 1% for all revenue producers and add a graduated rate structure for really huge revenue makers. Let's say in 2021, I made $40,000 at my job. At 1%, I would pay $400 to the government. Last year, Amazon grossed $470 billion in revenue. At a one percent tax rate, the company would have paid the government $4.7 billion. Amazon actually paid only $2.1 billion in income taxes, less than half of our imaginary one percent revenue tax rate.

In the idealsphere, every earner, whether a worker or an institution, pays a portion of their revenue to the government right off the top, just like entities in the ecosystem. It's more efficient, and its more fair. This method also allows people to choose how to use their resources once they pay their revenue tax. They don't have to worry if they'll receive a tax deduction by spending their funds on certain approved expenses. They can spend their resources on the things they choose to help their business without looking over their shoulder.

Now, let's return to the income tax system. The ecosystem doesn't have anything like the income tax system. Any excess resources are shared among all the cells. But people are people, and they are incented by profit. People are and ought to be free to pursue profits in any way that does not harm others. However, we know accumulation of *excess* resources can be used to thwart the smooth workings of the ideasphere.

In the ecosystem, most entities consuming excess resources immediately create offspring. If the equivalent existed in the ideasphere, every person and institution with excess profits would plow them back into their resource-producing efforts, thus keeping the resources flowing throughout the system. This doesn't always happen here in the ideasphere. Resources are accumulated and often used to prevent others from exerting their ability to choose. The body would never allow the heart to make life worse for the kidneys or the stomach. No way. So, why do we allow people to do so?

We can avoid this problem by charging income taxes only on highly profitable institutions and highly paid individuals. Unlike the revenue tax, this tax rate would be set quite high. Let's consider how this might work by using an example. Let's say you run an imaginary company named Big Mamazon and this year your net income is $35 billion dollars on revenue of $500 billion. From 2009 to 2021, your Big Mamazon had revenue of over $2 trillion and earned a net profit of about $90 billion. In the existing ideasphere, your company paid taxes of $16 billion in those 13 years. That's a tax rate of about 18% on its net income but less than 1% of its revenue. That's not much, considering it's the Big Mamazon of companies.

Now, let's throw Big Mamazon into the idealsphere. This year, the company will pay $5 billion in revenue tax, which leaves it with $30 billion in profit. How much of the profit should go back to the government to keep resources flowing and prevent Big Mamazon from using them to keep others from participating? It's a good question. On the one hand, we want the company to be incented to continue to operate by keeping some of its profits. On the other hand, if it accumulates too many resources it can disrupt the competitive and collaborative environment for others. The income tax rate must balance these two competing goals. Maybe the tax doesn't kick in until the company has accumulated a significant level of resources, but when it does, the rate would be quite high, perhaps 75% or 80%?

This is just an example, so don't go apoplectic on me. After all, there's no company called Big Mamazon. The idea is that most small individual companies would not pay any income tax at all. Most small businesses would never accumulate massive profits, so they would not pay any income taxes at all. Remember, they are already paying a revenue tax, so they are contributing to society already.

If a company amasses a huge amount of wealth from their operations, the chances are high the resources are going to be locked up and not utilized by the members of society. The goal of income taxation is twofold: one to keep resources flowing, and two, to dilute the concentration of resources so they cannot be used to subvert the smooth working of society's survival mechanisms.

Concentration of power is death to a society designed to make sure all citizens have maximum opportunities to choose their future without harming others. The revenue tax is the price we would all pay for participating in the society. The income tax is the vehicle to keep resources moving in a way to maximize human choice and prevent concentration of power.

Okay. There's a quick picture of how the coordinating system would work in the idealsphere. I hope the ideas presented here make you think.

Chapter 33

NOT PEOPLE

Institutions

In the ecosystem, organs exist solely to serve the body. Organs are a group of cells organized in such a way to perform a specific function. If cells could have figured out a more efficient structure than using organs to make a body survive, they would have. But they didn't. Organs are the solution cells and evolution together produced. They are complex, collaborative structures that perform in a way individual cells cannot.

Institutions in the ideasphere are both similar and dissimilar to organs in a body. Institutions are also efficient, collaborative structures that work to achieve a goal an individual alone cannot. The difference of course, is that institutions strive to achieve their own goals and not necessarily the goals of the society or the people within it. Which is fine. Institutions are composed of people and people like to choose how to act, so institutions are a perfectly adequate vehicle for making choices. If institutions are free to achieve their goals, though, they ought to be subject to the same rules any individual person who is free to maximize their choices would be: they can't harm other people and they can't harm the society in which they operate. This is not new; it's the same rule every cell and every person must adhere to in their relative ecosystems.

Institutions in the ideasphere come in all shapes and sizes just as entities in the ecosystem. Any group of people joining together to achieve a goal is an institution. That's a wide net. There are a few major types of institutions humans have developed. They all use similar structures, but what differentiates them is their goals. Let's consider some examples. Commercial institutions produce products or services for the purpose of maximizing profits. Non-profit institutions attempt to solve or lessen a societal problem. Religious institutions operate to maximize its number of believers. Government institutions exist to optimize the well-being of members in the society, at least in our imaginary idealsphere.

Collaborative institutions compete in the wider environment just as multi-celled entities do in the ecosystem. The major benefits of the survival mechanisms, innovation and productivity from competition, and complexity and specialization from collaboration, benefit institutions just as they benefit organs in a body. This is why institutions are so important in the ideasphere. Progress. Achievement. Technological advances. Complex ideas. These are things that result when people collaborate and produce outcomes a single person alone cannot achieve.

When excess profits skew the ideasphere environment, not only do the coordination efforts become less effective, but so do collaborative and competitive mechanisms. In other words, not only do people work less effectively, so do our institutions.

I know I made a big deal about businesses and profits when I talked to Dina. I was kind of an asshole, I know. So, I need to make something clear here. Profits are useful. They are a helpful feedback mechanism indicating an institution is meeting its goals. The ability to make profit is a proven incentive for people to start businesses and take risks. And they should benefit when they succeed at their risky activity of choice.

It's 'profit maximization' that is a problem. When profits are sought above all else, pressure to perform can become so intense that corners are cut. Let's say a company's product blows up and kills a customer. It may be cheaper to pay the customer's family a settlement and hide the problem rather than change the product itself. If this option maximizes profits, the company may see this as the 'right' action to take. But, if a company existed in the idealsphere, it would act differently. Let's say the government requires companies to disclose information about customer fatalities in its annual report. Let's say this company is so profitable that it is also subject to a high income tax rate on its excess profits. In this case, the incentive may shift. It might be better for the company to use some of its excess profits to fix its products. When it reports its fatality information, it can also report how it resolved the problem. And the money the company spent on the product fix protected some resources from taxation. Also, the likelihood of more consumer fatalities has gone down. This is a win-win. With other considerations besides profit now in play, the institution is incented to move toward the 'no harm' goal.

Curbing profit maximization through targeted income taxation also curbs another problem that concentrates resources: maximizing shareholder wealth. When a company is public, it means lots of institutions like university foundations and individuals, like you and me, can buy shares. Shares are little indivisible pieces of the company. The share value is generally equal to the sum of a company's profits over its lifetime. Let's say a company has made a profit of $1000 over its three-year lifespan and has issued 1000 shares to its investors. If we do the math and divide $1000 of accumulated profit by 1000 shares, we calculate each share to be worth $1. Let's say in the fourth year, the company makes a whopping $10,000. Now each share is worth $11 ($11,000 accumulated profit divided by 1000 shares). When the company earns more profits, the share value rises.

Now, the problem exists because the owners of the company, the shareholders, don't care squat about the other goals of the company. They only care about the profits and the share price. They can sell a share and convert it to resources, so the higher the share price, the better for the owner/investor. The managers of the company know this. If they don't maximize profits for their 'don't-care-squat-owners,' they may lose their jobs. This could not be a worse setup if the goal of society is to keep institutions from harming others. All the pressure is focused on making profit and nothing else. If government puts a high tax rate on excess earnings and requires reporting to disclose how well an institution does in not harming people or the society, the profit is modified, and so is the share price. The share price still varies, but on a less volatile basis and more in line with how well an institution meets all its goals, not only its profit goals. The result of these policies, of course, is a reduction in resource concentration among the investors. This is important because today, wealthy people and institutions own the majority of all shares in this country.

Taxation of income at high levels is required to keep resources flowing. Tax revenues aren't kept in a government's vault and hidden away. They are injected back into the economy in areas that serve the goals of society: education, infrastructure, health care, housing. People benefiting from these resources not only are able to live and thrive and maximize their ability to choose their best future, but they also keep the resources moving in a way that maximizes the production of ideas in the ideasphere.

Concentration of resources puts pressure on democratic systems and pushes them toward oligopoly and autocracy. People who have gained many resources and the power that comes with them, rarely give up their power without a fight. Once a country starts moving towards autocracy by concentrating resources and power, it can be hard to stop the slide. If resources are widely distributed, however, many citizens participate, and opportunities are maximized. Democracies have a much better chance for success if these conditions are present. As long as resources flow, participation follows, and choice flourishes. The promise of participation and profit moderation are conditions that favor democracies and republics.

Chapter 34

WORLD PEACE?

The Earth as Idealsphere

In the ecosystem, little coordination exists above the entity level. All non-humans exist as units that must fight for their own survival. They have no choice. Literally. Through long years of diversification and refinement of their own unique survival techniques, they have no ability to communicate. Therefore, they have no ability to collaborate except in the most rudimentary of ways.

If our idealsphere worked like the ecosystem, we would create a worldwide coordination function that strives to do what any country's government strives to do: optimize choice with minimal harm to others and the global society. It is possible. We all function the same way. Though we have many different languages and ways of living, we can communicate. If we can communicate, we can collaborate.

It is possible to develop a global structure to solve worldwide issues. It is also possible to create one single coordinating government in which we all live and strive to optimize choices while minimizing harm to each other and the global society. Possible? Yes. Likely? No.

Why? The ecosystem moves slowly towards collaboration, and so do we. Each time entities change; it takes a long period to sort out which collaborative structures work in the environment. People also take time to assess after they develop new collaborative structures. Tribes turned into towns. Slowly. Towns turned into cities. Slowly. Cities turned into states. Slowly. States turned into countries. Slowly. We haven't created a singular coordinated structure above this level. But we have made some efforts in that direction. Today countries make treaties with each other to achieve specific goals. Think of the SALT treaty between Russia and the U.S. They also make broader alliances to promote trade or mutual protection. NATO is an example of this. Most of the world's countries have joined the United Nations.

The United Nations is the first real effort people all over the globe have taken to join together in a coordinated way. It is not a worldwide government, but it seeks to provide worldwide coordination when problems arise that affect all of us. It is reactive, much like a neural system in the ecosystem, rather than proactive, like individual governments can be. Here are the goals of the UN, paraphrased for brevity:

- Prevent war
- Believe in fundamental human rights
- Establish conditions for treaties to be maintained
- Promote social progress.

These are the actions the UN prioritizes:

- Practice tolerance
- Live together in peace
- Unite to preserve peace
- Ensure armed force won't be used
- Promote economic achievement

Just by looking at these words, we can catch the parallel to the idealsphere:

> *People ought to be free to create any idea and convert it to any action they choose, as long as their actions do not harm other people or the world as a whole.*

It's always rough going at the start collaboration. We make a lot of mistakes when we begin a new activity. By observing how the UN works, we can deduce it is less than perfect. Countries with veto power work to enhance their self-interest rather than to benefit the global community as a whole. Countries boycott participation when they don't like the actions the UN takes.

The fact we have made the effort to form the United Nations is a huge evolutionary leap. We now have one place where all countries can come together to disclose issues, seek assistance, and discuss solutions. It does not carry the weight of a government which is expressly granted coordinating authority by its people. It is more loosely structured. The UN is more like a pack of lions than a bee colony. But this is where we are on the evolutionary track. There is no requirement we must strive for a unified government.

That being said, we do have a pressing problem today that needs more coordination than a pack of lions might provide: climate change.

We are only a few years away from the catastrophic breakdown of our Earth's climate systems.

The irony is rich here. People have created an ideasphere that produces ideas and structures outside our bodies. We've learned how to delay accountability or even avoid it when our actions don't turn out so well. We have become more interconnected and interdependent as we have grown more complex. Our actions have changed the environment in which we live and on which we depend for survival. We are now at an inflection point.

The two sides of humanity are meeting in a head-on collision. On the one side is the idea-producing, structure-storing, accountability-averse people who want to be free to choose to live however we want with the only constraint being not to harm others. On the other side is the insidious potential harm coming to all of us by our own actions, the potential destruction of our ecosystem, our physical home.

Here is the situation as I see it: to deal with a global catastrophic issue, we need the strength of a global coordinating mechanism to enable us to take actions or distribute resources in such a way as to reduce or eliminate our potential destruction while both inflicting as little harm as possible on each world citizen and providing as many choices as possible given the dire situation. We don't have such a global mechanism. Though the UN is not ideally designed for this task, it has taken on the effort to produce treaties to mitigate the problem. The UN is doing a good job, probably the best job it can, given its structure. The quality of the UN's effort is not the key issue, here. This is a race with deadly consequences. Can the UN coordinate enough change before catastrophe occurs?

How can we optimize our chances of survival given these circumstances? I do not know the answer to this question, but I know a little better how to view the problem. We have three basic choices:

- Do nothing. The result is inevitable destruction of much if not all the human species.
- Use our current treaty approach where we painstakingly try to resolve the issue while each country resists chipping in resources to aid in the solution or refuses to take concrete action that might damage its economic future. The result is a risky, uncertain survival outcome.
- Temporarily authorize the UN to act as a coordination unit with more powers to deal with this problem. This means they would make rules to distribute resources from wealthier countries who cause climate change to those with fewer resources who are harmed by climate change. They would also identify climate harming behaviors and work with countries to mitigate those behaviors with as little negative impact as possible. The result is a better chance of survival.

Of course, different flavors of each option exist, except maybe the first one. The point is, this is a hard problem that threatens our survival, and the only way to deal with it is for all of us interdependent, interconnected, idea-producing people to collaborate and coordinate as a single unit to avert our destruction.

What will we do? How will we respond? It would be one thing if we couldn't solve this problem, and we died out as a species, like so many others that went extinct before us. This is not the case with people. We can think. We can produce ideas. We can discover problems. We can solve them.

I would hate if we died out as a species because we were unwilling to solve this problem.

Chapter 35

FROM HERE. TO THERE

My last few notes outlined what I think the most critical parts of the idealsphere might look like. I needed to construct a clear view of that vision before I could conceive of how to put it into action. I think the idealsphere is a goal worth achieving. I have no illusions; it's a difficult task. But nobody achieves a goal without effort, so here I go . . .

To translate our current ideasphere into an idealsphere is a little more complicated than adding an 'L.' I've been thinking about how to do this ever since Dina left. I think I've come up with a process that makes sense to me. I use it to keep me on track. It's easy for me to get mixed up and lose my train of thought and wander into nonsensical 'idea overload,' when I think about our current issues. I admit, it's fun to go down all those rabbit holes, but it's not too helpful when it comes to solving a problem. Here is my handy reference guide:

First, I start with the primary goal:

> *People ought to be free to create any idea and convert it to any action they choose, as long as their actions do not harm other people or the society as a whole.*

Now, the thought process:

- Remember the primary goal
- Define the problem in the ideasphere
- Determine how the ecosystem would handle the equivalent issue
- Redefine the ideasphere goal in terms of the issue, if necessary
- Outline how to use competition, collaboration, and coordination to achieve the goal

- For coordination changes, add rules and enforcement actions that might be needed
- Add any other thoughts about the issue to address.

I thought I would test out the model on one problem we have in society today. We have a lot of problems to choose from, but I selected social media as my test subject.

SOCIAL MEDIA

The primary goal (again): People can implement any idea as long as they do not harm others or society.

The issue: Some posts may cause harm while others are benign, and they are mixed up together.

Ecosystem solution: All 'posts' made by cells are sent to the neural system. The neural system determines the potential for harm and responds to the harmful behavior by mitigating its spread.

Redefine the goal: People should be free to post whatever they want on social media as long as they don't harm others or the society.

Solution using the three 'C's': The Social Media companies are doing fine in the competition department, but they require more coordination to mitigate harm. In social media, the customers are the advertisers who pay the social media company to promote their products. The social media company's *products* are the people who post messages or more precisely, their posts that carry ads. Neither the advertising customers nor the company care about the posts people make; they only care how many people see their ads. So, the more viral a post is, the better, regardless of the nature of the post. The problem is the post itself may be benign or harmful. Coordination is needed to unsnarl this jumbled mess.

Rules and Enforcement: The key here is to understand what 'speech' is. Speech is an action. It is the expression of an idea. If the speech is spoken out into the world, it is not stored anywhere outside the body. But if speech is written or recorded, as in a social media post, it becomes a tangible physical object that has an existence of its own. That tangible physical product can be stored and shared by many people. It can influence others. It can benefit others. Think of a new cure for a disease or a study on how bees benefit agriculture. But that tangible physical product can also harm others. Think of online bullying of a child or fomenting armed rebellion against the country. Online posts can run the gamut between the two extremes of benefit and harm. On the one hand, a person can share a picture of a cute puppy. On the other, a person can promote a bleach cocktail COVID treatment. One of these two is harmful, and I'm pretty sure it's not the puppy gif.

We need to make the distinction between freedom of thought and freedom of speech. Freedom of thought should always be one hundred percent free. People should have the unlimited ability to think whatever ideas they want to think. This is the nature of people. This is the superpower of people. It should be unfettered and protected. But speech that is turned into physical form, whether it is a social media post or a newspaper or a talk show or a podcast, crosses into a new territory where it is an action that can have a life of its own. This action may be harmless, but it can just as easily be harmful. This switch from intangible to tangible form that may be harmful must be addressed, especially as technology continues to change and influence the way we live.

Add any other thoughts about the issue we might want to address: Boy do I have thoughts.

Social Media is a new phenomenon that is early in its own evolutionary development. It's equivalent to a new creature evolving in the ecosystem. Since social media is a new twist on the age-old function of speech, let's go back in time and examine a brief history of how we communicate with each other.

When people began to communicate, they talked. The farthest extent of speech depended on how well a person could be heard. Literally. People stood on soapboxes or in town squares or yakked to each while warming themselves over the communal fire. Whichever method, their reach was understandably limited. Later, when people printed documents and distributed ideas via writing, this reach expanded. Why? When things are written down, they take on an existence of their own. Pamphlets are passed around. Books are borrowed from the library. People read newspapers.

Soon, societies distinguished between two types of writing. Journalism is defined as the publication of news. News is defined as new or previously unknown information. To be 'news' in journalism, the information is required to be provably true.

Anything that is not true is defined as 'not news.' The 'not news' writing is a catchall. It's informally referred to as 'all the other crap people write.' Ironically, many offerings that are enjoyable and popular fall into this category: fiction, movies, books about how life can work better if it works like the ecosystem. To journalists, these other writings are all 'non-journalism.'

Let's focus on journalism for a minute. Journalists came up with their own rules by which to operate, which is interesting in and of itself. It turns out that, back in the day, no one would give up their hard-earned money to read a newspaper they couldn't trust to tell the truth. The five basic principles of journalism that have evolved are: 1-truth and accuracy, 2-independence, 3-fairness and impartiality, 4-humanity, and 5-accountability. Operating by these rules is a win-win situation for journalists and readers. Journalists are paid to tell the truth, and readers pay to obtain the truth.

It wasn't long before some newspapers discovered they could make decent money dressing up an idea as 'truth-like' to convince people to think a specific way. Swaying people to prefer one set of ideas over another wasn't new, but using journalism as a cover was. This change had significant impact. Newspapers and books, soon to be joined by radio and television, reached far more people than a soapbox. Nontruthful ideas that appeared to be truthful had the potential to harm a lot of people. So, governments stepped in to regulate journalism in the cases where they might cause harm. For example, if journalists lie about a person and cause financial or reputational harm, libel laws protect those victims.

In 1927, after radio became popular, the federal government created the 'Fairness Doctrine,' which covered news reported over the airwaves. This law required broadcasters to cover controversial issues of public importance and to fairly reflect opposing views of these issues. This law promoted the goal of an informed society. Lawmakers wanted to make sure citizens obtained a full and fair view of important issues. Congress abolished this law in 1987. Cable television had made the law moot, anyway. Cable TV does not go over the airwaves, so cable companies were not required to comply with the law.

When in effect, the Fairness Doctrine forced broadcasters to be honest in their coverage. After the law was abolished, media companies were free to return to the much murkier world of putting out 'news' that was neither true nor independent nor fair nor human nor accountable. In other words, they produced 'not-journalism' dressed up as journalism.

Social media is the next generation of information producer and it's a doozy. It mixes all types of information into one big pot, and it mixes people who post information into another big pot. Let's take a closer peek into this mysterious amalgamation of communication.

Who can post on social media? Anyone. What can they post? Pretty much anything. I can post a picture of my dog on my account and tell the world how cute he is. I can also post on my very same account that our recent election was corrupt, and the government should be overthrown. And I can post anything in between. Some posts are harmless. Some are harmful. Others are somewhere in the middle.

So, here's where we stand right now. We have a forum in which a person can act as a journalist, and that very same person can act as a producer of 'all the other crap people write.' It's all mixed up in a giant messy blob of posts streaming out of millions of social media accounts every minute of every day.

How would the ecosystem handle this? Well, first, cells can only talk to other cells when it is useful to the survival of the other cells or the body as a whole. If we took this tactic, we could wipe out 99.9% of all social media posts in that they are mostly useless crap. But given that people are people, and they should be allowed to share useless crap on social media if they want, then this option is out. Besides, DNA instructs cells to only perform useful actions. Cells don't do useless. But people? We're really good at useless.

So, the job of sorting out the mess that is social media is left to our coordinating function. Yup, it's another area where government should regulate. There's a catch. Some of the content on social media is harmless and therefore the right to post should be maximized. On the other hand, some of it is harmful and should be subject to regulation. So, the answer is easy-peasy; just separate the harmless from the harmful and regulate the harmful. Sure, there are only twelve gazillion posts coming out every minute. No problem!

So, let's try this. Let's do it the ecosystem way. We make the cells, I mean, people, report on their truthfulness status. The government makes a simple set of rules and requires the social media companies to enforce those rules.

Rule #1: Every person who posts must describe their post as either:

- true, that is, based on evidence that has been proven
  OR
- not true; that is, it is not proven or is the opinion of the posting person

Rule #2: Every person must take responsibility for their post.

It might work something like this: after a person writes a post, but before they send it out into the social media world, a screen pops up and requires information about the post. Let's consider some examples.

POST: My classmate, Bobby Jones is fat and ugly
NOTE: This is not a true statement. This is my opinion. I am responsible for posting this message as non-truthful.

You might as well post "I'm a dick." But I digress. The point is the person who posts is required to describe the truthfulness of the post and their accountability for posting it. The speech itself is not prohibited. The required disclosure, however, makes clear that the person who posts it is accountable for the content of the statement.

Let's take another example. Let's say I share a post from someone who already identified this post as a non-true opinion:

POST: A certain politician is a pedophile.
NOTE: This is not a true statement. This is my opinion. I am responsible for re-posting this message as non-truthful.

In this case, you might want to post, "I'm a super dick." But don't worry, super dicks out there; at least it would be noted as an opinion. The point is, free speech is not infringed, but the untruthfulness and responsibility of the post is noted.

Now, for a true post:

POST: This dog in Seattle rode the bus to the dog park by himself

NOTE: This is a true statement, and I have accurately stated the facts. I am responsible for this post and agree I am posting this information as truthful.

Now, you might say that people unable to distinguish truth from a lie will not describe their posts honestly. I agree with you. So, our regulators need to anticipate this and put some teeth into enforcement.

Enforcement Rule #1: People who respond to a post may dispute an inaccurate note and report their dispute to the social media company.

Enforcement Rule #2: If a person posts a false note, they're warned by the social media company.

Enforcement Rule #3: Those who inaccurately note their posts are warned. Those who inaccurately dispute a note are also warned.

Enforcement Rule #4: If the 'inaccurate poster' or an 'inaccurate disputer' posts false notes multiple times, they are removed from the social media platform.

Let me anticipate the screams of indignation. Free Speech! I can hear it now. You are violating a person's right to free speech! And I would respond, absolutely not. People are still free to speak, but the speech is now injected with accountability. You can still speak, but now you are required to define your speech. You, as a person who posts on social media, are allowed to speak as both a journalist and an 'other crap' person. Therefore, it is up to you to tell the people who read your post in which capacity you are speaking. That's not a violation of your free speech; that is a protection for others from your potentially dangerous, physically existing, free-flowing action you call speech, but which is actually a tangible social media post. This is a way for people who read your posts to distinguish whether your post is truthful or just a bunch of crap. You, the person who posts, knows this best, so it is your job to say.

Then there's the 'what is truth, really?' argument. Without going all existential on you, journalists deal with this problem every day. Journalists must prove what they say is true. If they're wrong, they are required to correct it. Now, here's where you argue you're not a journalist, you're just a simple citizen making a post. To which I would respond, if you are going to act like a journalist, you need to follow the rules of a journalist. If you are not going to act like a journalist, call your post an opinion and move on. What you shouldn't be allowed to do is act like a journalist without the rigor of a journalist. Sorry, no can do. If you can't prove what you say, you belong in the 'other crap' department. I mean the 'opinion department.'

The last argument, promoted by the social media companies themselves, is: 'This is too hard.' Nope. If you can set up a vehicle by which people can post, you can set up a vehicle by which people can describe the nature of their posts. Furthermore, you might need a few disclosure rules yourself. How many people posted harmful content? How many people lied about their posts? How many people got kicked off the platform? How many people should have been kicked off but weren't kicked off? How many of those who post are actually people? This kind of information would be useful to all of us who suffer from disinformation.

Hmm, it might be useful if some 'news' organizations or politicians were required to describe their content in this way, too. Any speech that is recorded into a physical form should be subject to this rule. Hmm. I'd like to hear Donald Trump say, "This is my opinion and not a proven fact. The election was rigged. Also, I am responsible for making this shit up.' That might be refreshing.

Chapter 36

ONLY SHADOW KNOWS

Hi Dina,

I'm not sure what happened to me after the last bit I wrote about social media. Maybe I got too worked up about the problems we face today. Or maybe I got too overwhelmed by how far we need to go to make this world a better place. At any rate, I'm exhausted. I feel pretty crappy. I think I'll stop and rest for a few days.

I've tried to work on a list of some of our current issues that we might be able to address by using what you have taught me about the ecosystem and ideasphere. I think I need more people to help with this. I need experts who understand these issues in depth, to help find new solutions given this new understanding of how humans operate in the ideasphere. We need everyone to help make the world a better place.

I need to think about the best way to encourage people to participate in this whole 'change the world project.' But right now, I'm too exhausted. I need to rest for a bit. Bye for now.

Sara

"Hello, Sara."

"Dina! You're back! How are you? Oh, have I missed you, dear friend."

"I have not been far, Sara, but I have missed our discussions. I am well. We . . . are well."

"Hold on a second, Dina. Josh told me if I talked with you again, I should turn on my phone and record our conversation."

"I do not believe it will be helpful to him, Sara."

"Why not? I thought it was a good idea."

"Your son will be able to hear your side of the conversation, but remember, I operate inside your memories. He will not to be able to 'hear' me."

"Josh is a smart cookie. He'll figure it out."

"Once again, I fail to understand your usage. Josh is a human, not a baked wafer."

"Ha! Oh Dina, it is so good to talk with you again."

How are you feeling, Sara?"

"Tired. I don't understand why I feel so drained all the time. There's nothing wrong with me. The doctors keep telling me I'm fine, and I believe them. Still, I'm so dang beat all the time."

"I am afraid that is my fault, Sara. As I explained earlier, it is exhausting for your body to continue this work while awake. I am concerned it is endangering your life."

"What? No, no, no, Dina. I'm just tired."

"No, Sara. Your energy will not last. Your job is done. It is time to pass on this task to others who can help. I asked you to do one thing. To learn. To understand the way the Earth operates, and to share your understanding. You have completed this task. You now understand how this Earth operates, how the ecosystem and ideasphere intertwine in an interdependent and chaotic dance, and how people can use this understanding to build a better world to benefit both people and the non-human beings inside it."

"No, Dina. I haven't written everything down. All I have written is some notes about the idealsphere. I can't tell if I hit the mark or I'm a million miles off. I haven't written about the three C's. Or how the ecosystems work together. Or about information. I haven't done at all what you asked me to do."

"There is no need to do more, Sara. You have explained our talks to your loved ones, and they have recorded your talks. The ideas are now stored outside your body. They will help you with the next steps."

"Randy and Josh?"

"Yes. It is by sharing ideas that their meaning comes into existence. They understand you. They believe you. This is what is important."

"Believe me? I think they suspect I've lost my marbles."

"Do you own any small round glass orbs, Sara?"

She let out a weak chuckle. "Oh, Dina. You can still make me laugh."

"What I am about to tell you is very important, Sara. I asked you to share what you have learned. Remember how the universe works? Slowly at first. Step by step. You have taken that first step. It is now the job of your fellow humans to choose to change the ideasphere. To take the next steps. Together.

"People have learned much in their time on earth. For every issue you face, many specialists exist who are equipped to study the problem and determine workable solutions. Up to now, they have been missing the overall picture of how our ecosystems work together. They need to blend their expertise with this new understanding of the ideasphere to take the next steps. They are the best people to determine how humans should use the three survival mechanisms to optimize both the chances of survival and choice. You have done your part. Now it is time for you to rest and let others carry on. I . . . We . . . are more than grateful for all you have done. If you continue to pursue this effort, you will not survive. We have detected a steady deterioration in your condition since you have awakened. We cannot permit you to risk your life by going further down this path. We . . . my compatriots and I, have discussed the options, and have determined your life is too valuable to risk any further effort. We want you to remain here as a living contributing member of the ideasphere. We want you to guide others in their thinking. Therefore, we have concluded we must end our interaction with you."

"What does that mean?"

"For you? That you will live."

"And for you?"

"That is of less importance, Sara."

"Tell me, Dina."

"We will expire."

"Expire. You mean die?"

"Yes, but not all of us. Others of our kind exist all over the Earth who are working to help people understand this world and the ideasphere in which they live."

"But no one has made it as far as you, have they, Dina?"

"True. However, this is only a temporary setback. Others will soon evolve the same abilities we have evolved."

"Come on. Don't expect me to fall for that line. Not after everything you taught me. You know very well the characteristics you've developed might not come along again. All the various conditions that brought you, as you are, to me, as I am, may never happen again. If the world can't lose me, how much more important is it that we not lose you? You are the one that needs to produce a new generation to interact with people without causing such drastic side effects. Only you can make the small changes to improve our co-existence over time. No, Dina. I think it's a bad idea for you to up and 'expire' without causing significant consequences to all of us in both the ecosystem and the ideasphere."

"Perhaps there is another way."

"Tell me."

"I will need you and your canine companion to assist me."

"Shadow? Shadow gets to help? Excellent, we're in."

"I warn you this method is risky and not optimal. It may require a long time for you to recover. Your energy may not return for months, perhaps years. You might lose your ability to remember the topics we discussed."

"Then how will I help other people learn about the ideasphere?"

"You may not be able to, Sara. This is the risk we must take if we are both to survive."

"Will I remember after a while? When my strength comes back?"

"I do not know."

"But neither of us will die, right?"

"I cannot be sure. We both may live. We both may die."

She thought for a long moment. "It's worth the risk, Dina."

"Are you sure? This is a moment in which you might want to slumber on the topic."

"Sleep on it?" She laughed. "I don't need to, Dina. This is the right choice for everyone."

"All right. We will proceed."

"Before we do, I have one question."

"What is that?"

Way back, when we first started talking, I asked you a question and you said I knew the answer."

"Yes. You asked, 'Where did the energy come from that created the universe?'"

"That's it. Dina, will you tell me?"

"Sara, you do not need me to tell you. You know the answer. It is right in front of you, in front of all people. It has been for years."

"Give me a hint. Please."

Dina paused for a long moment. "You are tired, Sara, and you have continued on this path of learning with perseverance. I will provide you a clue, but once I do, you will be disappointed you did not detect the obvious nature of the answer."

"Probably, but my biases are in the way. I can't clear my mind enough to think objectively. I have been loaded with all kinds of theories about gods and creation and big bangs ever since I was a tot. They are all mixed up and stopping my brain from seeing what you say is obvious. Just one teensy-weensy itty-bitty little hint."

"All right, but I warn you, you will strike yourself with your foot when I tell you."

"Kick myself? Ha! I probably will."

"What is the first law of thermodynamics, Sara?"

"Hmm, let me think. The first law. 'Energy cannot be created or destroyed.'

"Correct."

"And. . . ? Dina waited and said nothing. "Dina?"

"This is your answer, Sara."

Sara thought for a long moment. "Oh. Ohhhhhhhhhhh. Damn, Dina. You. Are. Right. It's been under our noses all along. For years. Ever since the scientist wrote down those laws in the mid 1800's."

"This law is attributed to both Rudolf Clausius and William Thomson."

"Of course. I don't remember the authors, but I remember the law. Every kid learns it in junior high science, but we blow by its importance. Energy is never created or destroyed. So, by definition, it has always existed. It has always been here. Always. It just exists, flowing and changing and blowing up into different universes and folding back into space, making whatever it can make along the way. Oh man. . . the statement is so simple and obvious it completely understates its true significance."

"Indeed. People have viewed this law as a mere tool of physics, rather than the ultimate reality of life."

"See what bias can do, Dina? I'm gobsmacked."

"No, Sara, you are still a human."

"Okay, a gobsmacked human."

"Now you understand. Everything around us is a form of energy that moves and forms into different shapes over time. Energy will continue to move and produce different shapes long after you and I have gone, long after the Earth is gone, long after the universe is gone. The bit of energy of which we are made will continue to exist in other forms, just as it has existed in many other forms prior to this moment. The cycle is never ending because energy is never ending."

Sara remained quiet for a moment. "But Dina, what existed before energy?"

For the first time she sensed a note of disparagement in Dina's tone. "Only a human who lives in an ideasphere could ask this question, Sara. You know this answer."

She laughed. "I'm teasing you. I know the answer."

"Energy has always been here," they said in unison.

Tears welled up in Sara's eyes. "Thank you, Dina. I'm going to miss you. When I was little, I used to think heaven was knowing everything in the world. I know now that is never going to happen. That was just a childhood fantasy. Even after all we have talked about, I won't pretend to understand all the secrets of the universe. But because of you, it all makes so much more sense. Thank you for giving me a slice of my own personal heaven."

"It was my pleasure. I hope we will meet again someday." Dina paused again. "Sara?"

"Yes?"

"Before I go, I am compelled to . . . share something with you."

"Oh? What is that?'

"I . . . we, wish to arrange ourselves in a configuration that is meaningful to you. We have never done this before, but our experience with you has enabled us to glean the utility of this action."

"Sweet Dina, I haven't the foggiest idea what you are trying to say."

Suddenly the pink mist drifted through her entire body. She was enveloped in a deep sense of contentment. It cradled her gently as she seemed to float through the air. "We have developed a strong regard for you, Sara. We wanted to give you this human sense of peace as an expression of our gratitude."

Warmth flooded through her hazy tranquility. "Oh, Dina, I see now." She closed her eyes and gave into the sense of contentment. "We're all connected. Underneath it all, we are all one." She exhaled hazily. "We're all the same energy."

Sara and Shadow sat beside the garden where this whole crazy event started months ago. The tomato vines were crinkly with their drying leaves. A few red orbs still languished on the branches, but the cool autumn nights had taken their toll. The bright green leaves were now a dusty sage, and the sharp tang of overripe tomatoes hung in the air. She took in a deep breath.

Okay," she said to Shadow as she leaned down and picked up a handful of soil. "Go ahead, boy." At once, Shadow began to dig. He stopped and looked up at Sara, as if to ask if she was sure it was acceptable to do something she never ever let him do before. "Go ahead, boy. Dig away."

Sara smiled as she watched Shadow burrow and tunnel the dirt with enthusiasm. When the hole reached about a foot in depth, she reached for his collar. "Good boy, Shadow. Now, come here. Sit down right here, by my side."

Now came the hard part. Dina told her she needed to think of her saddest memory. It wasn't difficult to select that horrible moment, but she was damned unwilling to conjure it up again. How long had it taken her to come to an uneasy truce with her own mind about the loss of her beautiful Maggy? She lay down, hugged Shadow close, and began to remember.

The pain was so sharp that at first she almost stopped. Then she thought of Dina, and what would happen if the world lost her sweet little prokaryote friend. So, she let the pain come. It did not take long for the tears to follow. She wept deeply. Wept for the husband Maggy would never have. For the grandchildren not born. For the holidays and birthdays that now carried a tinge of sadness at her absence. Of she and Randy growing old without their precious daughter. She bawled like a baby. She couldn't help it. The hurt went so deep. As the tears poured from her eyes, she was not aware of the pink mist emanating from her eyes. Shadow sensed it. He whined and cuddled closer to Sara, but she was lost in her grief and did not see the pink haze form into an orb and float over the newly dug hole.

The wind picked up sharply and Shadow stood up in alarm. The pinkish globe wobbled erratically. It wavered near Sara's prone figure and then moved back toward the hole. Shadow began to bark wildly as the orb broke into a myriad of sparkling droplets following a strong gust. A few of the drops drifted back over Sara's face. Most of them floated into Shadow's eyes and nose and ears. A tiny glowing mass shimmered across the yard before it lifted toward the trees and scattered into the afternoon light.

It was nearly dark when she felt a gentle shake on her shoulder. "Sara, Sara, are you all right?"

Randy's voice held a tinge of panic. She opened her eyes. She and Shadow still snuggled together in the darkening light of evening. "I'm fine, Randy. Really. Don't worry. I'm fine."

"Are you sure? What are you doing out here?"

"Remembering," she said softly. "Just remembering." She stared down at the hole Shadow had dug. It appeared untouched. Did it work? Did Dina make it back into the hole? She couldn't tell.

"Would you like to go in the house?"

"Yes. I would. It's getting cool out here."

Randy helped her up, but her legs couldn't carry her weight, and she collapsed to the ground. "Sara, Sara."

"It's okay. Dina said this would happen."

"Dina? Dina came back?"

"Yes. To say goodbye." She sighed. "Dina needed to leave. I'll never see my little friend again."

He lifted her up. "Let's get you into the house." He brought her inside and laid her on the sofa in the living room. "Comfortable? I'll get you some tea."

Sara stopped him as he stood up. She grasped his hand tightly. "Randy?"

He smiled down at her. "Yes, sweetheart?"

"She kissed his hand. "I'm glad I chose you."

He kissed her gently. "I'm glad you chose me, too. I love you." He stood up. "I'll be right back."

When he returned, she was fast asleep. Randy set the cup on the table, then pressed his fingers against the pulse at her neck.

Weak. Too weak. He sat down in the chair next to her. When would this nightmare end?

The phone rang, breaking into his thoughts.

"Hello? Hey Josh." He paused a moment, then a large smile broke across his face. "That's great, son. Congratulations! Seven pounds, nine ounces? Twenty inches? She has all her fingers and toes? Fantastic." He listened to Josh for a few more minutes. "I'm not sure we can make it tonight. Your mother's pretty tired today. I'll call you back in a bit and let you know."

Randy hung up and smiled down at his sleeping wife. "Well, sweetheart, your son is a father. Baby Maggy arrived a few minutes ago."

He leaned over and peered at his wife. "Sara. Sara?" He felt her pulse. "Saraaaaaaa!"

Chapter 37

NOW WHAT?

Dear Reader,

My mom was always a person who wanted to learn how things fit together. She had a certain sense when an idea didn't cut it. Her bullshit meter, she used to call it. She learned to heed that meter; it never let her down. I always thought it was a backwards tool at best. It told her when something didn't make sense, but it never gave her any answers about what did make sense.

That is why this book, which I transcribed from our talks about what happened when she was in her coma, resonates with me. She never caught on, but I did. In all the time she related these events, she never noticed her bullshit meter had gone quiet.

Dad and I recorded her experiences during long conversations after she woke up, when her memories were fresh and clear. I'm glad we did. It's all we have of her now. We miss her so much. It's kind of hard to explain how good it is to hear her voice on those recordings, her laugh, and, I can't believe I'm writing this, even her stupid jokes.

My dad is busted up. I worry about him and hope he will be able to recover and live the life Mom would want for him. He says he's lost his two favorite girls. How the hell am I supposed to respond to that? I can't begin to comprehend his pain.

I put together this book in the hopes that what turned out to be my mom's final wishes can be made available to anyone who might be interested in understanding them and, even better, act on them. She would say I have done what she couldn't do; store these ideas outside her body so they can be shared across space and time by other folks who live in the ideasphere. I think she would like this as her legacy. Otherwise, it will be her bullshit meter. Or her bad jokes.

As with any idea, the concepts can be simple, but the difficulty is always in the details. I think she'd be the first to say the ideas she and Dina discussed are clear, self-evident, even obvious, once they are explained. Mom lived her life learning these disparate ideas but was never able to assemble them into a cohesive whole until Dina came along. She would also say that life does make sense if we could only grasp the entire picture, not just the separate bits and pieces we manage to pick up as we live our lives.

Sensible as it is, getting from where we are today to a better future, what Mom called the 'idealsphere;' that's going to take some doing. That's another reason why I put this book together.

We are here, living in an imperfect ideasphere. If we continue to go on the way we have been, we face a real risk we won't survive. Other entities will continue, so life will go on. But humans won't. And when we go, we'll take the ideasphere with us.

Maybe Dina's clan is still out somewhere finding better ways to communicate with people. I'm not sure if Dina survived or died along with my mom. But I do know this. We have these ideas now. They survive now, without Dina or my mom. As long as they can be accepted and shared and acted on, they can grow. That's where you come in.

To travel from here to there, from our present status to a healthy future, it will take all of us to do our part to re-think how we exist in this interdependent world we live in. It is up to each of us to decide whether to put these ideas into action. How? That's a fair question.

First, we can focus on living our individual lives by acting on the ideas we value without harming others. Imagine how much better the world would be by just doing that. Right now. Wouldn't the world be better if each of us put this one clear idea into practice every day from here forward? Hell, at least we'd be a lot happier.

We have a lot of problems in our country and in our world. My Mom planned to write a 'Big List of Issues' she thought we should address. She thought if we viewed these issues using the perspective of the idealsphere, we might come up with more efficient and effective solutions. I think her approach is a valid one. It's a difficult task and will take a lot of time and knowledge from experts who understand the details about each of these problems. But she never got around to starting the list.

I recently had the fantastic fortune of welcoming a new baby into my life, so I won't be sleeping for the next 18 years. My Dad is not ready to take on any projects right now, least of all one that would bring more heartbreak than joy. So, I am doing what any creature in the ecosystem would do: I'm throwing this task out for collaboration. Let's test this concept that a new idea can survive though the originators are no longer here.

If you are an expert in any area, try this experiment: re-examine your area of expertise in the context of the idealsphere. Study the issue from the perspective of a cell or body in the ecosystem. Adjust for human choice. Create new solutions using this perspective. Then tell the world about them. Write a book. Make a podcast. Give a TED talk. Share your ideas! Bonus points if you add humor to the mix. My mom would love that.

Don't stop there. If you find possible solutions, and you believe in them, widen your net. Start an institution to put your ideas into action. Become a public servant where you can act on these ideas. Run for office with the idea of transforming your ideas into law.

I know my mom would be grateful. She'd be the first one to admit that experts will do a much better job of coming up with solutions and conveying them than she ever could. She always said she was a mile wide and an inch deep. But it was this exact mix that enabled her to combine the disparate ideas that have been running around the ideasphere for thousands of years into a cohesive whole. Personally, I think it was this mix that caused Dina to choose her.

If we are to survive for any length of time, if we are to have a future in which my daughter and her daughter and her daughter will survive and thrive, we need to align our ideas and actions in a way that fits the world we live in. The ecosystem has provided the roadmap. Our minds have given us the means to read the map. It is now time to go down the road and make a better world. Remember . . .

> *People ought to be free to create any idea and convert it to any action they choose, as long as their actions do not harm other people or the society as a whole.*

Thanks for reading,

Josh

Josh saved the file and closed his laptop. He leaned down to ruffle Shadow's ears. "Well, boy, let's see what happens now. I think Mom would be happy."

Shadow wagged his tail. Josh didn't notice the pink tinge around Shadow's eyes and nose as the dog plopped down and let out a deep sigh.

~~~~~~
~~~~~~

SOURCES

Chapter 1

1. "Tomato, Mortgage Lifter," *Burpee*, As of December 20, 2023. https://www.burpee.com/tomato-mortgage-lifter-prod000998.html.
2. "Tomato, Early Girl Hybrid," *Burpee*, As of December 20, 2023. https://www.burpee.com/tomato-early-girl-hybrid-prod000986.html.
3. "Amana Orange Tomato Seed," *Victory Gardeners*, As of December 20, 2023. https://victorygardeners.com/product/amana-orange-tomato-seed.

Chapter 2

1. Larry M. Silverberg, "Physicists Suggest All Matter May Be Made Up of Energy 'Fragments'." *Science Alert*, 11 December 2020. https://www.sciencealert.com/physicists-suggest-energy-fragments-is-the-best-way-to-describe-matter
2. James Trefil, *1001 Things Everyone Should Know About Science* (New York, Doubleday, 1992), p. 265.
3. Yoshinari Hayato (Associate Professor) and Yumiko Takenaga, "Were All Forces, Electromagnetic Force And Gravity, Originally One?" *Kamioka Observatory, ICRR, The University of Tokyo,* As of December 20, 2023, https://www-sk.icrr.u-tokyo.ac.jp/en/hk/special/yonde04/
4. Swinburne University of Technology, "Laws Of Physics Vary Throughout The Universe, New Study Suggests." *ScienceDaily*, 9 September 2010. www.sciencedaily.com/releases/2010/09/100909004112.htm.
5. CERN, "The Standard Model," *Origins Project*, September 16, 2022, https://www.exploratorium.edu/origins/cern/ideas/standard3.html
6. NASA Science Universe Exploration, "The Universe's History," *NASA*, At December 20, 2023, https://universe.nasa.gov/universe/basics/#:~:text=Around%20380%2C000%20years%20after%20the,call%20the%20epoch%20of%20recombination
7. Michael D. Lemonick, "Cosmic Nothing," *Scientific American*, January 2024, Volume 330, Number 1, p.20.
8. Paul Scott Anderson, "A New Look At The Universe's Oldest Light," *EarthSky.org*, January 12, 2021, https://earthsky.org/space/a-new-look-at-the-universes-oldest-light/.
9. Iqbal Pittalwala, "Scientists Precisely Measure Total Amount Of Matter In The Universe," *Phys.Org*, September 28, 2020, https://phys.org/news/2020-09-scientists-precisely-total-amount-universe.html.

Chapter 3

1. Las Cumbres Observatory, "The Early Universe," *Spacebook*, As of September 17, 2022, https://lco.global/spacebook/cosmology/early-universe/.

2. Cecille De Jesus, "The Edge of Physics: Do Gravitons Really Exist?" *Futurism*, December 24, 2016, https://futurism.com/the-edge-of-physics-do-gravitons-really-exist.
3. Andrew Zimmerman Jones, "Stellar Nucleosynthesis: How Stars Make All Of The Elements," *ThoughtCo.*, May 30, 2019, https://www.thoughtco.com/stellar-nucleosynthesis-2699311.
4. Peter Hoppe, Wataru Fujiya and Ernst Zinner, "The Chemistry Of Exploding Stars," *Max-Planck Gesellschaft*, January 19, 2012, https://www.mpg.de/4992099/meteorite_murchison_supernova.
5. Ethan Siegel, "Ask Ethan: How Many Generations Of Stars Formed Before Our Sun?" *Forbes*, October 26, 2019, https://www.forbes.com/sites/startswithabang/2019/10/26/ask-ethan-how-many-generations-of-stars-formed-before-our-sun-did/?sh=6789ffd51b84.
6. Jonathan B. Losos, *Improbable Destinies, Fate, Chance, and the Future of Evolution* (New York, Riverhead Books, 2017), p. 14.
7. PBS, "Punctuated Equilibrium," *PBS.org*, As of September 17, 2022, https://www.pbs.org/wgbh/evolution/library/03/5/l_035_01.html.
8. Shobhit Jain and Lindsay M. Iverson, "Glasgow Coma Scale," *National Library of Medicine, National Center for Biotechnology Information*, June 12, 2023, https://www.ncbi.nlm.nih.gov/books/NBK513298/

Chapter 4

1. John Wenz, "Scientists Find Strong Evidence That the Earth Was Hit Head-On by a Mars-Sized Planet," *Popular Mechanics*, January 28, 2016, https://www.popularmechanics.com/space/a19143/earth-moon-theia-collision/.
2. Richard Meckian, "When A Day Lasted Only 4 Hours," *Institute of Advanced Studies of the University of Sao Paulo*, March 16, 2016, http://www.iea.usp.br/en/news/when-a-day-lasted-only-four-hours.
3. NASA, "How Far Away Is The Moon?" *NASA Science Space Place*, As of September 20, 2022, https://spaceplace.nasa.gov/moon-distance/en/.
4. Marina Koren, "The Moon Is Leaving Us," *The Atlantic*, September 30, 2021, https://www.theatlantic.com/science/archive/2021/09/moon-moving-away-earth/620254/.
5. Dr. Alfredo Carpinetti, "Fragments Of The Planet That Formed The Moon May Be Buried By The Earth's Core," *IFL Science*, March 26, 2021, https://www.iflscience.com/fragments-of-the-planet-that-formed-the-moon-may-be-buried-by-the-earths-core-59172.
6. John Baez, "The Earth – For Physicists," *Physics World*, August 2009, https://math.ucr.edu/home/baez/physics_world_earth.pdf.
7. Alan Buis, "Earth's Magnetosphere: Protecting Our Planet from Harmful Space Energy," *NASA Jet Propulsion Laboratory*, August 3, 2021, https://climate.nasa.gov/news/3105/earths-magnetosphere-protecting-our-planet-from-harmful-space-energy/.

8. Jim Erickson, "Study Suggests Ground-Dwelling Mammals Survived Mass Extinction 66 Million Years Ago," *Michigan News – University of Michigan*, October 11, 2021, https://news.umich.edu/study-suggests-ground-dwelling-mammals-survived-mass-extinction-66-million-years-ago/.
9. Four Peaks Technologies, "Early Earth - Volcanoes Were Everywhere," *Early Earth Central*, As of September 24, 2022, http://www.earlyearthcentral.com/early_earth_page.html#volcanoes.
10. Hannah Hickey, "Volcanic eruptions may have spurred first 'whiffs' of oxygen in Earth's atmosphere," *UW News*, August 25, 2021, https://www.washington.edu/news/2021/08/25/volcanic-eruptions-may-have-spurred-first-whiffs-of-oxygen-in-earths-atmosphere/.
11. Catherine Brahic, "Volcanic mayhem drove major burst of evolution," *New Scientist,* January 15, 2014, https://www.newscientist.com/article/mg22129522-600-volcanic-mayhem-drove-major-burst-of-evolution/.
12. Michael Rothschild, *Bionomics, Economy as System* (New York, Henry Holt and Company, 1990), p. 260
13. Stephen Jay Gould, *Wonderful Life*, The Burgess Shale and the Nature of History (New York, W.W. Norton and Co., 1989), pp. 300-323.
14. Losos, *Improbable Destinies, Fate, Chance, and the Future of Evolution*, p. 7.
15. BBC History Magazine, "Did Charles Darwin Coin The Phrase 'Survival Of The Fittest'?, *History Extra*, January 28, 2022, https://www.historyextra.com/period/victorian/survival-of-the-fittest-meaning-origin/.

Chapter 5

1. Regents of the University of Michigan, "Evolution Of The Atmosphere: Composition, Structure And Energy," *Introduction to Global Change I – Lecture Notes*, As of September 23, 2022, https://www.globalchange.umich.edu/globalchange1/current/lectures/Perry_Samson_lectures/evolution_atm/
2. Paul Webb, "Chapter 5.2 Origin of the Oceans," *Introduction to Oceanography*, (Pressbooks.pub, 2021), https://rwu.pressbooks.pub/webboceanography/chapter/5-2-origin-of-the-oceans/.
3. Water Science School, "Water, the Universal Solvent," *USGS Science for a Changing World*, June 9, 2018, https://www.usgs.gov/special-topics/water-science-school/science/water-universal-solvent.
4. The Editors of Encyclopaedia Britannica, "Electric Charge – Physics," *Brittanica.com*, September 9. 2022, https://www.britannica.com/science/electric-charge.
5. Evrim Yazgin, "Scientists Narrow Down The Theory About The Origins Of Life – Volcanoes Or Meteors," *Cosmos Magazine*, As of December 30, 2023, https://cosmosmagazine.com/earth/meteor-volcano-origin-life/#:~:text=New%20research%20shows%20how%20iron,of%20life%20on%20our%20planet

6. Joseph A. Resing and Francis J. Sansone, "The Chemistry Of Lava–Seawater Interactions: The Generation Of Acidity," *Geochimica et Cosmochimica Acta*, May 17, 1999, http://www.soest.hawaii.edu/oceanography/faculty/sansone/Resing%20&%20Sansone%201999%20GCA.pdf
7. Trefil, *1001 Things Everyone Should Know About Science,* p. 254
8. Jason P. Schrum, Ting F. Zhu, Jack W. Szostak, "The Origins Of Cellular Life," *Cold Spring Harb Perspect Biol. 2010;2(9):a002212. doi:10.1101/cshperspect.a002212,* https://www.ncbi.nlm.nih.gov/pmc/articles/PMC2926753/
9. Jennifer Newton, "Carbohydrates Promoted In New Prebiotic Theory," *Chemistry World*, September 11, 2017, https://www.chemistryworld.com/news/carbohydrates-promoted-in-new-prebiotic-theory/3007970.article
10. Michael K Reddy, "Amino Acid – Chemical Compound," *Brittanica.com*, August 22, 2022, https://www.britannica.com/science/amino-acid
11. Harvard TH Chan School of Public Health, "The Nutrition Source - Protein," *Harvard TH Chan School of Public Health*, As of September 23, 2022, https://www.hsph.harvard.edu/nutritionsource/what-should-you-eat/protein/
12. Ana Gutiérrez-Preciado and Hector Romero and Mariana Peimbert, "An Evolutionary Perspective on Amino Acids," *Nature Education*, 2010, https://www.nature.com/scitable/topicpage/an-evolutionary-perspective-on-amino-acids-14568445/
13. Uma Shanker, Brij Bhushan, G. Bhattacharjee, and Kamaluddin, "Formation of Nucleobases from Formamide in the Presence of Iron Oxides: Implication in Chemical Evolution and Origin of Life," *Astrobiology*, April 19, 2011, https://www.liebertpub.com/doi/abs/10.1089/ast.2010.0530
14. Irene A Chen and Peter Walde, "From Self-Assembled Vesicles to Protocells," *Cold Spring Harb Perspect Biol. 2010;2(7):a002170. doi:10.1101/cshperspect.a002170,* https://www.ncbi.nlm.nih.gov/pmc/articles/PMC2890201/
15. Annabelle Biscans, "Exploring the Emergence of RNA Nucleosides and Nucleotides on the Early Earth." *Life (Basel)*. 2018;8(4):57. Published 2018 Nov 6. doi:10.3390/life8040057 https://www.ncbi.nlm.nih.gov/pmc/articles/PMC6316623/
16. Trefil, *1001 Things Everyone Should Know About Science,* p. 80
17. Ying-Wei Yang, Su Zhang, Elizabeth O McCullum, John C Chaput, "Experimental Evidence That GNA And TNA Were Not Sequential Polymers In The Prebiotic Evolution of RNA," *J Mol Evol*, 2007, https://pubmed.ncbi.nlm.nih.gov/17828568/.
18. David Wang and Aisha Farhana, "Biochemistry, RNA Structure," [Updated 2022 May 8]. In: *StatPearls* (Treasure Island (FL), StatPearls Publishing; 2022 Jan.) https://www.ncbi.nlm.nih.gov/books/NBK558999/

19. Ruairi J Mackenzie, "DNA vs. RNA – 5 Key Differences and Comparison," *Technology Networks – Genomic Research*, March 31, 2022, https://www.technologynetworks.com/genomics/lists/what-are-the-key-differences-between-dna-and-rna-296719
20. B Alberts, A Johnson, J Lewis, et al., "The RNA World and the Origins of Life," *Molecular Biology of the Cell. 4th edition.* (New York: Garland Science; 2002), https://www.ncbi.nlm.nih.gov/books/NBK26876/
21. B Alberts, A Johnson, J Lewis, et al., "From DNA to RNA," *Molecular Biology of the Cell. 4th edition.* (New York: Garland Science; 2002), https://www.ncbi.nlm.nih.gov/books/NBK26887/
22. Reginald Davey, "What Is A Protocell?" *AZO Life Sciences*, As of September 23, 2022, https://www.azolifesciences.com/article/What-is-a-Protocell.aspx
23. Norio Kitadai and Shigenori Maryuama, "Origins Of Building Blocks of Life: A Review," *Geoscience Frontiers*, July 2018, https://www.sciencedirect.com/science/article/pii/S1674987117301305

Chapter 6

1. John Hopkins Medicine, "What is an EEG?" *John Hopkins University*, As of September 26, 2022, https://www.hopkinsmedicine.org/health/treatment-tests-and-therapies/electroencephalogram-eeg
2. Sarah Lewis, "Understanding Your EEG Results," Healthgrades, August 3, 2020, https://www.healthgrades.com/right-care/electroencephalogram-eeg/understanding-your-eeg-results

Chapter 7

1. Tina Batista Napotnik, Tamara Polajžer, Damijan Miklavčič, "Cell Death Due To Electroporation – A review," *Bioelectrochemistry, Volume 141, 2021, 107871, ISSN 1567-5394*, https://doi.org/10.1016/j.bioelechem.2021.107871, https://www.sciencedirect.com/science/article/pii/S1567539421001341
2. Jason P. Schrum, Ting F. Zhu, and Jack W. Szostak, "The origins of cellular life," *Cold Spring Harb Perspect Biol. 2010;2(9):a002212. doi:10.1101/cshperspect.a002212*, https://www.ncbi.nlm.nih.gov/pmc/articles/PMC2926753/
3. GM Cooper, "The Origin and Evolution of Cells," *The Cell: A Molecular Approach. 2nd edition.* (Sunderland (MA): Sinauer Associates; 2000), https://www.ncbi.nlm.nih.gov/books/NBK9841/
4. Google Classroom, "Prokaryote Structure," *Khan Academy*, As of September 28, 2022, https://www.khanacademy.org/science/ap-biology/gene-expression-and-regulation/dna-and-rna-structure/a/prokaryote-structure

5. Google Classroom, "Prokaryote Metabolism," *Khan Academy*, As of September 28, 2022, https://www.khanacademy.org/science/biology/bacteria-archaea/prokaryote-metabolism-ecology/a/prokaryote-metabolism-nutrition
6. Sally Warring, "Cells Living In Cells,": *Ask A Biologist, Arizona State University*, As of September 28, 2022, https://askabiologist.asu.edu/explore/cells-living-in-cells
7. CK-12 Life Science for Middle School, "4.12 Geologic Time Scale," *CK-12*, June 1, 2020, https://flexbooks.ck12.org/cbook/ck-12-middle-school-life-science-2.0/section/4.12/primary/lesson/timeline-of-evolution-ms-ls/
8. Nicole Gleichmann, "Prokaryotes vs Eukaryotes: What Are the Key Differences?" *Technology Networks – Cell Science*, July 8, 2021, https://www.technologynetworks.com/cell-science/articles/prokaryotes-vs-eukaryotes-what-are-the-key-differences-336095
9. Georgia Tech Biological Sciences, "Organismal Biology," *Georgia Tech University*, As of September 28, 2022, https://organismalbio.biosci.gatech.edu/biodiversity/prokaryotes-bacteria-archaea-2/

Chapter 8

1. Science Council, "Our Definition of Science," *The Science Council*, As of September 29, 2022, https://sciencecouncil.org/about-science/our-definition-of-science/
2. Norio Kitadia and Shigenori Maryuama, "Origins Of Building Blocks of Life: A Review," *Geoscience Frontiers*, July 2018, https://www.sciencedirect.com/science/article/pii/S1674987117301305
3. John Staughton, "How Long Did It Take For Multicellular Life To Evolve From Unicellular Life?" *Science ABC*, January 6, 2022, https://www.scienceabc.com/pure-sciences/how-long-did-it-take-for-multicellular-life-to-evolve-from-unicellular-life.html
4. Palomar College, "The First Primates," *Palomar College*, As of September 20, 2022, https://www.palomar.edu/anthro/earlyprimates/early_2.htm
5. CK-12 Life Science for Middle School, "4.12 Geologic Time Scale," *CK-12*, June 1, 2020, https://flexbooks.ck12.org/cbook/ck-12-middle-school-life-science-2.0/section/4.12/primary/lesson/timeline-of-evolution-ms-ls/
6. Michael Dhar, "What is RNA?" *Live Science*, October 15, 2020, https://www.livescience.com/what-is-RNA.html
7. B Alberts, A Johnson, J Lewis, et al., *Molecular Biology of the Cell. 4th edition*, (New York: Garland Science; 2002), "From DNA to RNA," https://www.ncbi.nlm.nih.gov/books/NBK26887/
8. B Alberts, A Johnson, J Lewis, et al., *Molecular Biology of the Cell. 4th edition*, (New York: Garland Science; 2002), "DNA Replication Mechanisms." https://www.ncbi.nlm.nih.gov/books/NBK26850/

9. Duke University, "Tree Of Life' for 2.3 million species released," *Phys.Org*, September 19, 2015, https://phys.org/news/2015-09-tree-life-million-species.html
10. Your Genome.org, "What Is A Gene?", *Your Genome.org*, October 6, 2016, https://www.yourgenome.org/facts/what-is-a-gene/
11. Biology for Majors I, "Expression of Genes," *Lumen Learning*, As of September 30, 2022, https://courses.lumenlearning.com/wm-biology1/chapter/reading-expression-of-genes/

Chapter 9

1. Marshall Brain, "How Gene Pools Work," *How Stuff Works.com*, As of September 30, 2022, https://science.howstuffworks.com/life/genetic/gene-pool1.htm
2. Leslie Pray, "DNA Replication and Causes of Mutation," *Nature Education*, As of September 30, 2022, https://www.nature.com/scitable/topicpage/dna-replication-and-causes-of-mutation-409/
3. Charles Darwin, *From So Simple A Beginning – The Four Great Books of Charles Darwin, On the Origin of Species By Means of Natural Selection*, (1859), p. 502.
4. Hannah Ritchie and Max Roser, "Biodiversity - Extinctions," *Our World in Data*, As of September 30, 2022, https://ourworldindata.org/extinctions
5. USGS, "When Did Dinosaurs Become Extinct?" *USGS*, As of September 30, 2022, https://www.usgs.gov/faqs/when-did-dinosaurs-become-extinct
6. Carl Zimmer, 'How Many Cells are In Your Body?" *National Geographic*, October 23, 2013, https://www.nationalgeographic.com/science/article/how-many-cells-are-in-your-body

Chapter 10

1. Luis Villazon, "When We Die, Does Our Whole Body Die At The Same Time?" *BBC Science Focus*, As of October 1, 2022, https://www.sciencefocus.com/the-human-body/when-we-die-does-our-whole-body-die-at-the-same-time/
2. Nabeeha Khalid; Mahzad Azimpouran, "Necrosis," [Updated 2022 Mar 9]. *StatPearls Publishing* (Treasure Island (FL), 2022) March 9, 2022, https://www.ncbi.nlm.nih.gov/books/NBK557627/
3. B Alberts, A Johnson, J Lewis, et al., "Programmed Cell Death (Apoptosis)" *Molecular Biology of the Cell. 4th edition*, (New York: Garland Science; 2002), https://www.ncbi.nlm.nih.gov/books/NBK26873/
4. Rosa Ana Risques and Daniel E. L. Promislow, "All's Well That Ends Well: Why Large Species Have Short Telomeres," *Royal Society Publishing*, January 15, 2018, https://royalsocietypublishing.org/doi/10.1098/rstb.2016.0448

5. Gardner M, Bann D, Wiley L, Cooper R, Hardy R, Nitsch D, Martin-Ruiz C, Shiels P, Sayer AA, Barbieri M, Bekaert S, Bischoff C, Brooks-Wilson A, Chen W, Cooper C, Christensen K, De Meyer T, Deary I, Der G, Diez Roux A, Fitzpatrick A, Hajat A, Halaschek-Wiener J, Harris S, Hunt SC, Jagger C, Jeon HS, Kaplan R, Kimura M, Lansdorp P, Li C, Maeda T, Mangino M, Nawrot TS, Nilsson P, Nordfjall K, Paolisso G, Ren F, Riabowol K, Robertson T, Roos G, Staessen JA, Spector T, Tang N, Unryn B, van der Harst P, Woo J, Xing C, Yadegarfar ME, Park JY, Young N, Kuh D, von Zglinicki T, Ben-Shlomo Y; Halcyon study team. "Gender and telomere length: systematic review and meta-analysis," *Exp Gerontol*, March 2014;51:15-27. https://www.ncbi.nlm.nih.gov/pmc/articles/PMC4523138/
6. Nicholas Brealey, "Know Your Brain: Reticular Formation," *Neuroscientifically Challenged*, From J. Nolte, The Human Brain: An Introduction to its Functional Anatomy. 6th ed. (Philadelphia, PA. Elsevier; 2009), https://neuroscientificallychallenged.com/posts/know-your-brain-reticular-formation
7. Susan Tyree, Jeremy Borniger, and Luis de Lecea, "Hypocretin as a Hub for Arousal and Motivation," *Frontiers in Neurology*, June 6, 2018, https://www.frontiersin.org/articles/10.3389/fneur.2018.00413/full
8. Life XCHange, "Your Brain at Work: The Reticular Activating System (RAS) and Your Goals & Behaviour," *Life XCHange*, As of October 1, 2022, https://lifexchangesolutions.com/reticular-activating-system/

Chapter 11

1. Michael Pollen, *Food Rules, An Eaters Manual*, (New York, Penguin Books, 2009)
2. Gavin van De Walle, "9 Important Functions of Protein in Your Body," *HealthLine*, June 20, 2018, https://www.healthline.com/nutrition/functions-of-protein
3. Harvard TH Chan School of Public Health, "The Nutrition Source - Carbohydrates," *Harvard TH Chan School of Public Health*, As of September 23, 2022, https://www.hsph.harvard.edu/nutritionsource/carbohydrates/
4. Anne Helmenstine, "What Is the Most Abundant Element in the Universe?" *Science Notes Posts*, October 27, 2020, https://sciencenotes.org/what-is-the-most-abundant-element-in-the-universe/
5. Oxford University Museum of Natural History, "Bacterial World," *University of Oxford*, October 19, 2018, http://www.oum.ox.ac.uk/bacterialworld/
6. Hannah Ritchie, "Humans Make Up Just 0.01% of Earth's Life – What's The Rest?" *Our World in Data*, April 24, 2019, https://ourworldindata.org/life-on-earth

7. Bill Chappell, "Along with Humans, Who Else Is In The 7 Billion Club?" *The Two-Way, NPR,* November 3, 2011, https://www.npr.org/sections/thetwo-way/2011/11/03/141946751/along-with-humans-who-else-is-in-the-7-billion-club

Chapter 12

1. Bert Markgraf, "The Major Structural Advantage Eukaryotes Have Over Prokaryotes," *Sciencing,* July 31, 2019, https://sciencing.com/major-structural-advantage-eukaryotes-over-prokaryotes-14989.html
2. Jamie Ellis, "The Internal Anatomy of the Honey Bee," *American Bee Journal,* September 1, 2015, https://americanbeejournal.com/the-internal-anatomy-of-the-honey-bee/
3. Yella Hewings-Martin, PhD, "How many cells are in the human body?" *Medical News Today,* July 12, 2017, https://www.medicalnewstoday.com/articles/318342
4. Adam Smith, *The Wealth of Nations* (Penguin Books, London, 1999), p. 118.

Chapter 13

1. Darwin, *From So Simple A Beginning – The Four Great Books of Charles Darwin, On the Origin of Species By Means of Natural Selection*, pp. 489 – 502.
2. Rothschild, Bionomics pp. 40, 62, 207.
3. Press Association, "Ancestor Of Humans And Other Mammals Was Small Furry Insect Eater," *The Guardian,* February 7, 2013, https://www.theguardian.com/science/2013/feb/07/ancestor-humans-mammals-insect-eater
4. Galapagos Conservancy, "History of Galapagos," *Galapagos Conservancy,* As of October 6, 2022, https://www.galapagos.org/about_galapagos/history/
5. Marie Iannotti, 'What is Allelopathy?" *The Spruce,* May 11, 2022, https://www.thespruce.com/what-is-allelopathy-1402504

Chapter 14

1. Rothschild, *Bionomics*, p 336
2. Jason P. Schrum, Ting F. Zhu, Jack W. Szostak, "The Origins Of Cellular Life," *Cold Spring Harb Perspect Biol. 2010;2(9):a002212. doi:10.1101/cshperspect.a002212* https://www.ncbi.nlm.nih.gov/pmc/articles/PMC2926753/
3. B. Alberts, A Johnson, J Lewis, et al., "Fractionation of Cells," *Molecular Biology of the Cell. 4th edition,* (New York: Garland Science; 2002) https://www.ncbi.nlm.nih.gov/books/NBK26936/

4. Google Classroom, "Prokaryote Metabolism," *Khan Academy*, As of September 28, 2022, https://www.khanacademy.org/science/biology/bacteria-archaea/prokaryote-metabolism-ecology/a/prokaryote-metabolism-nutrition
5. Centers for Disease Control and Prevention, "SARS-CoV-2 Variant Classifications and Definitions," *CDC*, April 26, 2022, https://www.cdc.gov/coronavirus/2019-ncov/variants/variant-classifications.html
6. Marcelo Gleiser, "The Microbial Eve: Our Oldest Ancestors Were Single-Celled Organisms," *NPR*, January 31, 2018, https://www.npr.org/sections/13.7/2018/01/31/581874421/be-humbled-our-oldest-ancestors-were-single-celled-organisms
7. GM Cooper, "The Origin and Evolution of Cells," *The Cell: A Molecular Approach. 2nd edition.* (Sunderland MA, Sinauer Associates; 2000), https://www.ncbi.nlm.nih.gov/books/NBK9841/
8. William B Whitman, David C Coleman, William J Wiebe, "Prokaryotes: the unseen majority," *Proc Natl Acad Sci U S A. 1998 Jun 9;95(12):6578-83. doi: 10.1073/pnas.95.12.6578. PMID: 9618454; PMCID: PMC33863,* https://www.ncbi.nlm.nih.gov/pmc/articles/PMC33863/
9. Georgia Tech Biological Sciences, "Organismal Biology," *Georgia Tech University*, As of September 28, 2022, https://organismalbio.biosci.gatech.edu/biodiversity/prokaryotes-bacteria-archaea-2/
10. Jordi van Gestel, Martin A. Nowak, Corina E. Tarnita, "The Evolution of Cell-to-Cell Communication in a Sporulating Bacterium," *PLOS Computational Biology*, December 20, 2012, https://journals.plos.org/ploscompbiol/article?id=10.1371/journal.pcbi.1002818
11. Biology Online, "Dictionary: Sexual Reproduction," *Biology Online*, As of October 9, 2022, https://www.biologyonline.com/dictionary/sexual-reproduction
12. Carlos J Melián, David Alonso, Stefano Allesina, Richard S Condit, Rampal S Etienne, "Does Sex Speed Up Evolutionary Rate And Increase Biodiversity?" *PLoS Comput Biol. 2012;8(3):e1002414. doc: 10.1371/journal.pcbi.1002414. Epub 2012 Mar 8. PMID: 22412362; PMCID: PMC3297559.* https://www.ncbi.nlm.nih.gov/pmc/articles/PMC3297559/
13. George Washington University, "The Evolution of Early Animal Complexity," *George Washington University*, As of October 9, 2022, https://www2.gwu.edu/~darwin/BiSc151/NoCoelom/earlyanimal.html
14. Carl Zimmer, "To Bee," *Carl Zimmer.com*, October 25, 2006, https://carlzimmer.com/to-bee/

Chapter 15

1. Rusty Burlew, "Why Do Honey Bees Need Fur?" *Honey Bee Suite*, As of October 10, 2022, https://www.honeybeesuite.com/why-do-honey-bees-need-fur/
2. Massimo Nepi, Donato A. Grasso, Stefano Mancuso "Nectar in Plant–Insect Mutualistic Relationships: From Food Reward to Partner Manipulation," *Frontiers in Plant Science*, As of July 19, 2018, https://www.frontiersin.org/articles/10.3389/fpls.2018.01063/full
3. Bionumbers, "Cell Biology By The Numbers - How Many Cells Are There In An Organism?" *Bionumbers*, As Of October 10, 2022, http://book.bionumbers.org/how-many-cells-are-there-in-an-organism/
4. Rothschild, *Bionomics*, p 33
5. GM Cooper, "The Origin and Evolution of Cells," *The Cell: A Molecular Approach. 2nd edition.* (Sunderland (MA): Sinauer Associates; 2000), https://www.ncbi.nlm.nih.gov/books/NBK9841/
6. Andrea Stephenson, Justin W. Adams, Mauro Vaccarezza, "The Vertebrate Heart: An Evolutionary Perspective," *Journal Of Anatomy*, September 14, 2017, https://onlinelibrary.wiley.com/doi/10.1111/joa.12687
7. National Cancer Institute, "What Is Cancer?" *National Cancer Institute*, As of October 10, 2022, https://www.cancer.gov/about-cancer/understanding/what-is-cancer

Chapter 16

1. Rothschild, Bionomics, pp. 96, 161
2. AO Spakov, MN Pertseva, "Signal Transduction Systems Of Prokaryotes," *Zh Evol Biokhim Fiziol. 2008 Mar-Apr;44(2):113-30. Russian. PMID: 18669273,* https://pubmed.ncbi.nlm.nih.gov/18669273/
3. Pedro C. Marijuan, Jorge Navarro, Raquel del Moral, "How Prokaryotes 'Encode' Their Environment: Systemic Tools For Organizing The Information Flow," *ResearchGate*, October 2017, https://www.researchgate.net/publication/320320195_How_prokaryotes_%27encode%27_their_environment_Systemic_tools_for_organizing_the_information_flow
4. Gabe Buckley, "Cell Signalling," *Biology Dictionary*, January 15, 2021, https://biologydictionary.net/cell-signaling/
5. David Robson, "A Brief History of the Brain," *New Scientist*, September 21, 2011, https://www.newscientist.com/article/mg21128311-800-a-brief-history-of-the-brain/?ignored=irrelevant
6. Charlotte Swanson, "Bizarre Brains of the Animal Kingdom," *Science World*, February 1, 2022, https://www.scienceworld.ca/stories/bizarre-brains-animal-kingdom/
7. Jon Lieff M.D., "The Remarkable Bee Brain," *Jon Lieff, M.D.*, November 12, 2012, https://jonlieffmd.com/blog/the-remarkable-bee-brain-2

8. Jamie Ellis, "The Internal Anatomy of the Honey Bee," *American Bee Journal*, September 1, 2015, https://americanbeejournal.com/the-internal-anatomy-of-the-honey-bee/
9. DinoAnimals, "Number of Neurons in the Brain of Animals," *DinoAnimals*, As of October 12, 2022, https://dinoanimals.com/animals/number-of-neurons-in-the-brain-of-animals/
10. Bernd Heinrich, *Bumblebee Economics* (Harvard University Press, Cambridge, MA, 1979), p 8-16.
11. Cleveland Clinic, "Meninges," *Cleveland Clinic*, January 11, 2022, https://my.clevelandclinic.org/health/articles/22266-meninges
12. Ask the Scientists, "Cell Signaling: How Your Cells Talk To Each Other," Ask The Scientists, As of October 12, 2022, https://askthescientists.com/qa/what-is-cell-signaling/
13. Aaron Kandola, "What Is The Autonomic Nervous System?" *Medical News Today*, January 10, 2020, https://www.medicalnewstoday.com/articles/327450
14. DNA From the Beginning, "DNA Responds To Signals From Outside The Cell," *DNA From the Beginning*, As of October 12, 2022, http://www.dnaftb.org/35/
15. Kimberly Repp, "How the Body Repairs Itself," *Ask a Biologist - Arizona State University*, As of October 12, 2022, https://askabiologist.asu.edu/explore/when-body-attacked
16. Susan Elmore, "Apoptosis: A Review Of Programmed Cell Death," *Toxicol Pathol. 2007 Jun;35(4):495-516. doi: 10.1080/01926230701320337. PMID: 17562483; PMCID: PMC2117903,* https://www.ncbi.nlm.nih.gov/pmc/articles/PMC2117903/
15. Nabeeha Khalid, Mahzad Azimpouran, "Necrosis," [Updated 2022 Mar 9]. In: *StatPearls* [Internet]. Treasure Island (FL): StatPearls Publishing; 2022 Jan. Available from: https://www.ncbi.nlm.nih.gov/books/NBK557627/
16. NIH Research Matters, "How Your Body Senses the Urge to Urinate," *National Institute of Health website*, October 27, 2020. https://www.nih.gov/news-events/nih-research-matters/how-your-body-senses-urge-urinate#:~:text=Over%20time%2C%20your%20bladder%20fills,triggers%20the%20urge%20to%20urinate.

Chapter 17

1. Editorial Staff, "How Many Possible Moves Are There In Chess?" *Chess Journal*, As of October 13, 2022, https://www.chessjournal.com/how-many-possible-moves-are-there-in-chess/
2. Rothschild, Bionomics, pp. 96, 97, 160
3. Rob DeSalle and Ian Tattersall, "Do Plants Have Brains?" *Natural History*, As of October 13, 2022 (From: The Brain: Big Bangs, Behaviors, and Beliefs by Rob DeSalle and Ian Tattersall, published by Yale University

Press, 2012) https://www.naturalhistorymag.com/features/152208/do-plants-have-brains

Chapter 18

1. Michael Laub, "Keeping Signals Straight: How Cells Process Information and Make Decisions," *Plos Biology*, July 18, 2016, https://journals.plos.org/plosbiology/article?id=10.1371/journal.pbio.1002519
2. Scitable, "The Information in DNA Is Decoded by Transcription," *Nature Education*, As of October 13,2022, https://www.nature.com/scitable/topicpage/the-information-in-dna-is-decoded-by-6524808/
3. Diffen, "Data vs. Information," *Diffen*, As of October 13, 2022, https://www.diffen.com/difference/Data_vs_Information
4. Gazette Staff, "Why Onions Have More DNA Than You Do," *Harvard Gazette*, February 10, 2000, https://news.harvard.edu/gazette/story/2000/02/why-onions-have-more-dna-than-you-do/
5. Rothschild, Bionomics, p. 4, 156, 157

Chapter 19

1. Rothschild, Bionomics, p. 1-5, 71, 96, 157, 208, 258-259
2. Charles Darwin, *From So Simple A Beginning* p. 502
3. Andrea Stephenson, Justin W. Adams, Mauro Vaccarezza, "The Vertebrate Heart: An Evolutionary Perspective," *Journal Of Anatomy*, September 14, 2017, https://onlinelibrary.wiley.com/doi/10.1111/joa.12687
4. Neil Shubin, *Your Inner Fish*, (Vintage Books, New York, 2008), p 81- 96
5. Richard C. Mohs, "How Human Memory Works," *HowStuffWorks*, As of October 14, 2022, https://science.howstuffworks.com/life/inside-the-mind/human-brain/human-memory.htm
6. University of California – San Diego, : They Remember: Communities Of Microbes Found To Have Working Memory," *Science News*, April 28, 2020, https://www.sciencedaily.com/releases/2020/04/200428093506.htm
7. Hannah Ritchie and Max Roser, "Biodiversity - Extinctions," *Our World in Data*, As of September 30, 2022, https://ourworldindata.org/extinctions

Chapter 20

1. Dr. Dennis O'Neill, "The First Primates," *Behavioral Sciences Department, Palomar College*, 2014, https://www2.palomar.edu/anthro/earlyprimates/early_2.htm

2. Your Genome, "Evolution of Modern Humans," *Your Genome*, July 21, 2021, https://www.yourgenome.org/stories/evolution-of-modern-humans/
3. Rothschild, Bionomics, pp. 71, 160 -163

4. Alison George, "Code Hidden In Stone Age Art May Be The Root Of Human Writing," *New Scientist*, November 9, 2016, https://www.newscientist.com/article/mg23230990-700-in-search-of-the-very-first-coded-symbols/
5. Gwen DeWar, "Newborn Cognitive Development: What Are Babies Thinking And Learning?" *Parenting Science*, 2020, https://parentingscience.com/newborn-cognitive-development/

Chapter 21

1. Traci Watson, "Ancient Egyptian Cemetery Holds Proof of Hard Labor," *National Geographic*, March 14, 2013, https://www.nationalgeographic.com/history/article/130313-ancient-egypt-akhenaten-amarna-cemetery-archaeology-science-world
2. Jake Buehler, "The Complex Truth About 'Junk DNA'," Quanta Magazine, September 1, 2021, https://www.quantamagazine.org/the-complex-truth-about-junk-dna-20210901/
3. Charlotte Hsu, "Carnivorous Plant Throws Out 'Junk' DNA," *University at Buffalo*, May 13, 2013, https://www.buffalo.edu/news/releases/2013/05/023.html
4. History.com Editors, "Printing Press," *History.com*, October 10, 2019, https://www.history.com/topics/inventions/printing-press
5. Dr. Robert Pastore, "The Neurochemistry of Sleep," *Power On Power Off*, May 13, 2020, https://poweronpoweroff.com/blogs/longform/the-neurochemistry-of-sleep
6. Gabe Paoletti, "The Story Of Franz Reichelt, The Man Who Died Jumping Off The Eiffel Tower," *All That's Interesting*, October 4, 2021, https://allthatsinteresting.com/franz-reichelt
7. Michael Graw, "Why DNA Is the Most Favorable Molecule for Genetic Material & How RNA Compares to It in This Respect," *Sciencing*, March 13, 2018, https://sciencing.com/dna-favorable-molecule-genetic-material-rna-compares-respect-17806.html
4. Christian Christensen, "Why Are The Greek Gods No Longer Worshipped?" World History FAQ, S of October 16, 2022, https://worldhistoryfaq.com/why-greek-gods-no-longer-worshiped/

Chapter 22

1. Ishan Daftardar, "How Are Memories Stored And Retrieved?" *Science ABC*, July 8, 2022, https://www.scienceabc.com/humans/how-are-memory-stored-retrieved-forget-encode-retrieve-hippocampus-long-term-memory-short-term-memory.html

2. WorldData.info, "Life Expectancy," *WorldData.info*, As of October 17, 2022, https://www.worlddata.info/life-expectancy.php (Data is based on the year 2020)
3. Kendra Cherry, "Genie's Story," *Very Well Mind*, February 7, 2022, https://www.verywellmind.com/genie-the-story-of-the-wild-child-2795241

Chapter 23
1. History.com Editors, "Social Darwinism," *History.com*, August 21, 2018, https://www.history.com/topics/early-20th-century-us/social-darwinism
2. Catherine Wilson, "Darwinian Morality," *Evolution: Education and Outreach*, August 7, 2009, https://evolution-outreach.biomedcentral.com/articles/10.1007/s12052-009-0162-z
3. Nicholas Brealey, "Know Your Brain: Reticular Formation," *Neuroscientifically Challenged*, From J. Nolte, The Human Brain: An Introduction to its Functional Anatomy. 6th ed. (Philadelphia, PA. Elsevier; 2009), https://neuroscientificallychallenged.com/posts/know-your-brain-reticular-formation
4. World Population Review, "Most Christian Countries 2022," *World Population Review*, As of October 18, 2022, https://worldpopulationreview.com/country-rankings/most-christian-countries
5. Union of Concerned Scientists, "How Do We Know that Humans Are the Major Cause of Global Warming?" *Union of Concerned Scientists*, January 21, 2021, https://www.ucsusa.org/resources/are-humans-major-cause-global-warming
6. Max Roser, Joe Hasell, Bastian Herre and Bobbie Macdonald, "War and Peace," *OurWorldInData.org*, 2016, https://ourworldindata.org/war-and-peace#citation

Chapter 25
1. Ryan Patrick Hanley, "On Self Interest," *Princeton University Press*, June 30, 2021, https://press.princeton.edu/ideas/on-self-interest
2. Editors of Encyclopaedia Britannica, "A Social And Cultural History Of Smoking," *Encyclopaedia Britanica*, As of December 30, 2023, https://www.britannica.com/topic/smoking-tobacco/A-social-and-cultural-history-of-smoking
3. Truth Initiative,, "A Look At How Big Tobacco Infiltrated Baseball," *Truth Initiative*, March 29, 2018, https://truthinitiative.org/research-resources/tobacco-industry-marketing/look-how-big-tobacco-infiltrated-baseball
4. Phil Edwards, "What Everyone Gets Wrong About The History Of Cigarettes," *Vox*, April 6. 2015, https://www.vox.com/2015/3/18/8243707/cigarette-rolling-machines
5. Casey Ashlock, "A Brief History of Cigarette Cards," *GoCollect*, April 12, 2021, https://blog.gocollect.com/a-brief-history-of-cigarette-cards/

6. Office on Smoking and Health – National Center for Chronic Disease Prevention and Health Promotion, "2000 Surgeon General's Report Highlights: Tobacco Timeline," *CDC*, July 21, 2015, https://www.cdc.gov/tobacco/sgr/2000/highlights/historical/index.htm

7. North Carolina Historic Sites, "Cultivation of a Tobacco Empire," *North Carolina Historic Sites,* As of October 19, 2022, https://historicsites.nc.gov/all-sites/duke-homestead/history/cultivation-tobacco-empire
8. "Tobacco Litigation Trials: 1954-present, 'Great American Trials'," *Encyclopedia.com.* (October 19, 2022), https://www.encyclopedia.com/law/law-magazines/tobacco-litigation-trials-1954-present
9. Editors of Encyclopedia Britannica, "American Tobacco Company," *Encyclopedia Britannica,* As of October 19, 2022, https://www.britannica.com/topic/American-Tobacco-Company
10. From: "Long Record for Steady Work Held by Woman Who began Her Duties in Duke's First Factory." *Durham Morning Herald.* January 17, 1926, Anchor, As of October 19, 2022, https://www.ncpedia.org/anchor/working-tobacco-factory
11. Duke Homestead Education & History Corporation, "A New Product," *Duke Homestead Education and History Corporation,* As of October 19, 2022, https://dukehomestead.org/a-new-product.php

Chapter 26
1. M.T. Wroblewski, "What Is Puffery in Advertising?" *Chron*, September 17, 2020, https://smallbusiness.chron.com/puffery-advertising-24357.html
2. Tejvan Pettinger, "Profit Maximisation," *Economics Help*, July 16, 2019, https://www.economicshelp.org/blog/3201/economics/profit-maximisation/
3. Aaron O'Neill, "Estimated Global Population from 10,000 BCE to 2100," *Statista*, June 21, 2022, https://www.statista.com/statistics/1006502/global-population-ten-thousand-bc-to-2050/
4. Robert Reich, "How Capitalism Is Killing Democracy," *Foreign Policy*, October 12, 2009, https://foreignpolicy.com/2009/10/12/how-capitalism-is-killing-democracy/

Chapter 27
1. Bernd Heinrich, *Bumblebee Economics*, p. 142
2. Adam Smith, *The Wealth of Nations*, p.117

3. Hannah Ritchie, "Humans Make Up Just 0.01% Of Earth's Life – What's The Rest?" *Our World in Data*, April 24, 2019, https://ourworldindata.org/life-on-earth
4. Ron Elving, "The NRA Wasn't Always Against Gun Restrictions," *NPR*, October 10, 2017, https://www.npr.org/2017/10/10/556578593/the-nra-wasnt-always-against-gun-restrictions
5. Open Secrets, "What is a PAC?" *Open Secrets*, As of October 20, 2022, https://www.opensecrets.org/political-action-committees-pacs/what-is-a-pac
6. Elias Beck, "Working Conditions in the Industrial Revolution," *History Crunch*, November 26, 2016 (updated March 25, 2022), https://www.historycrunch.com/working-conditions-in-the-industrial-revolution.html#/
7. Mark Meredith, "Marble House," *House HisTree*, November 23, 2018, https://househistree.com/houses/marble-house
8. US Department of Labor, "5. Progressive Era Investigations," *US Department of Labor*, As of October 20, 2022, https://www.dol.gov/general/aboutdol/history/mono-regsafepart05

Chapter 28

1. nidirect, Government services, "What Happens When You Drink Alcohol?" *nidirect.gov*, As of October 20, 2022, https://www.nidirect.gov.uk/articles/what-happens-when-you-drink-alcohol
2. Niall Moyna, "What Happens To Your Body When You Run A Marathon?" *RTE*, October 18, 2022, https://www.rte.ie/brainstorm/2017/1024/914860-what-happens-to-your-body-when-you-run-a-marathon/
3. Miguel Douglas, "Native Americans Are Not All the Same: An Exploration of Indigenous Diversity," *American Indian Republic*, October 22, 2021, https://americanindianrepublic.com/native-americans-are-not-all-the-same-an-exploration-of-indigenous-diversity/

Chapter 29

1. Chad Stone, Danilo Trisi, Arloc Sherman And Jennifer Beltrán, "A Guide to Statistics on Historical Trends in Income Inequality," *Center on Budget and Policy Priorities*, January 13, 2020, https://www.cbpp.org/research/poverty-and-inequality/a-guide-to-statistics-on-historical-trends-in-income-inequality
2. The Editors of Encyclopaedia Britannica, "Property Law And The Western Concept Of Private Property," *Brittanica.com*, October 20, 2022, https://www.britannica.com/topic/property-law/Property-law-and-the-Western-concept-of-private-property
3. Maria S. Cox, "Taxation," *Britannica.com*, September 2, 2022, https://www.britannica.com/topic/taxation

4. National Archives, "The Constitution of the United States: A Transcription," *National Archives*, As of October 20, 2022, https://www.archives.gov/founding-docs/constitution-transcript
5. Wendy Sawyer and Peter Wagner, "Mass Incarceration: The Whole Pie 2022," *Prison Policy Initiative*, March 14, 2023, https://www.prisonpolicy.org/reports/pie2023.html

Chapter 32

1. elawtalk, "*9 of the Most Common Types of Government Systems Explained*," *elaw Talk*, January 10, 2022, https://elawtalk.com/types-of-government-systems/
2. Scott Kennedy, "Chinese State Capitalism," *Center For Strategic And International Studies*, October 7, 2021, https://www.csis.org/analysis/chinese-state-capitalism
3. National Archives, "The Constitution of the United States: A Transcription," *National Archives*, As of October 20, 2022, https://www.archives.gov/founding-docs/constitution-transcript
4. Wendy Sawyer and Peter Wagner, "Mass Incarceration: The Whole Pie 2022," *Prison Policy Initiative*, March 14, 2023, https://www.prisonpolicy.org/reports/pie2023.html
5. Macrotrends, "Amazon Revenue 2010-2023 | AMZN," *Macrotrends*, As of December 30, 2023, https://www.macrotrends.net/stocks/charts/AMZN/amazon/revenue

Chapter 33

1. Rothschild, Bionomics, p.324
2. Robin Hartill, CFP, "The Definitive Guide: How to Value a Stock," *The Motley Fool*, June 30, 2022, https://www.fool.com/investing/how-to-invest/stocks/how-to-value-stock/
3. Oren Levin-Waldman, "How Inequality Undermines Democracy," *E-International Relations*, December 10, 2016, https://www.e-ir.info/2016/12/10/how-inequality-undermines-democracy/

Chapter 34

1. NASA Global Climate Change, "The Effects of Climate Change," *NASA*, As of October 22, 2022, https://climate.nasa.gov/effects/

Chapter 35

1. Jonathan Edwards, "Bus-Riding Dog Who Took Herself To Park Remembered As 'Seattle Icon,'" *Washington Post*, October 18, 2022, https://www.washingtonpost.com/nation/2022/10/18/eclipse-bus-riding-dog-dies/

Chapter 36

1. Bahman Zohuri, "First Law of Thermodynamics," from Physics of Cryogenics, *Science Direct*, As of December 29, 2023,

https://www.sciencedirect.com/topics/chemistry/first-law-of-thermodynamics

2. Stephen Wolfram, "Some Historical Notes," from A New Kind of Science, *Wolfram Science*, 2002, https://www.wolframscience.com/reference/notes/1019b/

Afterword

Congratulations! If you've made it this far, I commend you. This was not an easy book to read for any number of reasons.

Though I have gone to great lengths to simplify the scientific ideas included in this book, they are still scientific ideas that can be hard to wrap your head around when you're not used to thinking of the world this way. Thank you for persevering.

I've blown up several foundational tenets of humanity's existence. Oops. I didn't intend to do that when I started writing. It's just that once I connected the dots between the ecosystem and the ideasphere, many of the key premises that underlie human life did themselves in. They couldn't stand up to the most basic tests that nature requires of even its smallest and simplest members.

I've told many bad jokes and awful puns. I can't help myself. I find that no matter how serious a topic is, there is always room for levity. Mark Twain said the only real effective weapon humans have is humor. I think he was right.

I would be remiss if I did not mention the most influential thinkers that enabled me to develop the Ideasphere theory.

First, Michael Rothschild who wrote *Bionomics - Economy as Ecosystem* in 1990. Rothschild thoughtfully detailed how economic systems mirror biological systems. It was a compelling, rich, and insightful book. I have been thinking about this book in one way or another since the year I first read it. Rothschild was brave. I think this was the first book to delve into the link between natural and human systems since the Social Darwinism fiasco of the late 1800's. People have been afraid to link humanity to biology for fear of repeating the tendency to give people license to do horrible things and blame it on nature. Rothschild showed that economic systems do indeed mirror natural behaviors. I am more than grateful for Rothschild's insights, but even reading Bionomics as a young adult, I couldn't help thinking that if economics mirrored biology, shouldn't all of humanity mirror biology? In fact, I was sure that within a year or two of the publication of *Bionomics*, someone would make this link. That book never came. Worse, the term bionomics was co-opted to measure the economic cost of climate change and other human activities. That is why I finally decided to write this book; the book I was waiting for.

Second, I would be remiss if I did not mention Charles Darwin. *The Origin of Species* is a tough read; Darwin wrote run on sentences on top of run on sentences. But what a thinker. I came away from each page astounded at his ability to reason through the ideas of evolution so clearly and compellingly using his experiments with beetles and bees and barnacles and discussions with others in every walk of life. In addition, *The Voyage of the Beagle* is as compelling. Darwin wrote this book while he served as naturalist companion to the captain on a five-year voyage beginning in 1831 when he was 22 years old. His observations and ability to work out how natural formations and living structures emerged are clearly evident in this book. It is apparent that this trip planted the idea of change in his mind early in his life and that evolution was the natural conclusion he drew from his observations.

Stephen Jay Gould's book, *Wonderful Life*, cemented the idea in my mind that change is the only constant in the universe, and that this Earth and the humans on it, are unlikely to even exist. That very improbability makes me, as it did to Gould, more appreciative of our existence; and ever more incented to make sure humans stick around.

Last, the writers of the Constitution have my deep gratitude. They had the ability to put into practical form ideas about government they could not prove as right but which they intuitively understood to be true. They have come the closest to mirroring the lessons nature would teach us if it could. It is my fervent hope that 200 plus years after the founding of our nation, we will heed these lessons and will not let the brilliant ideas upon which our country is founded perish from this earth.

I give my thanks to the many professionals in the world today. Scientists and thinkers, economists and philosophers, sociologists and leaders and too many others to count. Though I am not of your ilk, and am far more a generalist than a specialist, it is in you that I put my hope for a better future. You study and learn the details of your respective professions and build the tools that allow us to live better lives. I hope that this book might provide you with a different context, with a new lens, so to speak, by which you can continue your work in a way that improves life on Earth.

Anne Riley
February 1, 2024

ABOUT THE AUTHOR

Anne Riley is the Author of:

> Elusive- Journey to Happiness
> Aerie – A Romantic Suspense Story
> DINA – Nature's Case for Democracy

She lives in Illinois with her husband Tim and dog, Scout. She has three lovely children, two wonderful children-in-law, two fantastic grandchildren, and a myriad of grandpets. She lives her best life every day and hopes that we all will find ways to make life better for the generations that follow us.

Made in the USA
Las Vegas, NV
18 December 2024

14625905R00246